Board Review Series

Biochemistry
2nd edition

Board Review Series

Biochemistry
2nd edition

Dawn B. Marks, Ph.D.
Professor of Biochemistry
Department of Biochemistry
Temple University School of Medicine
Philadelphia, Pennsylvania

Harwal Publishing

Philadelphia • Baltimore • Hong Kong • London • Munich • Sydney • Tokyo

A Waverly Company

Harwal

Sponsoring Editor: Elizabeth A. Nieginski
Managing Editor: Susan E. Kelly
Production: Laurie Forsyth

Library of Congress Cataloging-in-Publication Data

Marks, Dawn B.
    Biochemistry / Dawn B. Marks. — 2nd ed.
      p.   cm. — (Board review series)
    Includes index.
    ISBN 0-683-05597-6 (alk. paper)
    1. Biochemistry—Outlines, syllabi, etc. I. Title. II. Series.
QP518.3.M37   1994
574.19'2'076—dc20
DNLM/DLC
for Library of Congress
                                          93-33778
                                            CIP

10 9 8 7 6 5 4 3 2 1

Contents

Comprehensive Examination   299

Index   321

Preface to the Second Edition

This book is intended to aid medical students in preparing for examinations, in particular the United States Medical Licensing Examination (USMLE) Step 1. The second edition has been developed with the aim of converting all books in the BOARD REVIEW SERIES to an outline format and to bring the questions into conformity with the new USMLE style.

The overviews at the start of each chapter as well as the summaries at the beginning of each section have been retained because of my strong conviction that the details of biochemistry are easier to comprehend if they are integrated into a physiologic understanding of the functioning of the human body. Therefore, the book has retained the first chapter summarizing fuel metabolism.

All of the material has been organized to eliminate redundancy. A chapter on molecular endocrinology has been added, the molecular biology section has been updated, many of the figures have been redrawn, and the questions have been revised, not only to conform with the USMLE style but also to reflect the current emphasis on clinical problem solving.

I hope that this edition will aid students not only with their immediate task of passing a set of examinations, but also help them with the more long-term objective of fitting the subject of biochemistry into the framework of basic and clinical sciences, so essential to the understanding of their future patients' problems.

Dawn B. Marks

1
Fuel Metabolism

Overview

- The major fuels of the body, carbohydrates, fats, and proteins, are obtained from the diet and stored in the body's fuel depots.
- In the fed state (after a meal), ingested fuel is used to meet the immediate energy needs of the body and excess fuel is stored.
- During fasting (e.g., between meals or overnight), stored fuels are used to derive the energy needed to survive until the next meal.
- In prolonged fasting (starvation), changes occur in the use of fuel stores that permit survival for extended periods of time.
- The level of insulin in the blood increases in the fed state and promotes fuel storage; the level of glucagon increases in the fasting state and promotes the release of stored fuel.

I. Metabolic Fuels and Dietary Components

- Carbohydrates, fats, and proteins serve as the major fuels of the body and are obtained from the diet. After digestion and absorption, these fuels may be oxidized for energy.
- Fuel consumed in excess of the body's immediate energy needs is stored, mainly as fat, but also as glycogen. To some extent, body protein also can be used as fuel.
- The daily energy expenditure of an individual includes the energy required for the basal metabolic rate (BMR) and the energy required for physical activity.
- In addition to providing energy, dietary components also produce precursors for the synthesis of structural compounds and supply the essential fatty acids and amino acids (which the body cannot synthesize) and the vitamins and minerals (which often serve as cofactors for enzymes).

A. Fuels
–When **fuels** are metabolized in the body, **heat** is generated and **ATP** (adenosine triphosphate) is synthesized.

1

1. Energy content of fuels

    a. The oxidation of **carbohydrates** to CO_2 and H_2O produces approximately 4 kcal/g.

    b. Protein also produces about 4 kcal/g.

    c. Fats produce more than twice as much energy (9 kcal/g).

    d. Alcohol, which is present in many diets, produces about 7 kcal/g.

2. Energy is often expressed by physicians and nutritionists with the term "calorie" in place of kilocalorie.

3. Heat generated by fuel oxidation is used to maintain **body temperature**.

4. ATP generated by fuel metabolism is used for biochemical reactions, muscle contraction, and other energy-requiring processes.

B. Composition of body fuel stores (Figure 1-1)

    1. Triacylglycerol (triglyceride)

        a. The major fuel store of the body is **adipose triacylglycerol**.

        b. Adipose tissue stores fuel very efficiently. It has more stored calories per gram and less water (15%) than other fuel stores. (Muscle tissue is about 80% water.)

    2. Glycogen stores, although small, are extremely important.

        a. Liver glycogen is used to maintain blood glucose during the early stages of fasting.

        b. Muscle glycogen is oxidized for muscle contraction.

    3. Protein does not serve solely as a source of fuel and can be degraded only to a limited extent.

        a. Approximately one-third of total body protein can be degraded.

        b. If too much protein is oxidized for energy, body functions can be severely compromised.

C. Daily energy expenditure is the amount of energy required each day.

    1. Basal metabolic rate (BMR) is the energy used by a person who has fasted for at least 12 hours and is awake but at rest.

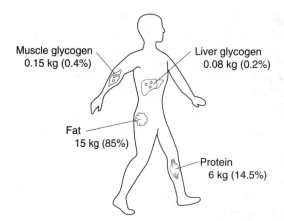

Muscle glycogen
0.15 kg (0.4%)

Liver glycogen
0.08 kg (0.2%)

Fat
15 kg (85%)

Protein
6 kg (14.5%)

Figure 1-1. Fuel composition of an average 70-kg man after an overnight fast (in kg and as percent of total calories).

2. Specific dynamic action (SDA) is the elevation in metabolic rate that occurs during digestion and absorption of foods. It is often ignored in calculations because its value is usually unknown and probably small (less than 10% of the total energy).

3. Physical activity

    a. The number of calories that physical activity adds to the daily energy expenditure varies considerably. A person can expend about 5 calories (kcal) each minute while walking but 20 calories while running.

    b. The daily energy requirement for an extremely sedentary person is about 30% of the BMR. For a more active person, it may be 50% or more of the BMR.

D. Other dietary requirements

  1. Essential fatty acids

    a. Linoleic and linolenic acids are required in the diet.

    b. Linoleic acid is necessary for the **synthesis of prostaglandins** and related compounds. (The function of linolenic acid is not firmly established.)

  2. Protein

    a. Essential amino acids

      (1) Nine amino acids cannot be synthesized in the body and, therefore, must be present in the diet in order for protein synthesis to occur. The essential amino acids are lysine, isoleucine, leucine, threonine, valine, tryptophan, phenylalanine, methionine, and histidine.

      (2) Only a small amount of **histidine** is required in the diet. However, larger amounts are required for **growth** (e.g., in children, pregnant women, and people recovering from injuries).

      (3) Because **arginine** can be synthesized only in limited amounts, it is required in the diet for **growth**.

    b. Nitrogen balance

      (1) The body's primary source of nitrogen is **dietary protein,** which is about 16% nitrogen.

      (2) Proteins are constantly being synthesized and degraded in the body.

      (3) As amino acids are oxidized, the nitrogen is converted to urea and excreted by the kidneys. Other nitrogen-containing compounds produced from amino acids are also excreted in the urine (**uric acid, creatinine,** and **NH_4^+**).

      (4) Nitrogen balance (the normal state in the adult) occurs when synthesis of body protein equals degradation. The amount of nitrogen excreted in the urine each day equals the amount of nitrogen ingested each day.

      (5) Negative nitrogen balance occurs when degradation of body protein exceeds synthesis. More nitrogen is excreted than ingested. It results from an inadequate amount of protein in the diet or the absence of one or more essential amino acids.

(6) Positive nitrogen balance occurs when synthesis of body protein exceeds degradation. Less nitrogen is excreted than ingested. It occurs during growth.

3. Vitamins and minerals

a. Vitamins and minerals are required in the diet. Many serve as **cofactors for enzymes**.

b. Minerals required in large amounts include **calcium and phosphate,** which serve as structural components of bone. Minerals required in trace amounts include **iron,** which is a component of heme.

II. The Fed or Absorptive State (Figure 1-2)

- Dietary carbohydrates are cleaved during digestion, forming monosaccharides (mainly glucose) that enter the blood. Glucose may be oxidized by various tissues for energy or stored as glycogen in the liver and in muscle. In the liver, glucose may be converted to triacylglycerols, which are packaged in very low-density lipoproteins (VLDL) and released into the blood. The fatty acids of the VLDL are stored in adipose tissue.
- Dietary fats (triacylglycerols) are digested to fatty acids and 2-monoglycerides. These digestive products are resynthesized to triacylglycerols by intestinal epithelial cells, packaged in chylomicrons, and secreted via the lymph into the blood. The fatty acids of chylomicrons may be stored in adipose triacylglycerols or oxidized by various tissues for energy.
- Dietary proteins are digested to amino acids and absorbed into the blood. The amino acids may be used by various tissues to synthesize proteins, to produce nitrogen-containing compounds (e.g., purines, heme, creatine, and epinephrine), or they may be oxidized to produce energy.

A. Digestion and absorption

1. Carbohydrates

a. The major dietary carbohydrate is **starch,** the storage form of carbohydrate in plants.

(1) Starch is cleaved by **salivary amylase** in the mouth and by **pancreatic amylase** in the intestine to disaccharides and oligosaccharides.

(2) Dextrinases, α-glucosidases, and **disaccharidases** located on the surface of the brush border of the intestinal epithelial cell complete the conversion of starch to glucose.

b. Ingested disaccharides are cleaved by disaccharidases on the surface of the intestinal epithelial cell.

(1) Sucrose (table sugar) is converted to **fructose** and **glucose** by sucrase.

(2) Lactose (milk sugar) is converted to **glucose** and **galactose** by lactase.

c. Some **free glucose** and **fructose** are consumed in the diet.

d. Monosaccharides (mainly **glucose** and some **fructose** and **galactose**), present in the diet or produced by the digestive process, are absorbed by the intestinal epithelial cells and pass into the blood.

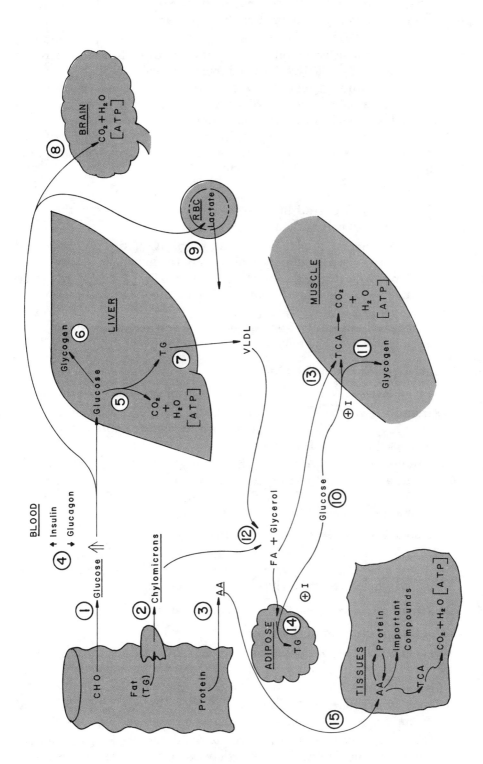

Figure 1-2. The fed state. The *circled numbers* serve as a guide, indicating the approximate order in which the processes begin to occur. CHO = carbohydrate; TG = triacylglycerols (triglycerides); FA = fatty acids; AA = amino acids; TCA = tricarboxylic acid cycle; RBC = red blood cells; VLDL = very low-density lipoprotein; I = insulin; ⊕ = stimulated by.

2. Fats

  a. The primary dietary fat is **triacylglycerol,** obtained from the fat stores of the plants and animals that serve as food.

    (1) Because they are water-insoluble, triacylglycerols are **emulsified** in the intestine by bile salts. They are **digested** by pancreatic lipase to 2-monoglycerides and free fatty acids.

    (2) The **2-monoglycerides and fatty acids** are packaged into micelles (solubilized by bile salts) and absorbed into intestinal epithelial cells, where they are reconverted to triacylglycerols.

  b. After digestion and resynthesis, the triacylglycerols are packaged as **chylomicrons** (lipoproteins) that first enter the lymph and then the blood.

3. Proteins

  a. Proteins are **digested** first by pepsin in the stomach and then by a series of enzymes in the intestine.

    (1) The pancreas produces **trypsin, chymotrypsin, elastase,** and the **carboxypeptidases,** which act in the lumen of the intestine.

    (2) **Aminopeptidases, dipeptidases,** and **tripeptidases** are associated with the intestinal epithelial cells.

  b. Proteins are ultimately **degraded** to a mixture of **amino acids,** which then enter intestinal epithelial cells, where some amino acids are metabolized. The remainder pass into the blood.

B. Digestive products in the blood

  1. Hormone levels are affected when the products of digestion enter the blood.

    a. Insulin levels rise principally as a result of increased **blood glucose levels**. Insulin levels can also be increased to a lesser extent by increased blood levels of amino acids.

    b. Glucose causes levels of the hormone **glucagon** to **decrease** but amino acids cause them to increase; overall, after a mixed meal (containing carbohydrate, fat, and protein), glucagon levels remain fairly constant in the blood.

  2. The products of digestion enter cells from the blood.

    –**Glucose** and **amino acids** leave the intestinal epithelial cells and enter the hepatic portal vein. Therefore, the **liver** is the first tissue through which these products of digestion pass.

C. The fate of glucose in the fed (absorptive) state

  1. The fate of glucose in the liver

    –Liver cells either oxidize glucose or convert it to glycogen and triacylglycerols.

    a. Glucose is oxidized to CO_2 and H_2O to meet the immediate energy needs of the liver.

    b. Excess glucose is stored in the liver as **glycogen,** which is used during periods of fasting to maintain blood glucose.

    c. Excess glucose can be converted to fatty acids and a glycerol moiety, which combine to form triacylglycerols, which are released from the liver into the blood as VLDL.

2. The fate of glucose in other tissues

  a. The **brain,** which depends on glucose for its energy needs, **oxidizes glucose to CO_2 and H_2O,** producing ATP.

  b. **Red blood cells,** lacking mitochondria, oxidize glucose to **pyruvate and lactate,** which are released into the blood.

  c. **Muscle cells** take up glucose by a transport process that is stimulated by insulin.

   (1) Glucose is oxidized to CO_2 and H_2O to generate ATP for muscle contraction.

   (2) Muscles store glucose as **glycogen** for use during contraction.

  d. Cells of **adipose tissue** take up glucose by a transport process that is stimulated by insulin and use it to produce energy and to form the glycerol moiety of its triacylglycerol stores.

D. The fate of lipoproteins in the fed state

 1. The triacylglycerols of **chylomicrons** (produced from dietary fat) and **VLDL** (produced from glucose by the liver) are digested in capillaries by lipoprotein lipase to form fatty acids and glycerol.

 2. The **fatty acids** are taken up by **adipose** tissue, converted to **triacylglycerols,** and stored.

E. The fate of amino acids in the fed state

 1. Amino acids from dietary proteins enter cells. In **muscle,** the transport of amino acids into cells is stimulated by insulin.

 2. Amino acids may be:

  a. Used for **protein synthesis** (which occurs on ribosomes and requires mRNA). Proteins are constantly being synthesized and degraded.

  b. Used to make **nitrogenous compounds** such as heme, creatine phosphate, epinephrine, and the purine and pyrimidine bases found in DNA and RNA.

  c. Oxidized to generate ATP.

III. Fasting (Figure 1-3)

- As blood glucose levels decrease after a meal, insulin levels decrease and glucagon levels increase, stimulating the release of stored fuels into the blood.
- The liver supplies glucose and ketone bodies to the blood. Blood glucose levels are maintained by the liver, initially by the process of glycogenolysis and subsequently by gluconeogenesis. Ketone bodies are synthesized by the liver from fatty acids supplied by adipose tissue.
- Adipose triacylglycerols are released as free fatty acids and glycerol. The fatty acids are oxidized to CO_2 and H_2O by various tissues. In the liver, the fatty acids are converted to ketone bodies, and the glycerol serves as a source of carbon for gluconeogenesis.
- Muscle releases amino acids. The carbons are used by the liver for gluconeogenesis, and the nitrogen is converted to urea.

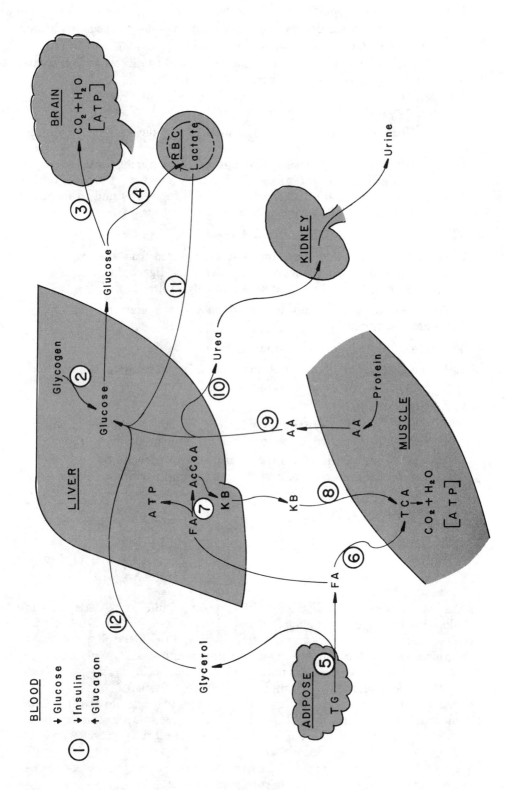

Figure 1-3. The fasting state. The *circled numbers* serve as a guide, indicating the approximate order in which the processes begin to occur. KB = ketone bodies. See legend for Figure 1-2 for explanation of abbreviations.

A. The liver during fasting

–The liver produces **glucose and ketone bodies** that are released into the blood and serve as sources of energy for other tissues.

1. Production of glucose by the liver

–The liver has the major responsibility for **maintaining blood glucose levels**. Glucose is required particularly by tissues such as the brain and red blood cells. The brain oxidizes glucose to CO_2 and H_2O, and red blood cells oxidize glucose to pyruvate and lactate.

a. Glycogenolysis

–About 2–3 hours after a meal, the liver begins to break down its glycogen stores by the process of **glycogenolysis,** and glucose is released into the blood. Glucose then can be taken up by various tissues and oxidized.

b. Gluconeogenesis

(1) After about 4–6 hours of fasting, the liver begins the process of **gluconeogenesis**. Within 30 hours, liver glycogen stores are depleted, leaving gluconeogenesis as the major process responsible for maintaining blood glucose.

(2) **Carbon sources** for gluconeogenesis

(a) **Lactate** produced by tissues such as red blood cells or exercising muscle

(b) **Glycerol** from breakdown of triacylglycerols in adipose tissue

(c) **Amino acids,** particularly alanine, from muscle protein

(d) **Propionate** from oxidation of odd-chain fatty acids (minor source)

2. Production of ketone bodies by the liver

a. As glucagon levels rise, adipose tissue breaks down its **triacylglycerol stores** into fatty acids and glycerol, which are released into the blood.

b. Through the process of **β-oxidation,** the liver converts the fatty acids to acetyl CoA.

c. **Acetyl CoA** is used by the liver for the synthesis of the ketone bodies, **acetoacetate** and **β-hydroxybutyrate**. The liver cannot oxidize ketone bodies. It releases them into the blood.

B. Adipose tissue during fasting

1. As glucagon levels rise, adipose **triacylglycerol** stores are mobilized. The liver converts the fatty acids to ketone bodies and the glycerol to glucose.

2. Tissues such as muscle oxidize the fatty acids to CO_2 and H_2O.

C. Muscle during fasting

1. Degradation of muscle protein

a. During fasting, muscle protein is degraded, producing amino acids.

b. These **amino acids** may be partially metabolized by muscle and released into the blood, mainly as **alanine and glutamine**.

c. Other tissues, such as the **gut and kidney,** further metabolize the glutamine.

d. The products (mainly **alanine**) travel to the **liver** where the carbons are converted to glucose or ketone bodies and the nitrogen is converted to urea.

2. Oxidation of fatty acids and ketone bodies

a. During **fasting,** muscle oxidizes fatty acids released from adipose tissue and ketone bodies produced by the liver.

b. During **exercise,** muscle can also use its own glycogen stores as well as glucose, fatty acids, and ketone bodies from the blood.

IV. Prolonged Fasting (Starvation)

- In starvation (prolonged fasting), muscle decreases its use of ketone bodies. As a result, ketone body levels rise in the blood, and the brain uses them for energy. Consequently, the brain needs less glucose, and gluconeogenesis slows, sparing muscle protein.
- These changes in the fuel utilization patterns of various tissues enable people to survive for extended periods of time without food.

A. Metabolic changes in starvation (Figure 1-4)

–When the body enters the **starved state,** after **3–5 days of fasting,** changes occur in the use of fuel stores.

1. Muscle decreases its use of ketone bodies and oxidizes fatty acids as its primary energy source.

2. Because of decreased use by muscle, **blood levels of ketone bodies rise**.

3. The **brain** then takes up and oxidizes ketone bodies to derive energy. Consequently, the brain decreases its use of glucose, although glucose is still a major fuel for the brain.

4. Liver **gluconeogenesis decreases**.

5. Muscle protein is spared (i.e., less muscle protein is degraded to provide amino acids for gluconeogenesis).

6. Because of decreased conversion of amino acids to glucose, **less urea is produced** from amino acid nitrogen in starvation than after an overnight fast.

B. Fat: the primary fuel

–The body uses its fat stores as its primary source of energy during starvation, conserving functional protein.

1. Overall, fats are quantitatively the most important fuel in the body.

2. The length of time a person can survive without food depends mainly on the amount of fat stored in adipose tissue.

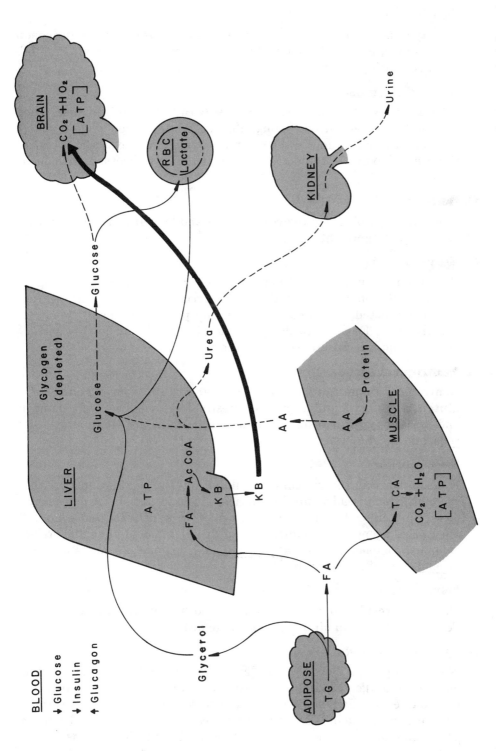

Figure 1-4. The starved state. *Dashed lines* indicate processes that have decreased, and the *heavy solid line* indicates a process that has increased relative to the fasting state. See legends for Figures 1-2 and 1-3 for explanation of abbreviations.

V. Clinical Correlations

A. Obesity

Obesity is associated with problems such as **hypertension, cardiovascular disease,** and **diabetes mellitus.** Treatment involves altering lifestyle, decreasing food intake, and increasing exercise.

B. Anorexia nervosa

Anorexia nervosa is characterized by **self-induced weight loss.** Those frequently affected include young, affluent, white women who, in spite of an emaciated appearance, often claim to be "fat." Anorexia nervosa is partially a behavioral problem: Those afflicted are obsessed with losing weight.

C. Bulimia

Bulimics suffer binges of **overeating** followed by **self-induced vomiting** to avoid gaining weight.

D. Cystic fibrosis

Cystic fibrosis is the most common lethal **genetic disease** among the white population of the United States. Eccrine and exocrine **gland** function are affected. Pulmonary disease and pancreatic insufficiency frequently occur. Food, particularly fats and proteins, may be only partially digested, and **nutritional deficiencies** result.

E. Nontropical sprue

Nontropical sprue (adult celiac disease) results from a **reaction to gluten,** a protein found in grains. Intestinal epithelial cells are damaged and **malabsorption** results. Common symptoms are steatorrhea, diarrhea, and weight loss.

F. Kwashiorkor

Kwashiorkor commonly occurs in children in third world countries where the diet, which is adequate in calories, is low in protein. A deficiency of dietary protein causes a **decrease in protein synthesis** that eventually affects the regeneration of intestinal epithelial cells and, thus, the problem is further compounded by **malabsorption.** Hepatomegaly and a distended abdomen are often observed.

G. Marasmus

Marasmus results from a diet deficient both in **protein and calories.** Persistent starvation ultimately results in death.

H. High-protein diets

When high-protein diets are very low in calories and the protein is of low biologic value (i.e., lacking in essential amino acids), **negative nitrogen balance** results. Body protein is degraded as amino acids are converted to glucose. A decrease in heart muscle may lead to death. Even if the protein is of high quality, ammonia and urea levels rise, putting increased stress on the kidneys. Vitamin deficiencies may occur due to a lack of fruits and vegetables.

I. Intravenous feeding

Solutions containing **5% glucose** are frequently infused into the veins of hospitalized patients. These solutions should be administered only for brief periods because they lack essential fatty acids and amino acids and because a high enough volume cannot be given per day to provide an adequate number of calories. More nutritionally complete solutions are available for parenteral administration.

J. Hyperthyroidism

One of the most common forms of hyperthyroidism, **Graves' disease,** results from excess thyroid hormone. It is characterized by an **elevated BMR,** an enlarged thyroid (goiter), protruding eyes, nervousness, tremors, palpitations, excessive perspiration, and weight loss.

K. Hypothyroidism

Hypothyroidism results from a **deficiency of thyroid hormone**. The BMR is decreased, and mucopolysaccharides accumulate on the vocal cords and in subcutaneous tissue. Common symptoms are lethargy, dry skin, a husky voice, decreased memory, and weight gain.

Review Test

Directions: Each of the numbered items or incomplete statements in this section is followed by answers or by completions of the statement. Select the **one** lettered answer or completion that is **best** in each case.

Questions 1–3

A young woman who has a sedentary job and does not exercise consulted a physician about her weight, which was 110 lb. A dietary history indicates that she eats approximately 100 g of carbohydrate, 20 g of protein, and 40 g of fat daily.

1. How many calories (kcal) does this woman consume each day?

(A) 1440
(B) 1340
(C) 940
(D) 840
(E) 640

2. What is the woman's approximate daily energy expenditure in calories (kcal) per day at this weight?

(A) 1200
(B) 1560
(C) 1800
(D) 2640
(E) 3432

3. Based on the woman's current weight, diet, and sedentary life-style, the physician correctly concludes that she should

(A) increase her exercise level
(B) decrease her protein intake
(C) increase her caloric intake
(D) decrease her fat intake to less than 30% of her total calories

4. In the normal adult, the fuel store that contains the fewest calories is

(A) adipose triacylglycerol
(B) liver glycogen
(C) muscle glycogen
(D) muscle protein

5. It is more advantageous for the human body to store fuel as triacylglycerol in adipose tissue than as protein in muscle because adipose triacylglycerol stores contain

(A) more calories and more water
(B) less calories and less water
(C) less calories and more water
(D) more calories and less water

6. Which one of the following amino acids is essential in the human diet?

(A) Serine
(B) Lysine
(C) Glutamate
(D) Tyrosine
(E) Cysteine

7. All of the following statements concerning fuel metabolism are true EXCEPT

(A) muscle protein can be a source of fuel for the body
(B) negative nitrogen balance occurs when the amount of nitrogen excreted in urine is greater than the amount of nitrogen ingested in the diet
(C) dietary protein is not required because amino acids can be synthesized from glucose
(D) a person who ingests 100 g of protein/day will excrete about 16 g of nitrogen in the urine

8. After fasting for 12 hours, a student consumes a large bag of pretzels. This meal will

(A) replenish liver glycogen stores
(B) increase the rate of gluconeogenesis
(C) reduce the rate at which fatty acids are converted to adipose triacylglycerols
(D) increase blood glucagon levels
(E) result in glucose being oxidized to lactate by the brain and to CO_2 and H_2O by red blood cells

9. Which of the following would be observed in a person who is resting after an overnight fast?

(A) Liver glycogen stores are completely depleted
(B) Liver gluconeogenesis is not an important process
(C) Muscle glycogen stores are used to maintain blood glucose
(D) Fatty acids are released from adipose triacylglycerol stores
(E) The liver is oxidizing ketone bodies to CO_2 and H_2O

10. Which of the following would be observed in a person after 1 week of starvation?

(A) The brain uses glucose and ketone bodies as fuel sources
(B) Liver glycogen stores are only partially depleted, due to an increase in gluconeogenesis
(C) Nitrogen balance is maintained because muscle protein releases amino acids to compensate for the lack of dietary protein
(D) Fatty acids from adipose stores are the major source of fuel for red blood cells

11. When compared to his state after an overnight fast, a person who fasts for 1 week will have

(A) higher levels of blood glucose
(B) less muscle protein
(C) more adipose tissue
(D) lower levels of ketone bodies in the blood

12. Patients with anorexia nervosa, untreated insulin-dependent diabetes mellitus, hyperthyroidism, and nontropical sprue all will

(A) have a high BMR
(B) have high insulin levels in the blood
(C) experience weight loss
(D) suffer from malabsorption
(E) have low levels of ketone bodies in the blood

Directions: Each group of items in this section consists of lettered options followed by a set of numbered items. For each item, select the **one** lettered option that is most closely associated with it. Each lettered option may be selected once, more than once, or not at all.

Questions 13–16

Match each of the characteristics below with the source of stored energy that it best describes.

(A) Protein
(B) Triacylglycerols
(C) Liver glycogen
(D) Muscle glycogen

13. The largest form of stored energy in the body

14. The energy source reserved for strenuous muscular activity

15. The primary source of carbon for maintaining blood glucose during an overnight fast

16. The major precursor of urea in the urine

Questions 17–21

Match each of the characteristics below with the tissue it best describes.

(A) Liver
(B) Brain
(C) Skeletal muscle
(D) Red blood cells

17. After a fast of a few days, ketone bodies become an important fuel

18. Ketone bodies are used as a fuel after an overnight fast

19. Fatty acids are not a significant fuel source at any time

20. During starvation, this tissue uses amino acids to maintain levels of blood glucose

21. This tissue converts lactate from muscle to a fuel for other tissues

Answers and Explanations

1–D. The woman consumes 400 calories (kcal) of carbohydrate (100 g × 4 kcal/g), 80 calories of protein (20 × 4), and 360 calories of fat (40 × 9) for a total of 840 calories daily.

2–B. This woman's daily energy expenditure is 1560 kcal. Daily energy expenditure equals BMR plus activity. Her weight is 110 lb/2.2 =50 kg. Her BMR (about 24 kcal/kg) is 50 kg × 24 = 1200 kcal/day. She is sedentary and needs only 360 additional kcal (30% of her BMR) to support her physical activity. Therefore, she needs 1200 + 360 = 1560 kcal each day.

3–C. Because her caloric intake (840 kcal/day) is less than her expenditure (1560 kcal/day), the woman is losing weight. She needs to increase her caloric intake. Exercise would cause her to lose weight. She is probably in negative nitrogen balance because her protein intake is low (0.8 g/kg/day is recommended). Although her fat intake is 43% of her total calories and recommended levels are less than 30%, she should increase her total calories rather than decreasing her fat intake.

4–B. In the average (70-kg) man, adipose tissue contains 15 kg of fat (135,000 calories). Liver glycogen contains about 0.08 kg carbohydrate (320 calories), and muscle glycogen contains about 0.15 kg carbohydrate (600 calories). In addition, about 6 kg of muscle protein (24,000 calories) can be used as fuel. Therefore, liver glycogen contains the fewest available calories.

5–D. Adipose tissue contains more calories and less water than muscle protein. Triacylglycerol stored in adipose tissue contains 9 kcal/g, and adipose tissue has about 15% water. Muscle protein contains 4 kcal/g and has about 80% water.

6–B. Lysine cannot be synthesized and must be obtained in the diet. Serine and glutamate can be synthesized from glucose. Cysteine can be synthesized from serine, obtaining its sulfur from methionine. Tyrosine is produced from phenylalanine.

7–C. Muscle protein provides amino acids for gluconeogenesis, which produces the glucose that is oxidized. Protein contains about 16% nitrogen; therefore, a person in nitrogen balance will excrete about 16 g of nitrogen for every 100 g of protein ingested. A person is in negative nitrogen balance if he excretes more nitrogen in his urine than he ingests in protein per day. Nine amino acids must be provided by the diet because they cannot be synthesized by the body.

8–A. After a meal of carbohydrates, glycogen is stored in the liver and in muscle, and triacylglycerols are stored in adipose tissue. The level of glucagon in the blood decreases, and gluconeogenesis decreases. The brain oxidizes glucose to CO_2 and H_2O while the red blood cells produce lactate.

9–D. Fatty acids are released from adipose tissue and oxidized by other cells. Liver glycogen is not depleted until about 30 hours of fasting. After an overnight fast, both glycogenolysis and gluconeogenesis by the liver help maintain blood glucose. Muscle glycogen stores are not used to maintain blood glucose. The liver produces ketone bodies but does not oxidize them.

10–A. After 3–5 days of starvation, the brain begins to use ketone bodies, in addition to glucose, as a fuel source. Glycogen stores in the liver are depleted during the first 30 hours of fasting. Inadequate protein in the diet results in negative nitrogen balance. Red blood cells cannot oxidize fatty acids because they do not have mitochondria.

11–B. If a person who has fasted overnight continues to fast for 1 week, muscle protein will decrease because it is being converted to blood glucose. His blood glucose levels will decrease only slightly because glycogenolysis and gluconeogenesis by the liver act to maintain blood glucose levels. Adipose tissue will decrease as triacylglycerol is mobilized. Fatty acids from adipose tissue will be converted to ketone bodies in the liver. Blood ketone body levels will rise.

12–C. All of these patients will lose weight; the anorexics because of insufficient calories in the diet, the diabetics because of low insulin levels, those with hyperthyroidism because of an increased BMR, and those with nontropical sprue because of decreased absorption of food from the gut. The untreated diabetics will have high ketone levels because of low insulin. Ketone levels may be elevated in anorexia and also in sprue.

13–B. Adipose triacylglycerols contain the largest amount of stored energy.

14–D. Muscle glycogen is used for energy during exercise.

15–C. Liver glycogenolysis is the major process for maintaining blood glucose after an overnight fast.

16–A. The nitrogen in amino acids derived from protein is converted to urea and excreted in the urine.

17–B. The brain begins to use ketone bodies when levels start to rise after 3 to 5 days of fasting.

18–C. Skeletal muscle oxidizes ketone bodies, which are synthesized in the liver from fatty acids derived from adipose tissue.

19–D. Oxidation of fatty acids occurs in mitochondria. Red blood cells lack mitochondria and therefore cannot use fatty acids.

20–A. The liver converts amino acids to blood glucose during gluconeogenesis.

21–A. Exercising muscle produces lactate, which the liver can convert to glucose by gluconeogenesis. Blood glucose is oxidized by red blood cells and other tissues.

2

Basic Aspects of Biochemistry

Overview

- Acids dissociate, releasing protons and producing their conjugate bases.
- Bases accept protons, producing their conjugate acids.
- Buffers consist of acid–base conjugate pairs that can donate and accept protons, thereby maintaining the pH of a solution.
- Proteins serve in many roles in the body (e.g., as enzymes, structural components, hormones, and antibodies).
- Interactions between amino acid residues produce the three-dimensional conformation of a protein.
- Enzymes are proteins that catalyze biochemical reactions.

I. Brief Review of Organic Chemistry

- Biochemical reactions involve the functional groups of molecules.

A. Functional groups in biochemistry

1. Types of groups

–Alcohols, amines, sulfhydryl groups, aldehydes, ketones, carboxyl groups, anhydrides, and esters are all important components of biochemical compounds (Figure 2-1).

2. Identification of carbon atoms (Figure 2-2)

a. The carbon atoms are **numbered** with the carbon of the most oxidized group designated as **carbon 1**.

b. The carbon atoms are assigned **Greek letters**. The α-carbon is the one next to the carbon atom of the most oxidized group.

B. Biochemical reactions

1. Reactions are classified according to the functional groups that react (e.g., **esterifications, hydroxylations, carboxylations,** and **decarboxylations**).

2. Oxidations of sulfhydryl groups to disulfides, of alcohols to aldehydes and ketones, and of aldehydes to carboxylic acids frequently occur.

a. Many of these oxidations are reversed by reductions.

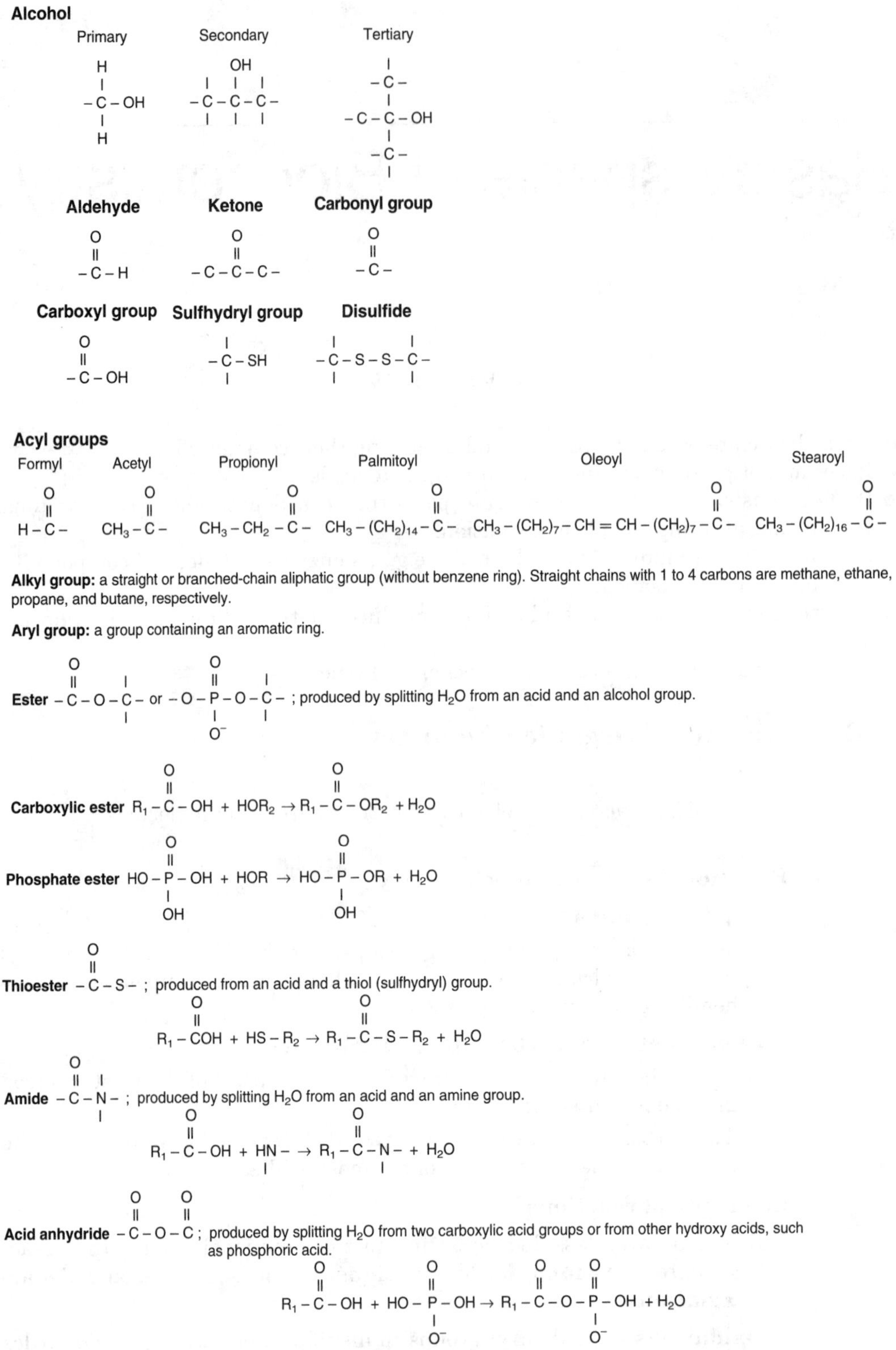

Alcohol

Primary

H
|
– C – OH
|
H

Secondary

OH
| | |
– C – C – C –
| | |

Tertiary

|
– C –
|
– C – C – OH
|
– C –
|

Aldehyde

O
‖
– C – H

Ketone

O
‖
– C – C – C –

Carbonyl group

O
‖
– C –

Carboxyl group

O
‖
– C – OH

Sulfhydryl group

|
– C – SH
|

Disulfide

| |
– C – S – S – C –
| |

Acyl groups

Formyl

O
‖
H – C –

Acetyl

O
‖
$CH_3 – C –$

Propionyl

O
‖
$CH_3 – CH_2 – C –$

Palmitoyl

O
‖
$CH_3 – (CH_2)_{14} – C –$

Oleoyl

O
‖
$CH_3 – (CH_2)_7 – CH = CH – (CH_2)_7 – C –$

Stearoyl

O
‖
$CH_3 – (CH_2)_{16} – C –$

Alkyl group: a straight or branched-chain aliphatic group (without benzene ring). Straight chains with 1 to 4 carbons are methane, ethane, propane, and butane, respectively.

Aryl group: a group containing an aromatic ring.

Ester
O |
‖ |
– C – O – C –
|

or

O |
‖ |
– O – P – O – C –
| |
O^-

; produced by splitting H_2O from an acid and an alcohol group.

Carboxylic ester
O
‖
$R_1 – C – OH + HOR_2 \rightarrow$
O
‖
$R_1 – C – OR_2 + H_2O$

Phosphate ester
O
‖
$HO – P – OH + HOR \rightarrow$
OH
O
‖
$HO – P – OR + H_2O$
OH

Thioester
O
‖
– C – S – ; produced from an acid and a thiol (sulfhydryl) group.
O
‖
$R_1 – COH + HS – R_2 \rightarrow$
O
‖
$R_1 – C – S – R_2 + H_2O$

Amide
O |
‖ |
– C – N – ; produced by splitting H_2O from an acid and an amine group.
O
‖
$R_1 – C – OH + HN – \rightarrow$
|
O
‖
$R_1 – C – N – + H_2O$
|

Acid anhydride
O O
‖ ‖
– C – O – C ; produced by splitting H_2O from two carboxylic acid groups or from other hydroxy acids, such as phosphoric acid.
O O O
‖ ‖ ‖
$R_1 – C – OH + HO – P – OH \rightarrow R_1 – C – O – P – OH + H_2O$
| |
O^- O^-

Figure 2-1. A brief review of organic chemistry. (Adapted with permission from Robert H. Hamilton.)

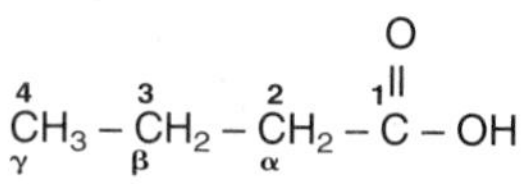

$$\overset{4}{\underset{\gamma}{CH_3}} - \overset{3}{\underset{\beta}{CH_2}} - \overset{2}{\underset{\alpha}{CH_2}} - \overset{1}{\overset{\overset{O}{\|}}{C}} - OH$$

Figure 2-2. Identification of carbon atoms in an organic compound. Carbons may be numbered starting with the most oxidized carbon-containing group, or they may be assigned Greek letters, with the carbon next to the most oxidized group designated as the α-carbon.

    b. In **oxidation** reactions, electrons are lost. In **reduction** reactions, electrons are gained.

    c. As foods are oxidized, electrons are released and passed through the electron transport chain. Adenosine triphosphate (**ATP**) is generated and supplies the energy to drive various functions of the body.

II. Acids, Bases, and Buffers

- Many biochemical compounds, ranging from small molecules to large polymers, are capable of releasing or accepting protons at physiologic pH and, as a consequence, may carry a charge.
- Most biochemical reactions occur in aqueous solutions.
- The pH of a solution is the negative $\log_{10}$ of its hydrogen ion concentration, H^+.
- Acids are proton donors, and bases are proton acceptors.
- The relationship among pH, pK (the negative log of the dissociation constant), and the concentrations of an acid and its conjugate base is described by the Henderson-Hasselbalch equation.
- Buffers consist of solutions of acid–base conjugate pairs that resist changes in pH when H^+ or OH^- are added.
- Acids that are ingested or produced by the body are buffered by bicarbonate and by proteins, particularly hemoglobin. These buffers help to maintain the pH in the body within the range compatible with life.

A. Water

    1. Water is the **solvent of life**. It dissociates

$$H_2O \rightleftharpoons H^+ + OH^-$$

with an equilibrium constant

$$K = \frac{[H^+][OH^-]}{[H_2O]}$$

    2. Because the extent of dissociation is not appreciable, H_2O remains constant at 55.5 M, and the ion product of H_2O is

$$K_w = [H^+][OH^-] = 1 \times 10^{-14}$$

    3. The **pH** of a solution is the negative $\log_{10}$ of its hydrogen ion concentration

$$pH = -\log_{10}[H^+]$$

  –For pure water,

$$[H^+] = [OH^-] = 1 \times 10^{-7}$$

Therefore, the **pH of pure water is 7**.

B. Acids and bases

—Acids are compounds that donate protons, and bases are compounds that accept protons.

1. Acids dissociate.

a. Strong acids, such as hydrochloric acid (HCl), dissociate completely.

b. Weak acids, such as acetic acid, dissociate only to a limited extent

$$HA \rightleftharpoons H^+ + A^-$$

where HA is the acid and A^- its conjugate base.

c. The **dissociation constant** for a weak acid is

$$K = \frac{[H^+][A^-]}{[HA]}$$

2. The **Henderson-Hasselbalch equation** is derived from the equation for the dissociation constant

$$pH = pK + \log_{10} \frac{[A^-]}{[HA]}$$

where pK is the negative $\log_{10}$ of K, the dissociation constant.

3. The **major acids** produced by the body include **phosphoric acid, sulfuric acid, lactic acid,** and the ketone bodies, **acetoacetic acid and β-hydroxybutyric acid**. CO_2 is also produced, which combines with H_2O to form **carbonic acid** in a reaction catalyzed by carbonic anhydrase

$$CO_2 + H_2O \rightleftharpoons H_2CO_3 \rightleftharpoons H^+ + HCO_3^-$$
$$\text{carbonic}$$
$$\text{anhydrase}$$

C. Buffers

1. Buffers consist of **solutions of acid–base conjugate pairs,** such as acetic acid and acetate.

a. Near its pK, a buffer maintains the pH of a solution, resisting changes due to addition of acids or bases (Figure 2-3). For a weak acid, the pK is often designated pK_a.

b. At the pK, $[A^-]$ and [HA] are equal, and the buffer has its maximal capacity.

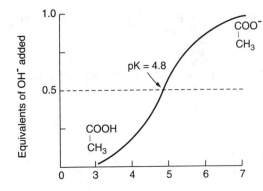

Figure 2-3. The titration curve of acetic acid. The molecular species that predominate at low and high pH are shown. At low pH (high $[H^+]$), the molecule is protonated and has zero charge. As alkali is added, the $[H^+]$ decreases ($H^+ + OH^- \rightarrow H_2O$), acetic acid dissociates, and the carboxyl group becomes negatively charged.

2. Buffering mechanisms in the body

–The **normal pH range** of arterial blood is **7.37 to 7.43**.

a. The major buffers of blood are **bicarbonate** (HCO_3^-/H_2CO_3) and **hemoglobin** (Hb/HHb).

b. These buffers act in conjunction with mechanisms in the kidneys for excreting protons and mechanisms in the lungs for exhaling CO_2 to maintain the pH within this range.

III. Amino Acids and Peptide Bonds

- An amino acid usually contains a carboxyl group, an amino group, and a side chain, all bonded to the α-carbon atom.
- Amino acids are usually of the L-configuration.
- At physiologic pH, amino acids carry a positive charge on their amino groups and a negative charge on their carboxyl groups.
- The side chains of the amino acids contain different chemical groups. Some side chains carry a charge.
- Peptide bonds link adjacent amino acid residues in a protein chain.

A. Amino acids (Figure 2-4)

–There are 20 amino acids used for the synthesis of proteins by the mRNA-directed process that occurs on ribosomes.

–Other amino acids exist for which there is no genetic code; for example, in the urea cycle or in proteins where they are generated by posttranslational modifications.

1. Structures of amino acids (Figure 2-5)

a. Most amino acids contain a **carboxyl group,** an **amino group,** and a **side chain** (R group), all attached to the α-carbon.

(1) Glycine does not have a side chain. Its α-carbon contains two hydrogens.

(2) The nitrogen of **proline** is part of a ring and, therefore, forms an imino group.

b. All of the 20 amino acids except glycine are of the L-**configuration** (see Figure 2-4). Glycine does not contain an asymmetric carbon atom and so is not optically active.

c. The **classification** of amino acids is based on their side chains.

(1) Hydrophobic amino acids have side chains that contain **aliphatic groups** (valine, leucine, and isoleucine) or **aromatic groups** (phenylalanine, tyrosine, and tryptophan) that may form hydrophobic interactions.

–**Tyrosine** has a phenolic group that carries a negative charge above its pK ($\approx$10), so it is not hydrophobic in this pH range.

(2) Hydroxyl groups found on serine and threonine can form hydrogen bonds.

(3) Sulfur is present in cysteine and methionine.

–The **sulfhydryl groups** of two cysteines may form a **disulfide,** producing cystine

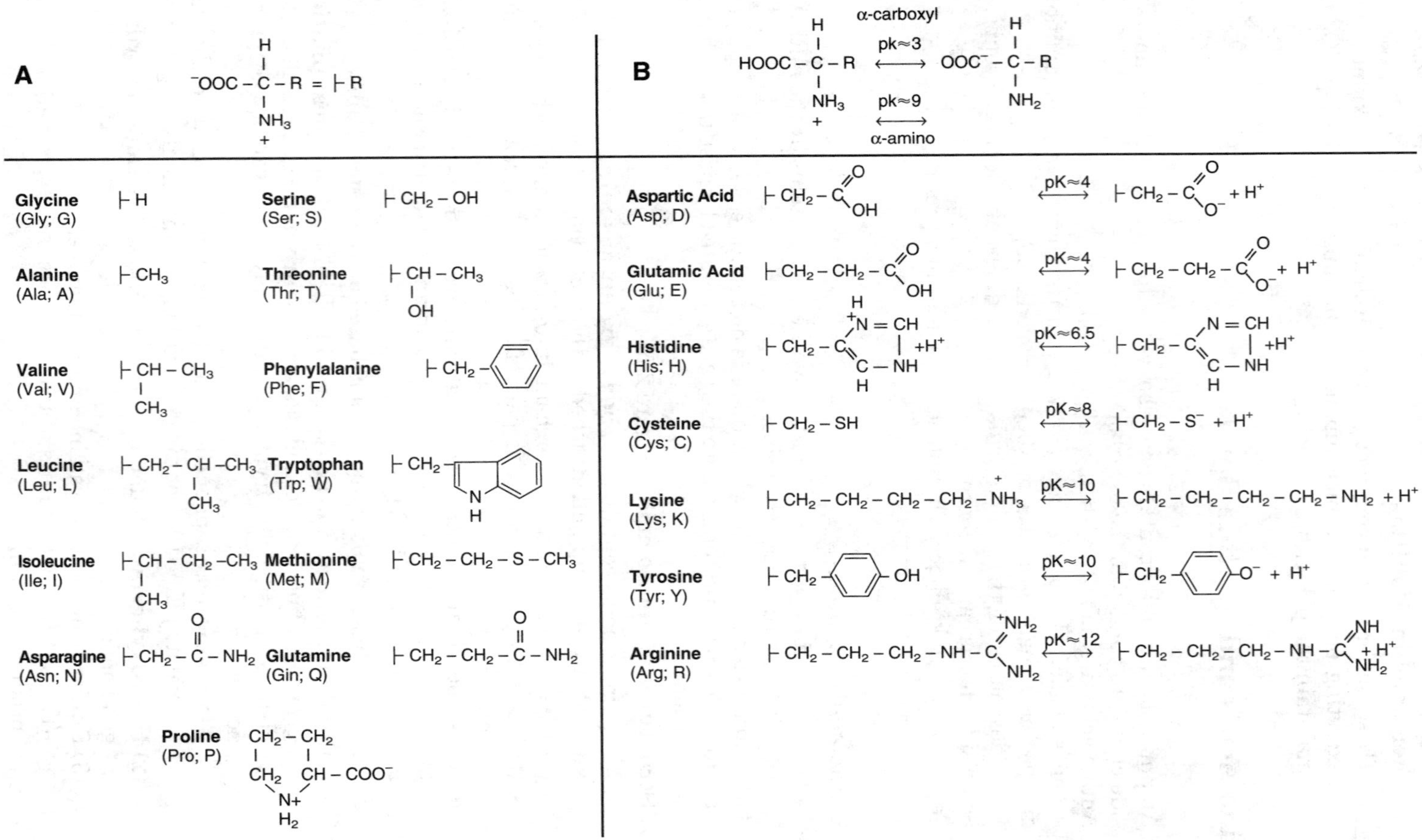

Figure 2-4. Structures of the amino acids. Abbreviations are given for all amino acids. (*A*) shows amino acids that do not have ionizable side chains. (*B*) shows side chains that are ionizable. In *B*, for each amino acid, the species that predominates at a pH below the pK is shown on the *left*; the species that predominates at a pH above the pK is shown on the *right* (*B*). Note that the charge changes from 0 to − or from + to 0. At the pK, equal amounts of both species are present.

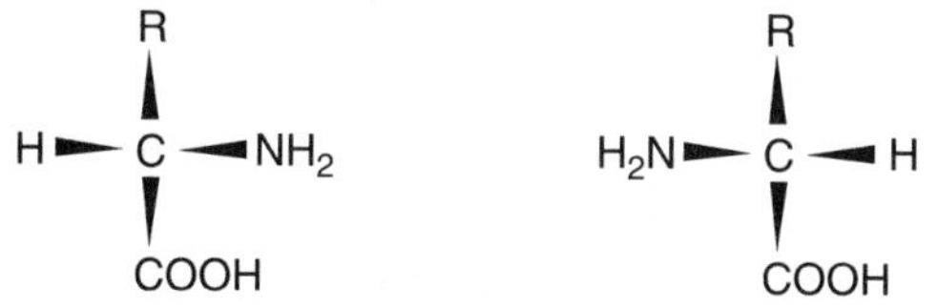

$$H \blacktriangleright \overset{\displaystyle R}{\underset{\displaystyle COOH}{C}} \blacktriangleleft NH_2 \qquad H_2N \blacktriangleright \overset{\displaystyle R}{\underset{\displaystyle COOH}{C}} \blacktriangleleft H$$

L-Amino acid   D-Amino acid

Figure 2-5. L- and D-amino acids. Groups at the broad parts of the arrows are closer to the viewer than groups at the narrow parts. R is the side chain. These forms are mirror images; they cannot be superimposed.

$$\underset{\text{Cysteine}}{\overset{\displaystyle COO^-}{\underset{\displaystyle NH_3^+}{HC}} - CH_2 - SH} \; + \; \underset{\text{Cysteine}}{HS - CH_2 - \overset{\displaystyle COO^-}{\underset{\displaystyle NH_3^+}{CH}}} \; \rightleftharpoons \; \underset{\text{Cystine}}{\overset{\displaystyle COO^-}{\underset{\displaystyle NH_3^+}{HC}} - CH_2 - S - S - CH_2 - \overset{\displaystyle COO^-}{\underset{\displaystyle NH_3^+}{CH}}}$$

    (4) Ionizable groups are present on the side chains of seven amino acids. They may carry a charge, depending on the pH. When charged, they may form **electrostatic interactions**.

    (5) Amides are present on the side chains of **asparagine and glutamine**.

    (6) The side chain of **proline forms a ring** with the nitrogen attached to the α-carbon.

  2. Charges on amino acids (see Figure 2-4*B*)

    a. Charges on α-amino and α-carboxyl groups

      –At physiologic pH, the **α-amino group** is protonated (pK ≈ 9) and carries a **positive charge,** and the **α-carboxyl group** is dissociated (pK ≈ 2) and carries a **negative charge**.

    b. Charges on side chains

      (1) Positive charges

        –The side chains of the basic amino acids, **arginine, lysine,** and **histidine,** are positively charged at pH 7.

        (a) The **guanidinium** group of arginine has a pK of about 12.

        (b) The **amino** group of lysine has a pK of about 10.

        (c) The **imidazole** group of histidine has a pK of about 6.5. Therefore, it titrates in the physiologic pH range.

      (2) Negative charges

        –The side chains of the acidic amino acids, **aspartate and glutamate,** are negatively charged at pH 7 (pK ≈ 4).

        –**Tyrosine** (pK ≈ 10) and **cysteine** (pK ≈ 8) also release protons from their side chains and may be negatively charged. However, dissociation does not usually occur in the physiologic pH range.

      (3) The **isoelectric point (pI)** is the pH at which the number of positive charges equals the number of negative charges.

  3. Titration of amino acids

    –Ionizable groups on amino acids carry protons at low pH (high $[H^+]$) that dissociate as the pH increases.

    a. For an amino acid that does not have an ionizable side chain, two pKs are observed during titration (Figure 2-6*A*).

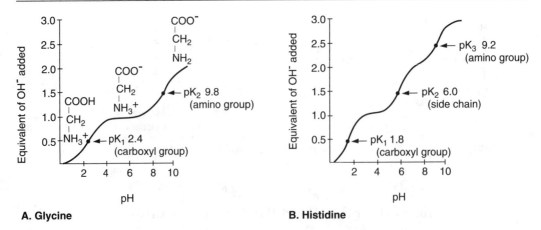

Figure 2-6. Titration curves for glycine (*A*) and histidine (*B*). The molecular species of glycine present at various pHs are indicated by the molecules above the curve. For histidine, pK_2 is the dissociation constant of the imidazole (side chain) group.

> **(1)** The first (**pK_1**) corresponds to the **α-carboxyl group** ($pK \approx 2$). As the proton dissociates, the carboxyl group goes from a zero to a minus charge.
>
> **(2)** The second (**pK_2**) corresponds to the **α-amino group** ($pK \approx 9$). As the proton dissociates, the amino group goes from a positive to a zero charge.

b. For an amino acid with an ionizable side chain, **three pKs** are observed during titration (Figure 2-6*B*).

> **(1)** The α-carboxyl and α-amino groups have pKs of about 2 and 9, respectively.
>
> **(2)** The **third pK varies** with the amino acid and depends on the pK of the side chain (see Figure 2-4*B*).

B. Peptide bonds

–**Amino acids** covalently join together to **form proteins** by means of peptide bonds, which are formed between the α-carboxyl group of one amino acid and the α-amino group of another (Figure 2-7).

1. Characteristics

a. The **atoms** involved in the peptide bond form a **rigid, planar unit**.

b. Because of its **partial double-bond** character, the peptide bond has **no freedom of rotation**.

c. However, the bonds involving the **α-carbon** can **rotate freely**.

2. Peptide bonds are extremely stable. Cleavage generally involves the action of proteolytic enzymes.

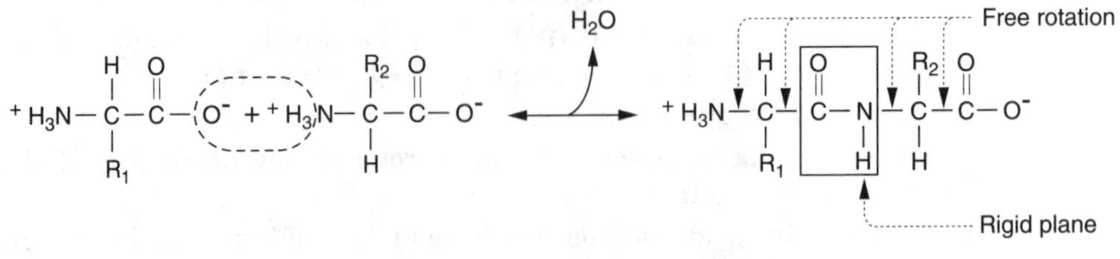

Figure 2-7. The peptide bond.

IV. Protein Structure

- The primary structure of a protein consists of the amino acid sequence along the chain.
- α-Helices, β-sheets, and other types of folding patterns provide secondary structure.
- Forces producing the three-dimensional conformation (tertiary structure) of a protein include electrostatic and hydrophobic interactions and hydrogen and disulfide bonds.
- Quaternary structure refers to the interaction of one or more subunits to form a protein.
- Proteins serve in many roles (e.g., as enzymes, hormones, receptors, antibodies, structural components, transporters of other compounds, and contractile elements in muscle).

A. General aspects of protein structure (Figure 2-8)

–The **linear sequence** of amino acid residues in a polypeptide chain determines the three-dimensional configuration of a protein, and the **structure** of a protein determines its function.

1. The **primary structure** is the sequence of amino acids along the polypeptide chain.

   a. By convention, the **sequence** is written from left to right, beginning with the **N-terminal** amino acid.

   b. Because there are no dissociable protons in peptide bonds, the **charges** on a polypeptide chain are due only to the N-terminal amino group, the C-terminal carboxyl group, and the side chains on amino acid residues.

   (1) The **pKs** of these groups in a polypeptide differ to some extent from those in the free amino acids, depending on the environment of the amino acid residues within the polymer (see Figure 2-4*B*).

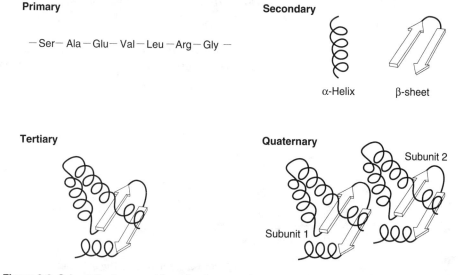

Figure 2-8. Schematic diagram of the primary, secondary, tertiary, and quaternary structure of a protein.

(2) A **protein** will **migrate** in an electric field, depending on the sum of its charges at a given pH (the net charge).

(a) **Positively charged proteins** are cations and migrate toward the cathode (−).

(b) **Negatively charged proteins** are anions and migrate toward the anode (+).

(c) At the **isoelectric pH** (pI), the net charge is zero, and the protein does not migrate.

2. **Secondary structure** includes various types of local conformations in which the atoms of the side chains are not involved.

a. In an **α-helix,** each carbonyl of a peptide bond forms a **hydrogen bond** with the −NH of a peptide bond four amino acid residues further along the chain (Figure 2-9).

(1) The formation of these hydrogen bonds causes a **helix** to be generated that contains 3.6 amino acid residues per turn.

(2) The side chains of the amino acid residues extend outward from the central axis of the **rod-like** structure.

(3) The **α-helix** is disrupted by proline residues, in which the ring imposes geometric constraints, and by regions in which numerous amino acid residues have charged groups or large, bulky side chains.

b. **β-Sheets** are formed by **hydrogen bonds** between two extended polypeptide chains or between two regions of a single chain (Figure 2-10).

(1) These **interactions** are between the **carbonyl** of one peptide bond and the **−NH** of another.

(2) The sheets are **parallel** if the chains run in the same direction; if the chains run in opposite directions, the sheets are **antiparallel**.

(3) **Globular proteins** usually contain regions of β-sheets that are formed in areas where the chain makes a β-turn and folds back on itself.

c. **Supersecondary structures**

(1) Certain **folding patterns** involving α-helices and β-sheets are frequently found and include the helix-turn-helix, the leucine zipper, and the zinc finger.

(2) Other types of **helices** or **loops** and **turns** can occur that differ from one type of protein to another and are sometimes called **random coils**.

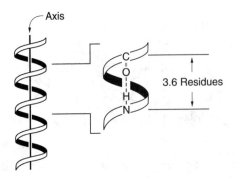

Axis

3.6 Residues

Figure 2-9. An α-helix. The enlarged segment shows hydrogen bonding.

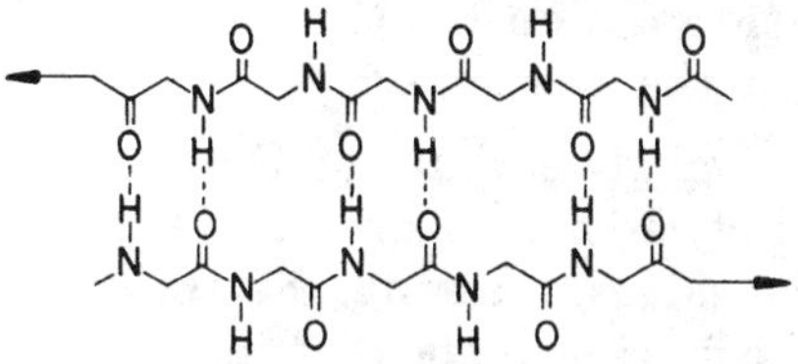

Figure 2-10. The structure of an antiparallel β-sheet. The orientation is indicated by *arrows*.

3. The **tertiary structure** of a protein refers to its overall **three-dimensional conformation**. It is produced by interactions between amino acid residues that may be located at a considerable distance from each other in the primary sequence of the polypeptide chain (Figure 2-11).

   a. **Hydrophobic amino acid residues** tend to collect in the interior of globular proteins, where they exclude water, while **hydrophilic residues** are usually found on the surface, where they interact with water.

   b. The types of **interactions** between amino acid residues that produce the three-dimensional shape of a protein include **hydrophobic** interactions, **electrostatic** interactions, and **hydrogen bonds,** all of which are noncovalent. Covalent disulfide bonds also occur.

4. **Quaternary structure** refers to the spatial arrangement of **subunits** in a protein that consists of more than one polypeptide chain (see Figure 2-8).

   –The subunits are joined together by the same types of **noncovalent interactions** that join various segments of a single chain to form its tertiary structure.

5. **Denaturation and renaturation**

   a. Proteins may be denatured by agents such as **heat** and **urea** that cause **unfolding** of polypeptide chains without causing hydrolysis of peptide bonds.

   b. If a denatured protein returns to its native state after the denaturing agent is removed, the process is called **renaturation**.

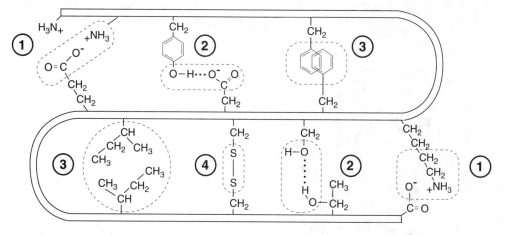

Figure 2-11. Interactions between amino acid residues in a polypeptide chain. ① Electrostatic interactions; ② hydrogen bonds; ③ hydrophobic interactions; ④ disulfide bonds.

6. Posttranslational modifications may occur after the protein has been synthesized on the ribosome.

–Phosphorylation, glycosylation, ADP-ribosylation, methylation, hydroxylation, and acetylation may affect the charge and the interactions between amino acid residues, **altering the three-dimensional configuration** and, thus, the function of the protein.

B. Hemoglobin (Figure 2-12)

1. Structure of hemoglobin

–Adult hemoglobin (HbA) consists of **four polypeptide chains** (two α and two β chains), each containing a molecule of **heme**.

a. The **α and β chains** of HbA are similar in three-dimensional configuration to each other and to the single chain of muscle myoglobin, although their amino acid sequences differ.

b. There are **eight regions of α-helix** in each chain.

c. The **nonpolar amino acids** are in the **interior** and the **charged amino acids** are on the **surface**.

d. **Heme** fits into a crevice in each globin chain and interacts with two histidine residues.

2. Function of hemoglobin

a. The **oxygen saturation curve** for hemoglobin is **sigmoidal** (Figure 2-13).

(1) Each heme binds one O_2 molecule, for a total of four O_2 molecules per HbA molecule. HbA changes from the taut or tense **(T) form** to the relaxed **(R) form** when oxygen binds.

(2) Binding of O_2 to one heme group in hemoglobin increases the affinity for O_2 of its other heme groups. This effect produces the sigmoidal oxygen saturation curve.

b. The binding of **protons** to HbA stimulates the release of O_2, a manifestation of the **Bohr effect** (see Figure 2-13).

(1) Thus, O_2 is readily released in the tissues where $[H^+]$ is high due to the production of CO_2 by metabolic processes.

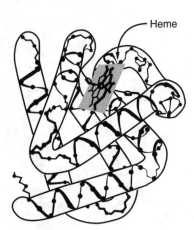

— Heme

Figure 2-12. The structure of the β chain of hemoglobin. Cylindrical regions contain α-helices. Heme fits into a space as indicated by the *arrow*.

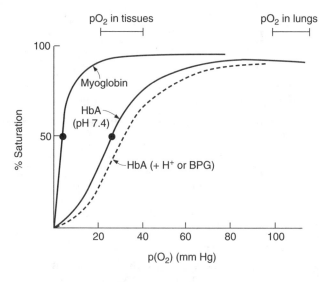

Figure 2-13. Oxygen saturation curves for myoglobin and adult hemoglobin (HbA). Myoglobin has a hyperbolic saturation curve. HbA has a sigmoidal curve. The HbA curve shifts to the right at lower pH, with higher concentrations of 2,3-bisphosphoglycerate (BPG), or as CO_2 binds in the tissues. Thus, O_2 is released more readily. P_{50} (●) is the partial pressure of O_2 at which HbA is half-saturated with O_2.

$$CO_2 + H_2O \rightleftharpoons H_2CO_3 \rightleftharpoons H^+ + HCO_3^-$$

Tissues →   ← Lung

$$H^+ + HbAO_2 \rightleftharpoons HHbA + O_2$$

(**2**) These reactions are reversed in the lung. O_2 binds to HbA, and CO_2 is exhaled.

   c. Covalent binding of CO_2 to HbA in the tissues also causes the **release of O_2**.

   d. Binding of **2,3-bisphosphoglycerate, or BPG** (formerly known as 2,3-diphosphoglycerate [DPG]), a side product of glycolysis in red blood cells, decreases the affinity of HbA for O_2. Consequently, O_2 is more readily released in tissues when BPG is bound to HbA.

     –**Fetal hemoglobin (HbF),** composed of two α subunits and two γ subunits, has a lower affinity for BPG than does HbA and, therefore, has a higher affinity for O_2.

C. Collagen refers to a group of similar structural proteins that are found, for example, in the **extracellular matrix,** the **vitreous humor** of the eye, and in **bone and cartilage**.

  1. Structure of collagen

   a. Collagen consists of **three chains** that wind around each other to form a **triple helix** (Figure 2-14).

   b. Collagen contains approximately 1000 amino acids, one-third of which are **glycine**. The sequence Gly-X-Y frequently occurs, in which X is often **proline** and Y is **hydroxyproline or hydroxylysine**.

Figure 2-14. The triple helix of collagen. Three polypeptide chains wrap around each other, as indicated by the *shading*. Cross links hold the chains together.

2. Synthesis of collagen

a. The polypeptide **chains of preprocollagen are synthesized** on the rough endoplasmic reticulum, and the signal (pre) sequence is cleaved.

b. Proline and lysine residues are hydroxylated by a reaction that requires O_2 and vitamin C.

c. Galactose and glucose are added to hydroxylysine residues.

d. The **triple helix forms;** procollagen is secreted from the cell and is cleaved to form collagen.

e. Cross links are produced. The side chains of lysine and hydroxylysine residues are oxidized to form aldehydes, which can undergo aldol condensation or form **Schiff bases** with the amino groups of lysine residues.

D. Insulin

1. Structure of insulin (Figure 2-15)

–Insulin is a polypeptide hormone that is produced by the **β cells** of the **pancreas**. It has 51 amino acids in two **polypeptide chains,** which are linked by two **disulfide bridges**.

2. Synthesis of insulin

a. Preproinsulin is synthesized on the rough endoplasmic reticulum and the pre-, or signal, sequence is removed to form proinsulin.

b. In secretory granules, **proinsulin** is cleaved, and the **C-peptide** is released. The remainder of the molecule forms the active hormone.

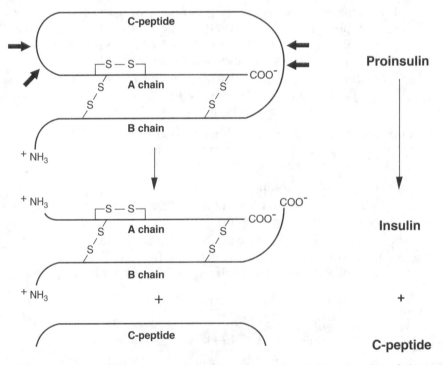

Figure 2-15. The cleavage of proinsulin to form insulin. *Heavy arrows* indicate the sites of cleavage, which releases the C-peptide.

V. Enzymes

- A major role of proteins is to serve as enzymes, catalysts of biochemical reactions.
- The active sites of enzymes are the regions where substrates bind, are converted to products, and are released.
- The rate (v) of many enzyme-catalyzed reactions can be described by the Michaelis-Menten equation. For enzymes that exhibit Michaelis-Menten kinetics, plots of velocity versus substrate concentration are hyperbolic.
- The Michaelis-Menten equation may be rearranged to give the Lineweaver-Burk equation.
- Competitive inhibitors compete with the substrate for binding at the active site of the enzyme.
- Noncompetitive inhibitors bind to the enzyme or the enzyme–substrate complex at a site different from the active site.
- Allosteric enzymes bind activators or inhibitors at sites other than the active site. Plots of the velocity versus substrate concentration for allosteric enzymes produce curves that are sigmoidal.

A. General properties of enzymes

1. The reactions of the cell would not occur rapidly enough to sustain life if enzyme catalysts were not present.

2. **Substrates** bind at the active sites of enzymes, where they are converted to products and released.

3. Enzymes are usually **highly specific** for their substrates and products.

   a. Many enzymes recognize only a single compound as a substrate.

   b. Some enzymes, such as those involved in digestion, are less specific.

4. Many enzymes require **cofactors** that frequently are **metal ions** or derivatives of **vitamins**.

5. Enzymes decrease the energy of activation for a reaction. They do not affect the equilibrium concentrations of the substrates and products.

B. Dependence of velocity on [E], [S], temperature, and pH

1. The **velocity** of a reaction, **v, increases with the enzyme concentration, [E],** if the substrate concentration, [S], is constant.

2. If the enzyme concentration is constant, the velocity increases with the substrate concentration until the maximum velocity, V_m, is attained.

   –The maximum velocity is attained when all the active sites of the enzyme are **saturated** with substrate.

3. The **velocity** of a reaction **increases with temperature** until a maximum is reached, after which the velocity decreases due to denaturation of the enzyme.

4. Each enzyme-catalyzed reaction has an **optimal pH** at which appropriate charges are present on both the enzyme and the substrate, and the velocity is at a maximum.

–Changes in the pH may alter these charges so that the reaction proceeds at a slower rate. If the pH is too high or too low, the enzyme may also undergo **denaturation.**

C. The Michaelis-Menten equation

1. Michaelis and Menten proposed that, during a reaction, an **enzyme–substrate complex** is formed that may **dissociate** (to reform the free enzyme and the substrate) or **react** (to release the product and regenerate the free enzyme)

$$E + S \underset{k_2}{\overset{k_1}{\rightleftharpoons}} ES \overset{k_3}{\rightarrow} E + P$$

where E is the enzyme, S the substrate, ES the enzyme–substrate complex, P the product, and k_1, k_2, and k_3 are rate constants.

2. From this concept, the **Michaelis-Menten equation** was derived

$$v = \frac{V_m[S]}{K_m + [S]}$$

where $K_m = (k_2 + k_3)/k_1$ and **V_m is the maximum velocity**.

3. The rate of formation of products (the velocity of the reaction) is related to the **concentration of the enzyme–substrate complex**

$$v = k_3[ES]$$

V_m is reached when all of the enzyme is in the enzyme–substrate complex.

4. **K_m** is the substrate concentration at which **$v = \frac{1}{2} V_m$**.

–When [S] = K_m, substitution of K_m for [S] in the Michaelis-Menten equation yields $v = \frac{1}{2} V_m$.

5. When the velocity is plotted versus [S], a **hyperbolic curve** is produced (Figure 2-16A).

D. The Lineweaver-Burk equation (see Figure 2-16B)

1. Because of the difficulty in determining V_m from a hyperbolic curve, the Michaelis-Menten equation was transformed by Lineweaver and Burk into an equation for a straight line

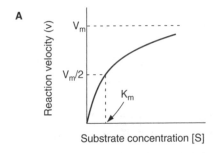
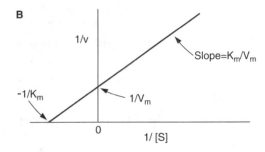

Figure 2-16. The velocity of an enzyme-catalyzed reaction. (*A*) Velocity (v) versus substrate concentration ([S]). (*B*) Lineweaver-Burk plot. Note the points on each plot from which V_m and K_m can be determined. V_m = maximum velocity, and K_m = substrate concentration at $\frac{1}{2} V_m$.

$$\frac{1}{v} = \frac{K_m}{V_m}\frac{1}{[S]} + \frac{1}{V_m}$$

2. The intercept on the $1/v$ axis equals $1/V_m$.

3. The intercept on the $1/[S]$ axis equals $-1/K_m$.

4. The slope of the line equals K_m/V_m.

E. Inhibitors of enzymes decrease the rate of enzymatic reactions.

   1. Competitive inhibitors compete with the substrate for the active site of the enzyme (Figure 2-17).

   a. An enzyme–inhibitor complex, EI, is formed.

   b. Competitive inhibition may be reversed by increasing the substrate concentration.

   c. V_m remains the same, but the **apparent K_m is increased**.

   d. For Lineweaver-Burk plots, **lines** for the inhibited reaction **intersect** on the **Y-axis** with those for the uninhibited reaction.

   2. Noncompetitive inhibitors do not compete with the substrate but **bind** to the enzyme or the enzyme–substrate complex at a site different from the active site, decreasing the activity of the enzyme (see Figure 2-17).

   a. V_m is decreased and, in the simplest case, **K_m is not affected**.

   b. For Lineweaver-Burk plots, **lines** for the inhibited reaction may **intersect** on the **X-axis** with those for the uninhibited reaction.

   3. Irreversible inhibitors bind tightly to the enzyme and **inactivate** it.

F. Allosteric enzymes

   1. Allosteric enzymes **bind activators or inhibitors** at sites other than the active site (Figure 2-18).

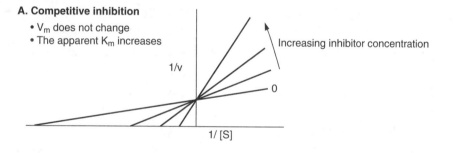

A. Competitive inhibition
 • V_m does not change
 • The apparent K_m increases

1/v

Increasing inhibitor concentration

0

1/ [S]

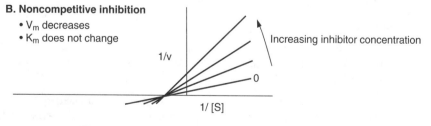

B. Noncompetitive inhibition
 • V_m decreases
 • K_m does not change

1/v

Increasing inhibitor concentration

0

1/ [S]

Figure 2-17. Effect of inhibitors on Lineweaver-Burk plots.

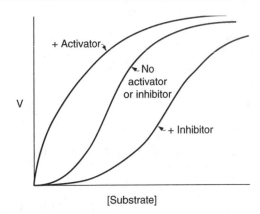

[Substrate]

Figure 2-18. Effect of activators and inhibitors on an allosteric enzyme.

2. **Sigmoidal curves** are generated by plots of the velocity versus the substrate concentration.

   a. An allosteric enzyme has two or more subunits, each with **substrate binding sites** that exhibit cooperativity.

   (1) Binding of a substrate molecule at one site facilitates binding of other substrate molecules at other sites. The enzyme shifts from a T form to an R form as substrate is bound.

   (2) Allosteric activators shift the enzyme toward the R form, which binds substrate more readily.

   (3) Allosteric inhibitors cause a shift toward the T form, which binds substrate less readily.

   b. Similar effects occur during O_2 binding to **hemoglobin**.

G. Regulation of enzyme activity by covalent modification

   1. Regulation of enzyme activity may occur by the **covalent addition** to an enzyme of chemical groups that increase or decrease its activity.

   2. Many enzymes may be activated or inhibited by **phosphorylation**.

   –**Pyruvate dehydrogenase** and **glycogen synthetase** are inhibited by phosphorylation, while glycogen phosphorylase is activated.

   3. Phosphatases that remove the phosphate groups alter the activities of the enzymes.

H. Regulation by protein–protein interactions

   –Proteins may bind to enzymes, altering their activity.

   1. Protein kinase A has subunits that inhibit its activity.

   2. When these subunits bind cyclic AMP (cAMP) and are released from the enzyme, the catalytic subunits become active.

I. Isoenzymes

   1. Isoenzymes (or isozymes) are enzymes that catalyze the same reaction but differ in their amino acid sequence and, therefore, in many of their properties.

   2. Tissues contain **characteristic isozymes** or mixtures of isozymes. Enzymes such as lactate dehydrogenase and creatine kinase differ from one tissue to another.

    a. Lactate dehydrogenase contains four subunits. Each subunit may be either of two types: heart (H) or muscle (M). Five isozymes exist: HHHH, HHHM, HHMM, HMMM, and MMMM.

    b. Creatine kinase, or creatine phosphokinase (**CPK**), contains two subunits. Each subunit may be either of two types: muscle (M) or brain (B). Three isozymes exist: MM, MB, and BB. The MB fraction is most prevalent in heart muscle.

VI. Clinical Correlations

A. Acid–base disturbances

Hypoventilation causes retention of CO_2 by the lungs, which can lead to a **respiratory acidosis**. Hyperventilation can cause a **respiratory alkalosis**. **Metabolic acidosis** can result from accumulation of metabolic acids (lactic acid or the ketone bodies β-hydroxybutyric acid and acetoacetic acid), or ingestion of acids or compounds that are metabolized to acids (methanol, ethylene glycol). **Metabolic alkalosis** is due to increased HCO_3^-, which is accompanied by an increased pH. Acid–base disturbances lead to compensatory responses that attempt to restore normal pH. For example, a metabolic acidosis causes hyperventilation and the release of CO_2, which tends to lower the pH. During metabolic acidosis, the kidneys excrete NH_4^+, which contains H^+ buffered by ammonia.

B. Hemoglobinopathies

There are many **mutations** that produce alterations in the structure of hemoglobin. One common mutation results in **sickle cell anemia,** in which the β chain of hemoglobin contains a valine rather than a glutamate at position 6. Thus, in the mutant hemoglobin (HbS), a hydrophobic amino acid replaces an amino acid with a negative charge. This change causes deoxygenated molecules of HbS to polymerize. Red blood cells that contain large complexes of HbS molecules assume a sickle shape. These cells undergo hemolysis and an anemia results. Painful **vasoocclusive crises** also occur, and end-organ damage may result.

C. Problems associated with connective tissue

Mutations in the synthesis and processing of collagen occur in **Ehlers-Danlos syndrome,** which is characterized by abnormalities of skin, ligaments, and internal organs. The skin is fragile and often stretches easily. Joint laxity occurs. Abnormalities in type I collagen genes have been found in **osteogenesis imperfecta**. In this condition, bones are fragile and are readily fractured. In **scurvy,** which is due to vitamin C deficiency, there is **decreased hydroxyproline synthesis,** which results in an unstable form of collagen. Bones, teeth, blood vessels, and other structures rich in collagen develop abnormally. **Bleeding gums** and **poor wound healing** are often observed.

D. Diabetes mellitus

Diabetes mellitus, which is due to a **deficiency of insulin** or the resistance of tissues to its action, results in **hyperglycemia**. Autoimmunity may play a role in the etiology of insulin-dependent diabetes mellitus. In this condition, the plasma contains antibodies to islet cells of the pancreas, including the β cells that produce insulin.

E. Deficient or defective enzymes

Thousands of diseases related to deficient or defective enzymes occur, many of which are **rare**. For example, in **phenylketonuria** (which has an incidence of 1 in 10,000 births in whites and Asians), the enzyme phenylalanine hydroxylase, which converts **phenylalanine** to tyrosine, is deficient. Phenylalanine accumulates, and tyrosine becomes an essential amino acid that is required in the diet. Mental retardation is a result of metabolic derangement. A more common problem is **lactase deficiency,** which occurs in 69% to 90% of American Indians, blacks, and Asians, and in 10% of whites. Lactose is not digested normally and accumulates in the gut where it is metabolized by bacteria. **Bloating, abdominal cramps, and watery diarrhea** result.

F. Therapeutic uses of enzyme kinetics

Drugs are frequently used therapeutically to **inhibit enzymes;** for example, 5-fluorouracil (5-FU) is used to inhibit the enzyme thymidylate synthetase. This enzyme converts dUMP to dTMP, which ultimately provides the thymine for DNA synthesis. **5-FU** is used as a **chemotherapeutic agent** to inhibit the proliferation of cancer cells.

G. Monitoring tissue damage by measuring enzyme levels in the blood

Enzymes, which are normally produced in cells, are released into the blood when cells are injured. For example, after a **heart attack,** there is an increase in blood levels of **creatine phosphokinase (CPK)** and other enzymes such as lactate dehydrogenase (LDH). The extent of damage and the rate of recovery can be estimated by periodically measuring the levels of these enzymes. Measurement of the **MB isozyme of CPK** is also used as an aid in diagnosis.

Review Test

Directions: Each of the numbered items or incomplete statements in this section is followed by answers or by completions of the statement. Select the **one** lettered answer or completion that is **best** in each case.

1. The most oxidized group in the compound shown below is

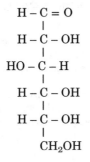

(A) an acid
(B) an aldehyde
(C) a ketone
(D) an alcohol

Questions 2 and 3

Use the following structure for questions 2 and 3.

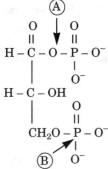

2. The bond labeled A in the compound shown is

(A) an anhydride
(B) an ether
(C) an ester
(D) a phosphodiester

3. The bond labeled B in the compound shown is

(A) an anhydride
(B) an ether
(C) an ester
(D) a phosphodiester

4. The conversion of β-hydroxybutyrate to acetoacetate

$$\underset{OH}{CH_3-\underset{|}{CH}-CH_2-COO^-} \rightleftharpoons \underset{O}{CH_3-\overset{||}{C}-CH_2-COO^-}$$

occurs by what type of reaction?

(A) Oxidation
(B) Reduction
(C) Dehydration
(D) Dehydroxylation

5. Physiologic pH is 7.4. What is the $[H^+]$ of a solution at physiologic pH?

(A) -7.4
(B) 0.6
(C) 0.6×10^{-8}
(D) 1.0×10^{-8}
(E) 4.0×10^{-8}

$-\log [H^+] = 7.4$

$10^{-7.4}$ $^{-7.4}$

Questions 6 and 7

6. The ionization of a weak acid, expressed as

$$HA \rightleftharpoons H^+ + A^-$$

has an apparent equilibrium constant (K) equal to

(A) $\log [A^-]/[HA]$
(B) $[HA]/[A^-]$
(C) $[H^+][A^-]/[HA]$
(D) $[A^-]/[HA]$
(E) $[HA]/[H^+][A^-]$

7. When the pH for a solution of this acid is equal to the pK, the ratio of the concentrations of the salt and the acid ($[A^-]/[HA]$) is

(A) 0
(B) 1
(C) 2
(D) 3
(E) 4

39

8. Proteins are effective buffers because they contain

(A) a large number of amino acids
(B) amino acid residues with different pKs
(C) N-terminal and C-terminal residues that can donate and accept protons
(D) peptide bonds that readily hydrolyze, consuming hydrogen and hydroxyl ions
(E) a large number of hydrogen bonds in α-helices

Questions 9 and 10

Use the structure to answer questions 9 and 10.

Asp-Ala-Ser-Glu-Val-Arg

9. The C-terminal amino acid of the hexapeptide shown is

(A) alanine
(B) asparagine
(C) aspartate
(D) arginine

10. At physiologic pH (7.4), this hexapeptide will contain a net charge of

(A) −2
(B) −1
(C) 0
(D) +1
(E) +2

11. Which one of the following types of bonds is covalent?

(A) Hydrophobic
(B) Hydrogen
(C) Disulfide
(D) Electrostatic

12. In sickle cell anemia, the hemoglobin molecule (HbS) is abnormal. If the β chains of normal hemoglobin (HbA) and HbS have the N-terminal sequences shown below and the chains otherwise are the same, which of the following statements is TRUE?

HbA Val-His-Leu-Thr-Pro-Glu-Glu-Lys-Ser-Ala-Val-Thr. . .

HbS Val-His-Leu-Thr-Pro-Val-Glu-Lys-Ser-Ala-Val-Thr. . .

(A) HbS contains one more hydrophobic amino acid than HbA
(B) HbS contains one more negative charge than HbA
(C) Neither HbS nor HbA contains an amino acid that has a side chain with a pK of 6
(D) The entire sequences shown for both HbA and HbS can form α-helices

13. Which one of the following conditions causes hemoglobin to release oxygen more readily?

(A) Metabolic alkalosis
(B) Increased production of 2,3-bisphosphoglycerate (BPG)
(C) Hyperventilation, leading to decreased levels of CO_2 in the blood
(D) Replacement of the β subunits with γ subunits

14. Production of which of the following proteins would be most directly affected in scurvy?

(A) Myoglobin
(B) Collagen
(C) Insulin
(D) Hemoglobin

15. The active site of an enzyme

(A) is formed only after addition of a specific substrate
(B) is directly involved in binding of allosteric inhibitors
(C) resides in a few adjacent amino acid residues in the primary sequence of the polypeptide chain
(D) binds competitive inhibitors

16. An enzyme catalyzing the reaction

$$E + A \rightleftharpoons EA \rightarrow E + P$$

was mixed with 4 mM substrate. The initial rate of product formation was 25% of V_m. The K_m for the enzyme is

$V = \frac{V_{max}}{4} = \frac{V_{max}[s]}{K_m+[s]}$

(A) 2 mM
(B) 4 mM
(C) 9 mM
(D) 12 mM
(E) 25 mM

$K_m+[s] = 4[s]$

17. The velocity (v) of an enzyme-catalyzed reaction

(A) decreases as the substrate concentration increases
(B) is lowest when the enzyme is saturated with substrate
(C) is related to the substrate concentration at ½ V_m
(D) is independent of the pH of the solution

Questions 18 and 19

Refer to the following reaction when answering questions 18 and 19.

$$\text{Fumarate} + H_2O \quad \underset{\text{fumarase}}{\rightleftharpoons} \quad \text{malate}$$

18. Fumarase catalyzes the conversion of fumarate to malate. It has a K_m of 5 μM for fumarate and a V_m of 50 μmol/min/ mg of protein when measured in the direction of malate formation. The concentration of fumarate required to give a velocity of 25 μmol/min/mg protein is

(A) 2 μM
(B) 5 μM
(C) 10 μM
(D) 20 μM
(E) 50 μM

$V = \frac{50[s]}{K_m+[s]}$ $B_m = \frac{50[s]}{}$

$25(5+s) = $
$125 + 25s = 50s$
$s = 5$

19. The K_m for fumarase is approximately 5 μM for fumarate. The fumarate concentration in mitochondria is approximately 2 mM. If the fumarate concentration dropped to 1 mM, the reaction rate would

(A) increase slightly
(B) decrease slightly
(C) decrease by one half
(D) stay exactly the same

20. Hexokinase and glucokinase both catalyze the phosphorylation of glucose to glucose 6-phosphate. The values of K_m for the enzymes are 10 μM and 0.02 M, respectively. If blood glucose is 5 mM under fasting conditions and 20 mM after a high-carbohydrate meal

(A) hexokinase will function near its V_m under fasting conditions
(B) glucokinase will function near its V_m under fasting conditions
(C) hexokinase will function at less than one-half V_m after a high-carbohydrate meal
(D) glucokinase will function at less than one-half V_m after a high-carbohydrate meal

21. A competitive inhibitor of an enzyme

(A) increases K_m but does not affect V_m
(B) decreases K_m but does not affect V_m
(C) increases V_m but does not affect K_m
(D) decreases V_m but does not affect K_m
(E) decreases both V_m and K_m

Questions 22–25

Refer to the graph when answering questions 22–25.

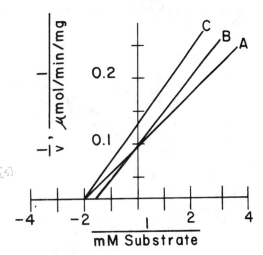

22. The value of K_m for the enzyme depicted by curve A is

(A) 0.5 mM
(B) 1 mM
(C) 2 mM
(D) 1 μmol/min/mg
(E) 10 μmol/min/mg

$-2 = \frac{-1}{K_m}$

0.5

23. The value of V_m for the enzyme depicted by curve A is

(A) 0.1 µmol/min/mg
(B) 1 µmol/min/mg
(C) 10 µmol/min/mg
(D) 0.5 mM
(E) 2 mM

24. Curve B depicts the effect of an inhibitor on the system described by curve A. This inhibitor

(A) is a competitive inhibitor
(B) is a noncompetitive inhibitor
(C) increases the V_m
(D) decreases the K_m

25. Curve C depicts the effect of a different inhibitor of the system described by curve A. This second inhibitor

(A) is a competitive inhibitor
(B) is a noncompetitive inhibitor
(C) increases the V_m
(D) decreases the K_m

Questions 26 and 27

Refer to the graph when answering questions 26 and 27.

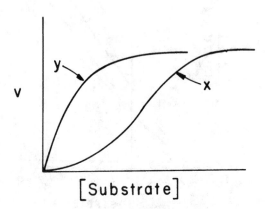

[Substrate]

The plot represents the relationship between substrate concentration and velocity for a single enzyme in the absence (curve x) and presence (curve y) of a compound that binds allosterically to the enzyme.

26. In the presence of the allosteric compound

(A) K_m and V_m both increase
(B) K_m and V_m both decrease
(C) K_m increases and V_m decreases
(D) K_m decreases and V_m increases
(E) K_m decreases and V_m stays the same

27. The allosteric compound is

(A) a competitive inhibitor
(B) a noncompetitive inhibitor
(C) an irreversible inhibitor
(D) an activator

28. Isocitrate dehydrogenase catalyzes the reaction

isocitrate + NAD^+ → α-ketoglutarate + CO_2 + NADH + H^+

The curves illustrated below are obtained when the initial velocity (v) of the reaction is plotted against isocitrate concentration in the presence of various levels of ADP and excess NAD^+. Which of the following statements about this system is correct?

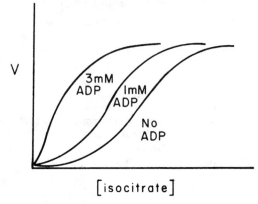

[isocitrate]

(A) Isocitrate dehydrogenase exhibits simple Michaelis-Menten kinetics in the absence of ADP
(B) ADP increases the K_m of the enzyme for isocitrate
(C) ADP increases the V_m of the enzyme
(D) ADP activates the enzyme

Directions: Each group of items in this section consists of lettered options followed by a set of numbered items. For each item, select the **one** lettered option that is most closely associated with it. Each lettered option may be selected once, more than once, or not at all.

Questions 29–32

Match each statement below with the compound it best describes.

(A) $^-OOC - CH_2 - CH_2 - CH - COO^-$
$\qquad\qquad\qquad\qquad\quad |$
$\qquad\qquad\qquad\qquad\; ^+NH_3$

(B) $^-OOC - CH - CH_2 - CH$ ⟨CH=CH / CH / CH-CH⟩
$\qquad\quad |$
$\qquad\; ^+NH_3$

(C) $^-OOC - CH - CH_2 - S - S - CH_2 - CH - COO^-$
$\qquad\quad |\qquad\qquad\qquad\qquad\qquad |$
$\qquad\; ^+NH_3\qquad\qquad\qquad\qquad\; ^+NH_3$

(D) $CH_3 - S - CH_2 - CH_2 - CH - COO^-$
$\qquad\qquad\qquad\qquad\qquad |$
$\qquad\qquad\qquad\qquad\; ^+NH_3$

29. Produced from two amino acids by an oxidation reaction *C*

30. Contains an aromatic side chain *B*

31. Contains a side chain that participates in electrostatic interactions *A*

32. Migrates toward the anode in an electric field

Questions 33–37

Match each characteristic below with the protein it best describes.

(A) Hemoglobin
(B) Myoglobin
(C) Collagen
(D) Insulin

33. Requires vitamin C for its synthesis

34. Has one oxygen binding site and one polypeptide chain

35. Contains four molecules of heme per molecule of protein

36. Is converted into a triple helix during its synthesis

37. Is composed of two polypeptide chains joined by disulfide bonds

Answers and Explanations

1–B. The most oxidized group in this molecule of glucose is an aldehyde (see carbon 1). The other functional groups are alcohols.

2–A. Bond A, an anhydride, is formed when a carboxylic acid and a phosphoric acid react, splitting out H_2O.

3–C. Bond B, a phosphate ester, is formed when phosphoric acid reacts with an alcohol, splitting out H_2O.

4–A. An alcohol is oxidized to a ketone when β-hydroxybutyrate is converted to acetoacetate. These compounds are ketone bodies.

5–E. If the pH is 7.4, the [H$^+$] is $10^{-7.4}$, or $10^{0.6} \times 10^{-8}$, or 4×10^{-8}.

6–C. The equilibrium constant equals the product of the concentrations of the products divided by the product of the concentrations of the reactants, or in this case, K = [H$^+$][A$^-$]/[HA].

7–B. The Henderson-Hasselbalch equation, pH = pK + $\log_{10}$ [A$^-$]/[HA], gives the relationship between these parameters. If pH = pK, $\log_{10}$ [A$^-$]/[HA] = 0, and [A$^-$]/[HA] = 1.

8–B. The side chains of the amino acid residues in proteins contain functional groups with different pKs. Therefore, they can donate and accept protons at various pH values and act as buffers over a broad pH spectrum. There is only one N-terminal amino group (pK≈9) and one C-terminal carboxyl group (pK≈3) per polypeptide chain. Peptide bonds are not readily hydrolyzed, and such hydrolysis would not provide buffering action. Hydrogen bonds have no buffering capacity.

9–D. By convention, peptides are drawn with the N-terminal amino acid on the left and the C-terminal amino acid on the right. Therefore, this peptide contains arginine at its C-terminus.

10–B. The N-terminal aspartate contains a positive charge on its N-terminal amino group and a negative charge on the carboxyl group of its side chain. Glutamate contains a negative charge on the carboxyl group of its side chain. The C-terminal arginine contains a negative charge on its C-terminal carboxyl group and a positive charge on its side chain. Thus, the overall charges are +2 and –3, which gives a net charge of –1.

11–C. Disulfide bonds are covalent.

12–A. The glutamate at position 6 in HbA is replaced by valine in HbS. Therefore, HbS has one more hydrophobic amino acid and one less negative charge than HbA. Both HbA and HbS contain histidine, which has a side chain with a pK of 6. Both sequences contain proline, which is cyclic and interrupts formation of an α-helix.

13–B. Increased [H$^+$], BPG, and CO_2 decrease the affinity of HbA for O_2. Fetal hemoglobin (HbF = $\alpha_2\gamma_2$) has a greater affinity for O_2 than HbA ($\alpha_2\beta_2$). Increased BPG would cause O_2 to be more readily released.

14–B. Scurvy is caused by a deficiency of vitamin C. The hydroxylation of proline and lysine residues in collagen requires vitamin C and oxygen. Globin synthesis might be indirectly affected because absorption of iron from the intestine is stimulated by vitamin C. Iron is involved in heme synthesis, which regulates globin synthesis.

15–D. The active site is formed when the enzyme folds into its three-dimensional configuration and may involve amino acid residues that are far apart in the primary sequence. Substrate molecules bind at the active site. Competitive inhibitors compete with the substrate. (Both bind at the active site.) Allosteric inhibitors bind at a site other than the active site.

16–D. In the Michaelis-Menten equation, v = (V$_m$ × [S])/(K$_m$ + [S]). In this case, $\frac{1}{4}$ V$_m$ = (V$_m$ × 4)/(K$_m$ + 4), or K$_m$ = 12 mM.

17–C. The velocity of an enzyme-catalyzed reaction increases as the substrate concentration increases. It is highest when the enzyme is saturated with substrate. Then, v equals V$_m$, the maximum velocity. The velocity depends on K$_m$. Enzymes have an optimal pH at which their activity is maximal.

18–B. A velocity of 25 is $\frac{1}{2}$ V$_m$, which is 50. K$_m$ = [S] at $\frac{1}{2}$ V$_m$. K$_m$ = 5 μM.

19–B. The velocity decreases slightly when the concentration of the substrate drops from 2 mM to 1 mM. At 2 mM, $v = (V_m \times 2,000 \; \mu M)/(5 \; \mu M + 2,000 \; \mu M) = 99.8\% \; V_m$. At 1 mM, $v = (V_m \times 1,000 \; \mu M)/(5 \; \mu M + 1,000 \; \mu M) = 99.5\% \; V_m$.

20–A. During fasting, for hexokinase, $v = (5 \times V_m)/(0.01 + 5) = 99.8\% \; V_m$; for glucokinase, $v = (5 \times V_m)/(20 + 5) = 20\% \; V_m$. In the fed state, for hexokinase, $v = (20 \times V_m)/(0.01 + 20) = 99.9\% \; V_m$; for glucokinase, $v = (20 \times V_m)/(20 + 20) = 50\% \; V_m$. Hexokinase will function near its V_m in both the fed and fasting states. Glucokinase (a liver enzyme) is more active in the fed than the fasting state. At 20 mM glucose, its velocity is 50% V_m.

21–A. A competitive inhibitor competes with the substrate for the active site of the enzyme, in effect increasing the K_m. As the substrate concentration is increased, the substrate, by competing with the inhibitor, can overcome its inhibitory effects, and eventually the normal V_m is reached.

22–A. The intercept on the x axis is $-1/K_m = -2$. Therefore, $K_m = 0.5$ mM.

23–C. The intercept on the y axis is $1/V_m = 0.1$. Therefore, $V_m = 10 \; \mu mol/min/mg$.

24–A. With this inhibitor, V_m is the same (the y intercept is the same), but K_m is larger (the x intercept is less negative). Therefore, this is a competitive inhibitor.

25–B. With this inhibitor, V_m is lower but K_m is the same. It is a noncompetitive inhibitor.

26–E. K_m, the substrate concentration at $\frac{1}{2} \; V_m$, is lower for curvey than for curve x. As the substrate concentration increases, both curves approach a plateau (V_m) that has the same velocity.

27–D. The allosteric compound increases the rate of the reaction at lower substrate concentrations, so it is an activator of the reaction.

28–D. Without ADP, the curve is sigmoidal, so Michaelis-Menten kinetics are not exhibited. V_m is the same at all ADP concentrations shown. The substrate concentration at $\frac{1}{2} \; V_m$ decreases as the ADP concentration increases; therefore, ADP decreases the K_m, activating the enzyme. (The velocity is higher at lower substrate concentrations in the presence of ADP.)

29–C. The sulfhydryl (-SH) groups of two cysteines react to form the disulfide bond (-S-S-) of cystine (compound C).

30–B. Phenylalanine (compound B) contains an aromatic phenyl ring.

31–A. The side chain of glutamate (compound A) contains a carboxyl group that carries a negative charge and participates in electrostatic interactions.

32–A. Glutamate is the only one of these amino acids that carries a net charge. It migrates toward the positive electrode (the anode).

33–C. Proline and lysine residues in collagen are hydroxylated in a reaction that requires vitamin C.

34–B. Each myoglobin molecule contains one polypeptide chain and one heme molecule that binds one O_2 molecule. Each hemoglobin molecule contains four polypeptide chains, four molecules of heme, and four molecules of oxygen.

35–A. Each molecule of hemoglobin contains four molecules of heme. Each myoglobin molecule contains one molecule of heme.

36–C. Collagen forms a triple helix during its synthesis.

37–D. Insulin is composed of an A chain and a B chain, which are linked by disulfide bonds.

3

Synthesis of Proteins

Overview

- Genetic information is encoded in DNA, which, in eukaryotes, is located mainly in nuclei with small amounts in mitochondria.
- Genetic information is inherited and expressed. Inheritance occurs by the process of replication. The strands of parental DNA serve as templates for synthesis of copies that are passed to daughter cells.
- Mutations that result from damage to DNA can lead to genetic alterations, including abnormal cell growth. Repair mechanisms can correct damaged DNA.
- Recombination of genes promotes genetic diversity.
- Expression of genes requires two steps, transcription and translation. DNA is transcribed to produce messenger RNA (mRNA), which is translated to produce proteins. Ribosomal RNA (rRNA) and transfer RNA (tRNA) participate in the process of translation.
- Proteins are involved in cell structure and function as enzymes, determining the reactions that occur in cells. Thus, proteins, the products of genes, determine what cells look like and how they behave.
- Gene expression is regulated. Only a small fraction of the genome is expressed in any one cell.

I. Nucleic Acid Structure

- The monomeric units of nucleic acids are nucleotides; each nucleotide contains a heterocyclic nitrogenous base, a sugar, and phosphate.
- DNA contains the bases adenine (A), guanine (G), cytosine (C), and thymine (T).
- RNA contains uracil (U) instead of thymine.
- Deoxyribose is present in DNA, whereas RNA contains ribose.
- Polynucleotides consist of nucleosides joined by 3',5'-phosphodiester bridges. The genetic message resides in the sequence of bases along the polynucleotide chain.
- In DNA, two antiparallel polynucleotide chains are joined by pairing between their bases and are twisted to form a double helix. Adenine pairs with thymine, and guanine pairs with cytosine. One chain runs in a 5' to 3' direction and the other runs 3' to 5'.

- DNA molecules interact with histones to form nucleosomes, and the strands are wound into more tightly coiled structures.
- RNA is single-stranded, but the strands loop back on themselves and the bases pair: guanine with cytosine and adenine with uracil.
- Messenger RNA (mRNA) has a cap at the 5' end and a poly(A) tail at the 3' end.
- Ribosomal RNA (rRNA) has extensive base-pairing.
- Transfer RNA (tRNA) forms a cloverleaf structure that contains many unusual nucleotides and an anticodon.

A. The structure of DNA

1. Chemical components of DNA

–Each polynucleotide chain of DNA contains nucleotides, which consist of a **nitrogenous base** (A, G, C, or T), **deoxyribose, and phosphate** (Figures 3-1, 3-2, and 3-3).

a. The bases are the **purines** adenine (A) and guanine (G), and the **pyrimidines** cytosine (C) and thymine (T).

b. Phosphodiester bonds join the 3'-carbon of one sugar to the 5'-carbon of the next (Figure 3-4).

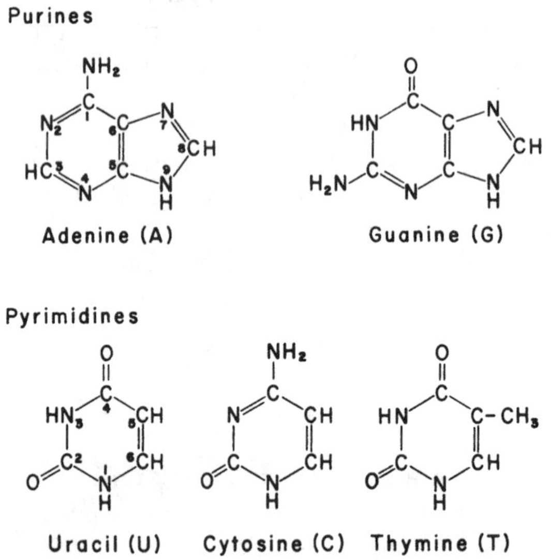

Figure 3-1. The nitrogenous bases of nucleic acids.

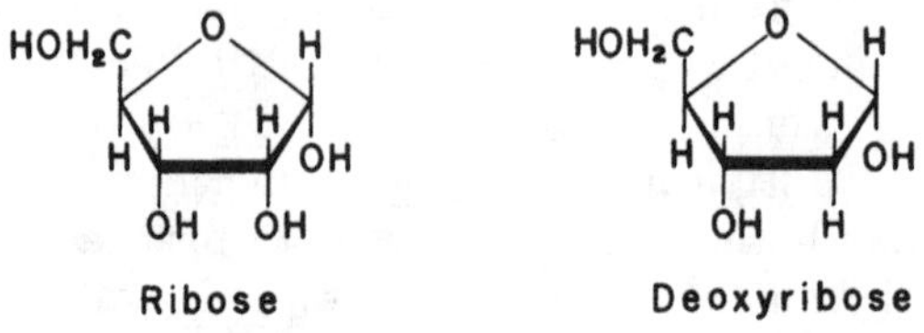

Figure 3-2. The sugars of the nucleic acids.

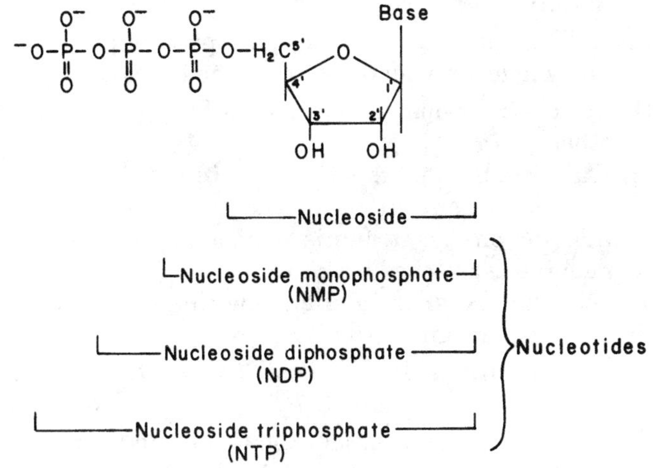

Figure 3-3. Nucleoside and nucleotide structures.

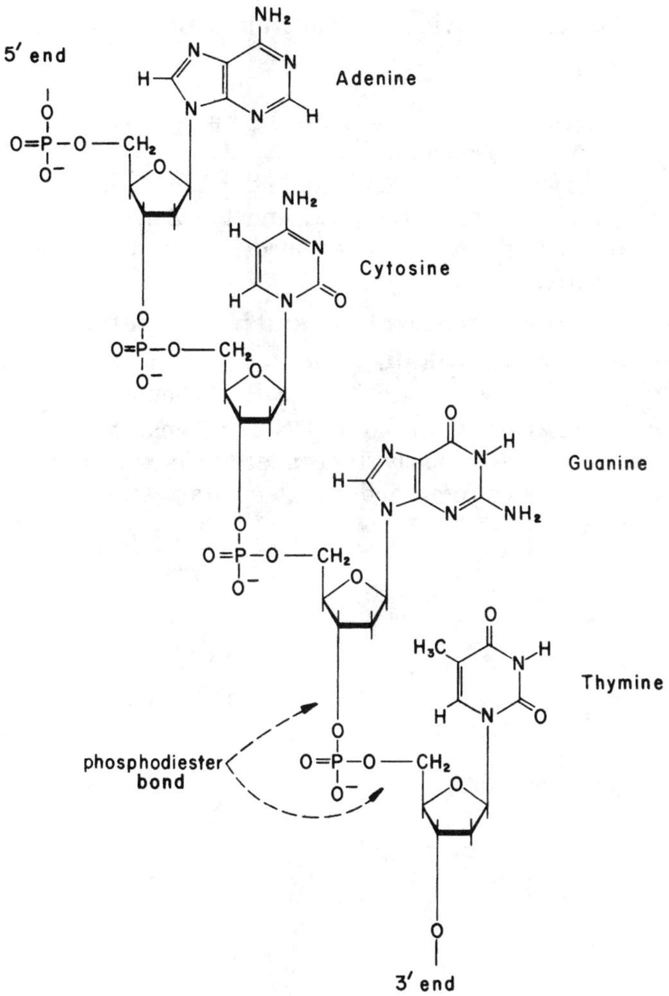

Figure 3-4. A polynucleotide.

2. DNA double helix

a. Each DNA molecule is composed of two **polynucleotide chains** joined by hydrogen bonds between the bases (Figure 3-5).

(1) **Adenine** on one chain forms a base pair with **thymine** on the other chain.

(2) **Guanine** base-pairs **with cytosine**.

(3) The **base sequences** of the two strands are complementary. Adenine on one strand is matched by thymine on the other, and guanine is matched by cytosine.

b. The **chains are antiparallel.** One chain runs in a **5' to 3'** direction; the other chain runs **3' to 5'** (Figure 3-6).

c. The double-stranded molecule is twisted to form a helix with major and minor grooves (Figure 3-7).

(1) The **base pairs** that join the two strands are **stacked** like a spiral staircase in the interior of the molecule.

(2) The **phosphate groups** are on the outside of the double helix. Two acidic groups of each phosphate are involved in phosphodiester bonds. The third is free and dissociates its proton at physiological pH, giving the molecule a **negative charge** (see Figure 3-4).

(3) The **B form of DNA,** first described by Watson and Crick, is right-handed and contains **10 base pairs per turn,** with 3.4Å between base pairs.

—Other forms of DNA include the A form, which is similar to the B form but more compact, and the Z form, which is left-handed and has its bases positioned more toward the periphery of the helix.

3. Denaturation, renaturation, and hybridization

a. Denaturation: Alkali, or heat, causes the strands of DNA to separate but does not break phosphodiester bonds.

b. Renaturation: If strands of DNA are separated by heat and then the **temperature** is slowly **decreased** under the appropriate conditions, **base pairs re-form** and complementary strands of DNA come back together.

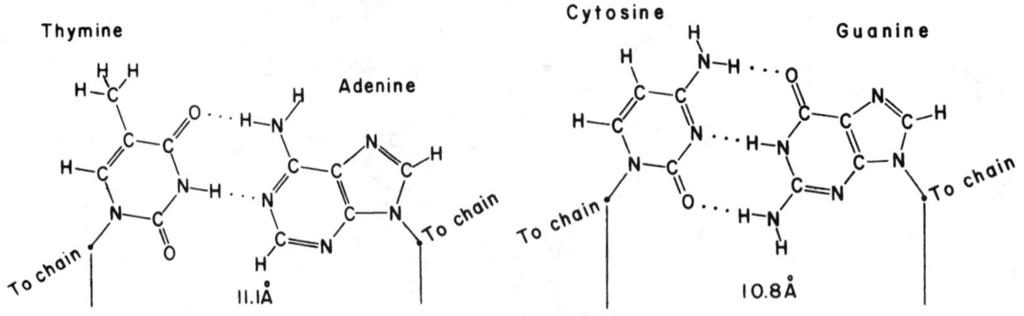

Figure 3-5. The base pairs of DNA.

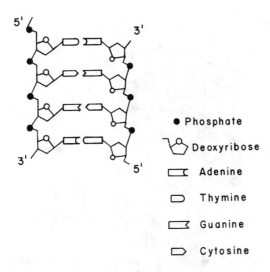

Figure 3-6. Antiparallel strands of DNA.

    c. Hybridization: A single strand of DNA pairs with complementary base sequences on another strand of DNA or RNA.

4. DNA molecules are extremely large.

    a. The entire chromosome of the bacterium *Escherichia coli* is circular and contains over 4×10^6 base pairs.

    b. The DNA molecule in the longest human chromosome is linear and is over 7.2 cm long.

5. Packing of DNA in the nucleus

    a. The **chromatin** of eukaryotic cells consists of DNA complexed with histones in **nucleosomes** (Figure 3-8).

        (1) Histones are relatively small, basic proteins with a high content of arginine and lysine. (Prokaryotes do not have histones.)

        (2) Two molecules each of histones H2A, H2B, H3, and H4 form a core around which approximately 140 base pairs of DNA are wound.

        (3) The DNA that joins one nucleosome core to the next is complexed with histone H1.

    b. The "beads on a string" nucleosomal structure of chromatin is further compacted to form solenoid structures (helical, tubular coils).

B. The structure of RNA

    1. RNA differs from DNA

        a. The polynucleotide structure of **RNA** is similar to DNA except that RNA contains the sugar **ribose** rather than deoxyribose and **uracil** rather than thymine. (A small amount of thymine is present in tRNA.)

        b. RNA is generally single-stranded (in contrast to DNA, which is double-stranded).

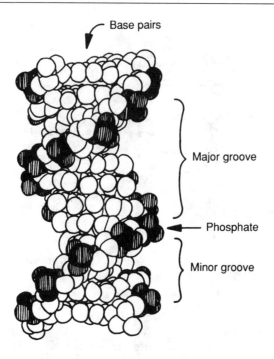

Figure 3-7. The DNA double helix.

(1) When **strands loop back** on themselves, the bases on opposite sides may pair: adenine with uracil and guanine with cytosine.

(2) **RNA molecules have extensive base-pairing,** which produces secondary and tertiary structures that are extremely important for RNA function.

(3) RNA molecules recognize DNA and other RNA molecules by base-pairing.

c. Some **RNA molecules** act as catalysts of reactions; thus, RNA, as well as protein, can have enzymatic activity.

(1) **Ribozymes,** usually precursors of rRNA, remove internal segments of themselves, splicing the ends together.

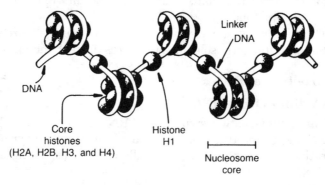

Figure 3-8. A polynucleosome. (Adapted from Olins DE and Olins AL: Nucleosomes: The structural quantum in chromosomes. *American Scientist 66* (1978), p 708.)

(2) RNAs also act as **ribonucleases,** cleaving other RNA molecules (e.g., RNase P cleaves tRNA precursors).

(3) Peptidyl transferase, an enzyme in protein synthesis, is composed of RNA.

2. Messenger RNA (mRNA) contains a cap structure and a poly(A) tail.

  a. The **cap** consists of **methylated guanine triphosphate** attached to the hydroxyl group on the ribose at the 5' end of the mRNA molecule.

    –The N7 in the guanine is methylated.

    –The 2'-hydroxyl groups of the first and second ribose moieties of the mRNA also may be methylated (Figure 3-9).

  b. The **poly(A) tail** contains up to 200 adenine (A) nucleotides attached to the hydroxyl group at the 3' end of the mRNA molecule.

3. Ribosomal RNA (rRNA) contains many loops and extensive base-pairing.

  –rRNA molecules differ in their sedimentation coefficients (S). They associate with proteins to form **ribosomes** (Figure 3-10).

  a. Prokaryotes have three types of rRNA: 16S, 23S, and 5S rRNA.

  b. Eukaryotes have four types of cytoplasmic rRNA: 18S, 28S, 5S, and 5.8S rRNA.

4. Transfer RNA (tRNA) has a cloverleaf structure and contains modified nucleotides. Transfer RNA molecules are relatively small, containing about 80 nucleotides.

  a. In eukaryotic cells, up to 10% of the nucleotides in tRNA are modified.

    –Modified nucleotides containing **pseudouridine (ψ), dihydrouridine (D),** and **ribothymidine (T)** are present in most tRNAs (Figure 3-11).

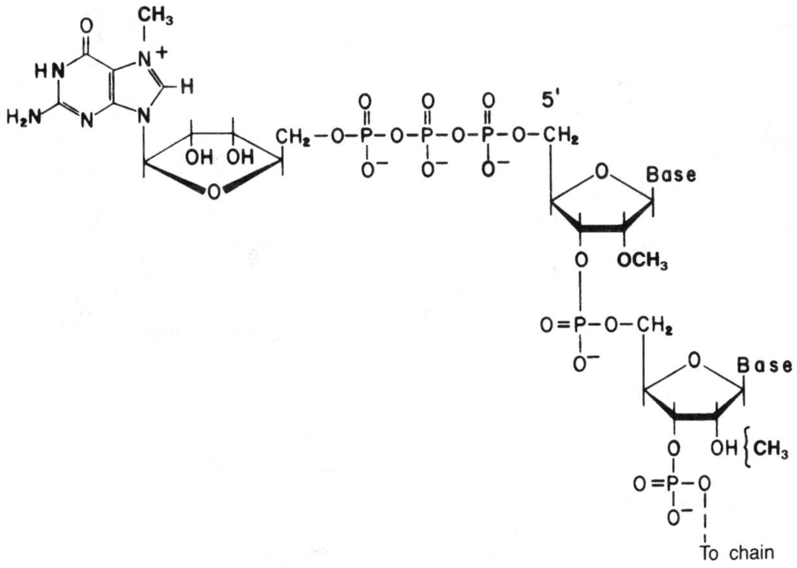

Figure 3-9. The cap structure.

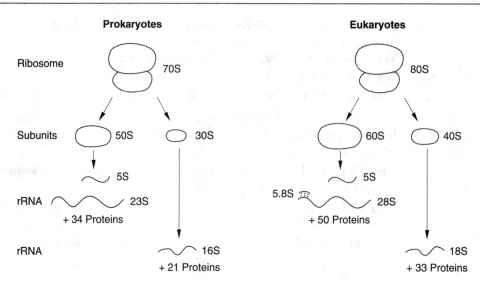

Figure 3-10. The composition of ribosomes.

b. All tRNA molecules have a similar **cloverleaf structure** even though their base sequences differ (Figure 3-12).

 (1) The first loop from the 5' end, the **D loop,** contains dihydrouridine.

 (2) The middle loop contains the **anticodon,** which base-pairs with the codon in mRNA.

 (3) The third loop, the **TψC loop,** contains both ribothymidine and pseudouridine.

 (4) The **CCA sequence** at the 3' end carries the amino acid.

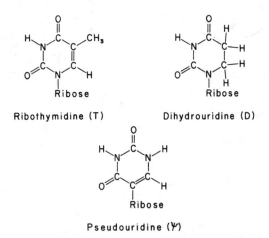

Figure 3-11. Three modified nucleosides found in most tRNAs.

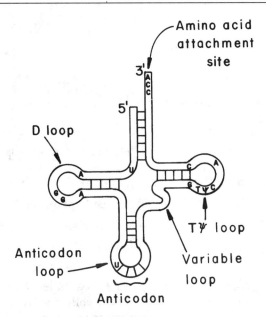

Figure 3-12. The cloverleaf structure of tRNA. Bases that commonly occur in a particular position are indicated by *letters*. Base-pairing in stem regions is indicated by *lines* between strands. ψ = pseudouridine; *T* = ribothymidine; *D* = dihydrouridine.

II. Synthesis of DNA (Replication)

- Replication, the process of DNA synthesis, occurs during the S phase of the cell cycle and is catalyzed by a complex of proteins that includes the enzyme DNA polymerase.
- Each strand of the parent DNA acts as a template for the synthesis of its complementary strand.
- DNA polymerase copies the template strand in the 3' to 5' direction and synthesizes the new strand in the 5' to 3' direction. Deoxyribonucleoside triphosphates serve as the precursors.
- DNA polymerase cannot initiate the synthesis of a new strand. A short stretch of RNA serves as a primer.
- Other proteins and enzymes are required to unwind the parental strands and allow both strands to be copied simultaneously.
- Errors that occur during replication are corrected by enzymes associated with the replication complex.
- Damage that occurs to DNA molecules may be corrected by **repair mechanisms**, which usually involve removal and replacement of the damaged region with the intact undamaged strand serving as a template.
- DNA molecules may recombine. A portion of a strand from one molecule may be exchanged for a portion of a strand from another molecule.
- Genes may be transposed (moved from one chromosomal site to another).

A. The cell cycle of eukaryotic cells (Figure 3-13)

  1. During the G_1 (first gap) phase, cells **prepare to duplicate** their chromosomes.

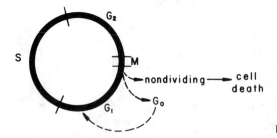

Figure 3-13. The cell cycle of eukaryotes.

2. During the **S** (synthesis) phase, **synthesis of DNA** (replication) occurs.

3. During the **G₂** (second gap) phase, cells **prepare to divide**.

4. During the **M** (mitosis) phase, **cell division** occurs.

5. Cells may traverse the cell cycle many times.

   –They may leave the cycle never to divide again.

   –They may enter a phase (sometimes called **G₀**) in which they may remain for extended periods. These cells may reenter the cell cycle and divide in response to an appropriate stimulus.

B. **Mechanism of replication**

   1. **Replication** is bidirectional and semiconservative (Figure 3-14).

      a. **Bidirectional** means that replication begins at a site of origin and simultaneously moves out in both directions from this point.

         –Prokaryotes have one site of origin on each chromosome.

         –Eukaryotes have multiple sites of origin on each chromosome.

      b. **Semiconservative** means that, following replication, each daughter molecule of DNA contains one intact parental strand and one newly synthesized strand joined by base pairs.

   2. **Replication forks** are the sites at which DNA synthesis is occurring (Figure 3-15).

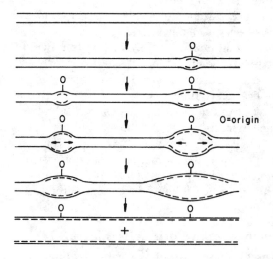

Figure 3-14. Replication of a eukaryotic chromosome. *Solid lines* are parental strands. *Dashed lines* are newly synthesized strands. Synthesis is bidirectional from each point of origin (*O*).

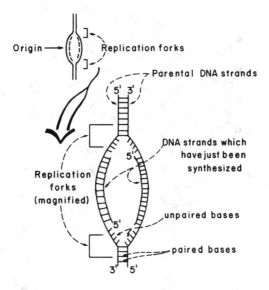

Figure 3-15. Replication forks.

–The **parental strands** of DNA separate and the helix unwinds ahead of a replication fork.

a. Helicases unwind the helix, and single-strand binding proteins hold it in a single-stranded conformation.

b. Topoisomerases act to prevent the extreme supercoiling of the parental helix that would result as a consequence of unwinding at a replication fork.

–Topoisomerases break and rejoin DNA chains.

–**DNA gyrase,** a topoisomerase inhibited by quinolones, is found only in prokaryotes.

3. DNA polymerases catalyze the synthesis of DNA.

a. Prokaryotes have three DNA polymerases: **pol I, pol II, and pol III**. Pol III is the replicative enzyme, and pol I is involved in repair.

b. Eukaryotes have four DNA polymerases: **α, β, γ, and δ.** DNA polymerase α is involved in replication, and β is involved in repair of nuclear DNA. Polymerase δ acts in conjunction with α, and γ functions in mitochondria.

c. DNA polymerases can only copy a DNA template in the 3' to 5' direction and produce the newly synthesized strand in the 5' to 3' direction.

d. Deoxyribonucleoside triphosphates (dATP, dGTP, dTTP, and dCTP) are the precursors for the DNA chain.

(1) Each one pairs with the corresponding base on the template strand and forms a phosphodiester bond with the hydroxyl group on the 3'-carbon of the sugar at the end of the growing chain (Figure 3-16).

(2) Pyrophosphate is produced and cleaved to two inorganic phosphates, releasing energy that drives the reaction.

4. DNA polymerase requires a **primer** (Figure 3-17).

a. DNA polymerases **cannot initiate** synthesis of new strands.

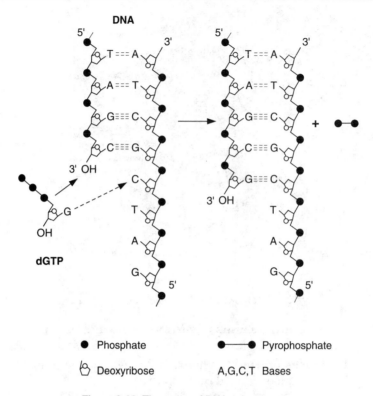

DNA

● Phosphate   ●———● Pyrophosphate

⬠ Deoxyribose   A,G,C,T Bases

Figure 3-16. The action of DNA polymerase.

b. RNA serves as the primer for DNA polymerase in vivo.

-The RNA primer, which contains 10 or fewer nucleotides, is formed by copying of the parental strand in a reaction catalyzed by **primase**.

c. DNA polymerase adds deoxyribonucleotides to the 3'-hydroxyls of the RNA primers and subsequently to the ends of the growing DNA strands.

d. DNA parental (template) strands are copied simultaneously at replication forks, although they run in opposite directions.

(1) The **leading strand** is formed by continuous copying of the parental strand that runs 3' to 5' toward the replication fork.

(2) The **lagging strand** is formed by discontinuous copying of the parental strand that runs 3' to 5' away from the replication fork.

-As more of the helix is unwound, synthesis of the lagging strand begins from another primer. The short fragments formed by this process are known as **Okazaki fragments**.

-The RNA **primers are removed by RNase H;** then the resulting **gaps are filled** with the appropriate deoxyribonucleotides by another DNA polymerase.

-Finally, the **Okazaki fragments are joined by polynucleotide ligase,** an enzyme that catalyzes formation of phosphodiester bonds between two polynucleotide chains.

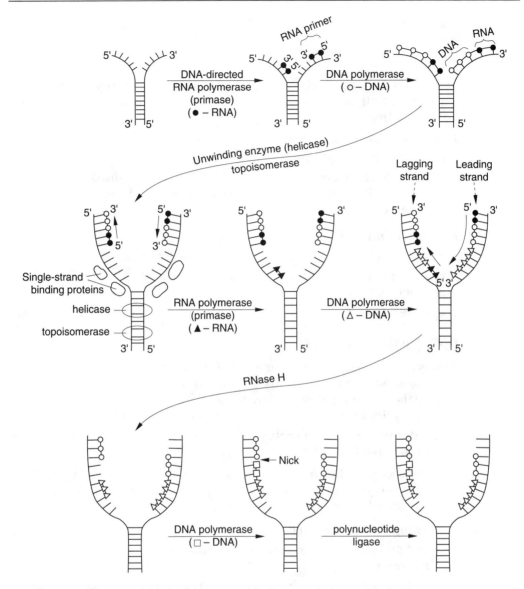

Figure 3-17. Mechanism of DNA synthesis at the replication fork. Two rounds of polymerase action are shown. The number of nucleotides added in each round is much larger than shown; in eukaryotes, about 10 ribonucleotides and 200 deoxyribonucleotides are polymerized on the lagging strand. Synthesis on the leading strand is continuous.

> **e.** In eukaryotic cells, about 200 deoxyribonucleotides are added to the lagging strand in each round of synthesis, whereas in prokaryotes about 2000 are added.

5. The **fidelity of replication** is very high with an overall error rate of 10^{-9} to 10^{-10}.

> **a.** **Errors** (insertion of an inappropriate nucleotide) that occur during replication **may be corrected by editing** during the replication process. This proofreading function is performed by a 3' to 5' exonuclease associated with pol III in prokaryotes and by DNA polymerase δ in eukaryotes.

      b. Postreplication repair processes also increase the fidelity of replication.

C. Mutations

    –**Changes in DNA** molecules cause mutations. After replication, these changes result in a permanent alteration of the base sequence in the daughter DNA.

    1. Changes causing mutations

        a. Uncorrected errors made during replication

        b. Damage that occurs **to nonreplicating DNA** caused by oxidative deamination, radiation, or chemicals, resulting in cleavage of DNA strands or chemical alteration or removal of bases

    2. Types of mutations

        a. Point mutation (substitution of one base for another)

        b. Insertion (addition of one or more nucleotides within a DNA sequence)

        c. Deletion (removal of one or more nucleotides from a DNA sequence)

D. DNA repair (Figure 3-18)

    1. In general, repair involves **removal** of the segment of DNA that contains a **damaged region,** filling in the gap by action of a DNA polymerase that uses the undamaged sister strand as a template, and **ligation** of the newly synthesized segment to the remainder of the chain.

      –**Endonucleases, exonucleases,** a **DNA polymerase,** and a **ligase** are required for repair.

    2. Specific **glycosylases** remove damaged bases by hydrolyzing N-glycosidic bonds, producing an apurinic or apyrimidinic site, which is cleaved and repaired.

E. Rearrangements of genes

    –Several processes produce new combinations of genes, thereby promoting genetic diversity.

    1. Recombination occurs between homologous DNA segments, that is, those that have very similar sequences.

    2. Transposition involves movement of a DNA segment from one site to a nonhomologous site.

      –Transposons are mobile genetic elements that facilitate movement of genes.

F. Reverse transcription

    1. Synthesis of DNA from an **RNA template** is catalyzed by reverse transcriptase.

    2. Retroviruses contain RNA as their genetic material.

        a. The retroviral RNA serves as a template for synthesis of DNA by reverse transcriptase.

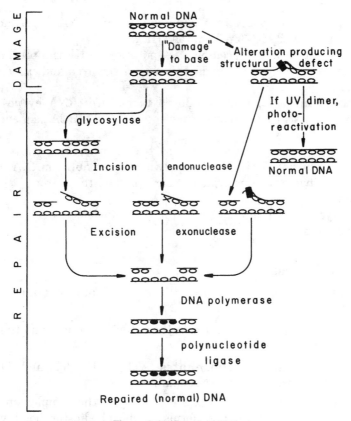

Figure 3-18. Repair of DNA.

   b. The DNA that is generated may be inserted into the genome (chromosomes) of the host cell and be expressed.

   3. Reverse transcriptase also may play a role in normal development.

III. Synthesis of RNA (Transcription)

● Transcription, the synthesis of RNA from a DNA template, is catalyzed by RNA polymerase. RNA polymerase copies a DNA template in the 3' to 5' direction and synthesizes a single-stranded RNA molecule in a 5' to 3' direction. Unlike DNA polymerase, RNA polymerase can initiate the synthesis of new strands.

● In eukaryotes, the primary product of transcription is modified and trimmed before it participates in protein synthesis.

● Messenger RNA, produced by RNA polymerase II, is capped at the 5' end and has a poly(A) tail added at the 3' end. Introns (segments that do not code for protein) are removed, and exons (segments that code for protein) are spliced together.

● Ribosomal RNA is produced by RNA polymerase I as a 45S precursor that is methylated and cleaved to form three of the rRNAs (18S, 28S,

and 5.8S) that appear in ribosomes. The 5S rRNA is produced from a separate gene by RNA polymerase III.
- Transfer RNA is produced by RNA polymerase III as a precursor that is trimmed at the 5' and 3' ends. Introns are removed and exons are spliced together.
- Unusual nucleotides are produced in mature tRNA by posttranscriptional modification of normal nucleotides, and a CCA sequence is added at the 3' end.
- Bacteria do not contain nuclei, so transcription and translation occur simultaneously. A single RNA polymerase produces mRNA, rRNA, and tRNA in bacteria. Bacterial transcripts (e.g., those from *E. coli*) do not contain introns.
- Eukaryotic RNA must travel from the nucleus to the cytoplasm for translation.

A. RNA polymerase

1. RNA polymerase can **initiate** the synthesis of **new chains**. A primer is not required.

2. The DNA template is copied in the 3' to 5' direction, and the RNA chain grows in the 5' to 3' direction.

3. **Ribonucleoside triphosphates** (ATP, GTP, UTP, and CTP) serve as the **precursors** for the RNA chain.

   –A ribonucleoside triphosphate pairs with the complementary base in the DNA template and forms a phosphodiester bond with the 3'-hydroxyl of the ribose at the end of the growing chain.

   –**Pyrophosphate** is produced and cleaved to two inorganic phosphates, releasing energy that drives the reaction.

B. Synthesis of RNA in bacteria

1. The RNA polymerase of *E. coli* contains **four subunits,** $\alpha_2\beta\beta'$, which form the core enzyme, and a fifth subunit, the **σ (sigma) factor,** which is required for initiation of RNA synthesis.

2. Genes contain a **promoter region** to which RNA polymerase binds.

   a. Promoters contain the consensus sequence **TATAAT** (called the Pribnow box) about 10 bases upstream from (before) the start point of transcription.

   –A **consensus sequence** consists of the most commonly found sequence of bases in a given region of all DNAs tested.

   b. A **second consensus sequence** (TTGACA) is usually located upstream from the Pribnow box, about 35 nucleotides (–35) from the start point of transcription.

3. When RNA polymerase binds to a **promoter,** local unwinding of the DNA occurs, so that the DNA strands partially separate. The polymerase then begins to transcribe the **template strand**.

   a. As the polymerase moves along the DNA, the next region of the double helix unwinds while the single-stranded region that has been transcribed rejoins its partner.

b. Termination occurs in a region in which the transcript forms a hairpin loop that precedes four U residues.

c. The ρ (rho) factor aids in the termination of some transcripts.

4. mRNA is often produced as a **polycistronic transcript** that is translated as it is being transcribed.

    a. A polycistronic mRNA produces several different proteins during translation, one from each cistron.

    b. *E. coli* mRNA has a short half-life. It is degraded in minutes.

5. rRNA is produced as a **large transcript** that is cleaved, producing the 16S rRNA that appears in the 30S ribosomal subunit and the 23S and 5S rRNAs that appear in the 50S ribosomal subunit.

–The 30S and 50S ribosomal subunits combine to form the 70S ribosome.

6. tRNA usually is produced from **larger transcripts** that are cleaved. One of the cleavage enzymes, RNase P, contains an essential RNA molecule that acts as a catalyst.

C. Synthesis of RNA in nuclei of eukaryotes

1. mRNA synthesis (Figure 3-19)

    a. Eukaryotic genes that produce mRNA contain a **promoter region** to which RNA polymerase II binds. Promoters contain a number of conserved sequences, often called elements.

      (1) A **TATA** (Hogness) **box,** containing the consensus sequence TATATAA, is located about 25 base pairs upstream (–25) from the transcription start site.

      (2) A **CAAT box** (consensus sequence GGTCAATCT) is frequently found about 70 base pairs upstream from the start site.

      (3) GC-rich regions (GC boxes) often occur between –40 and –110.

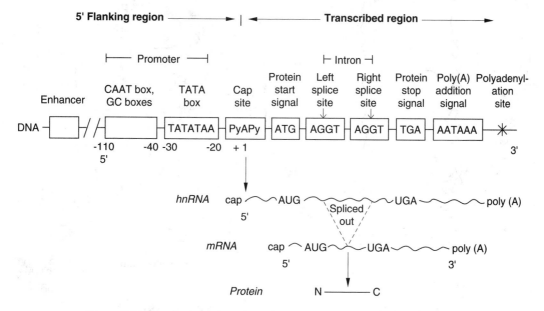

Figure 3-19. The structure of a eukaryotic gene and its products. *Py* = pyrimidine.

b. Enhancers are DNA sequences that modulate the transcription rate. They may be located thousands of base pairs upstream or downstream from the start site.

c. RNA polymerase II initially produces a large primary transcript called heterogeneous nuclear RNA (**hnRNA**), which contains exons and introns.

   (1) Exons are sequences within a transcript that **code for amino acids** in the protein product of a gene.

   (2) Introns are sequences within the coding region of a transcript that **do not code for amino acids** in the protein product.

d. Processing of hnRNA yields functional mRNA, which enters the cytoplasm through nuclear pores (Figure 3-20).

   (1) The **primary transcript** is capped at its 5' end as it is being transcribed.

   (2) A **poly(A) tail,** 20 to 200 nucleotides in length, is added to the 3' end of the transcript.

     –The sequence AAUAAA in hnRNA serves as a signal for cleavage and addition of the poly(A) tail by poly(A) polymerase. ATP serves as the precursor.

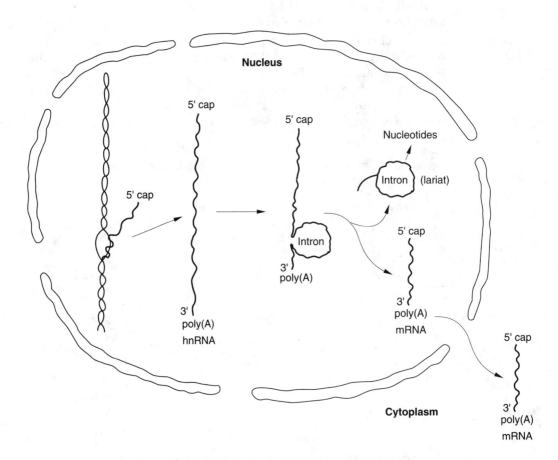

Figure 3-20. Synthesis of mRNA in eukaryotes.

(3) **Cleavage and splicing** remove introns and connect the exons.

–The splice point at the left flank of introns usually has the sequence AGGU. At the right flank, an invariant AG is frequently followed by GU.

–Small nuclear RNAs complexed with protein (snRNPs) [e.g., U1 and U2] are involved in the cleavage and splicing process. A lariat structure may be generated during the splicing reaction (see Figure 3-20).

(4) Some hnRNAs contain 50 or more exons that must be spliced correctly to produce functional mRNA. Other hnRNAs have no introns.

2. **rRNA synthesis** and assembly of ribosomes (Figure 3-21)

   a. A **45S precursor** is produced by RNA polymerase I from rRNA genes located in the fibrous region of the nucleolus. Many copies of the genes are present, linked together by spacer regions.

   b. Some of the nucleotides in the 45S precursor are methylated, mainly on the 2'-hydroxyl groups of their ribose moieties. These methyl groups are conserved in mature rRNA.

   c. The **45S precursor** undergoes a number of cleavages that ultimately produce 18S rRNA and 28S rRNA; the latter is hydrogen-bonded to a 5.8S rRNA.

   d. **18S rRNA** complexes with proteins and forms the 40S ribosomal subunit.

   e. The **28S, 5.8S, and 5S rRNAs** complex with proteins and form the 60S ribosomal subunit. 5S rRNA is produced by RNA polymerase III outside of the nucleolus.

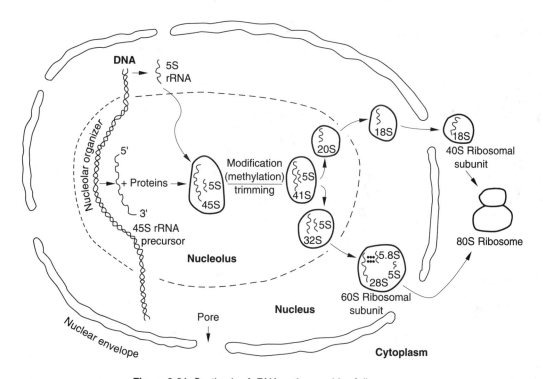

Figure 3-21. Synthesis of rRNA and assembly of ribosomes.

f. The **ribosomal subunits migrate** through the nuclear pores into the cytoplasm where they complex with mRNA, forming 80S ribosomes. Because sedimentation coefficients reflect both shape and particle weight, they are not additive.

g. rRNA precursors may contain introns that are removed during maturation. In some organisms, the enzymatic activity that removes rRNA introns resides in the rRNA precursor. No proteins are required. These autocatalytic RNAs are known as ribozymes.

3. tRNA synthesis (Figure 3-22)

a. RNA polymerase III is the enzyme that produces tRNA. The promoter is located within the coding region of the gene.

b. Primary transcripts for tRNA are cleaved at the 5' and 3' ends.

c. Some precursors contain **introns** that are removed.

d. During processing of tRNA precursors, **nucleotides** are modified. Posttranscriptional modification includes the conversion of uridine to pseudouridine (ψ), ribothymidine (T), and dihydrouridine (D). Other unusual nucleotides are also produced.

e. Addition of the sequence **CCA to the 3' end** is catalyzed by nucleotidyl transferase.

D. Inhibitors of RNA synthesis

1. Agents that bind to DNA

–**Actinomycin D** contains pentapeptide groups that fit into the grooves of DNA and a phenoxazone ring that intercalates (slips) between the

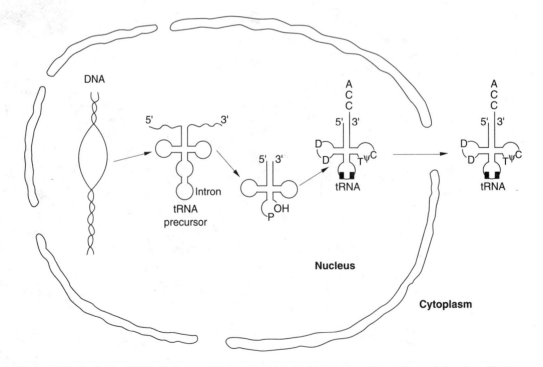

Figure 3-22. Synthesis of tRNA. D, T, ψ, and ■ are unusual nucleotides produced by posttranscriptional modifications.

DNA base pairs. When actinomycin D is present, DNA cannot act as a template for replication or transcription.

2. Agents that bind to RNA polymerase

    a. Rifampicin binds to bacterial RNA polymerase and prevents initiation of RNA synthesis.

    b. Streptolydigin binds to bacterial RNA polymerase and prevents elongation of RNA chains.

    c. α-Amanitin, derived from the poisonous mushroom *Amanita phalloides,* binds to eukaryotic RNA polymerases. RNA polymerase II is more readily inhibited than polymerase III. Polymerase I is insensitive.

IV. Protein synthesis (Translation of mRNA)

- During translation, messenger RNA (mRNA) determines the sequence of the amino acids in the protein that is produced.
- mRNA combines with ribosomes, which contain rRNA. Many ribosomes may be attached simultaneously to a single molecule of mRNA, forming a polysome.
- tRNA carries amino acids to the ribosomal site of protein synthesis. The anticodon in each aminoacyl-tRNA combines with the complementary codon in mRNA. A codon is the sequence of three nucleotides in mRNA that specifies a particular amino acid.
- Initiation of a polypeptide chain begins with the amino acid methionine (codon = AUG).
- Subsequently, amino acids are added to the growing polypeptide chain according to the codon sequence in the mRNA. Aminoacyl-tRNAs and GTP provide the energy for chain elongation.
- A protein is synthesized from its N- to its C-terminus, following the codons in the mRNA in a 5' to 3' direction.
- When synthesis of the polypeptide is completed, a termination codon (UGA, UAG, or UAA) causes the polypeptide chain to be released.

A. The genetic code (Table 3-1)

    –The genetic code is the collection of codons that specify all the amino acids found in proteins.

    –A **codon is a sequence of three bases** (triplet) in mRNA (5' to 3') that specifies, or corresponds to, a particular amino acid. During translation, the successive codons in an mRNA determine the sequence in which amino acids add to the growing polypeptide chain.

    1. The genetic code is degenerate (redundant). There is at least 1 codon for each of the 20 common amino acids; many amino acids have numerous codons.

    2. The genetic code is **nonoverlapping** (i.e., each nucleotide is used only once), beginning with a start codon (**AUG**) near the 5' end of the mRNA and ending with a termination (stop) codon (**UGA, UAG,** or **UAA**) near the 3' end.

Table 3-1. The Genetic Code

First Base (5')	Second Base U	C	A	G	Third Base (3')
U	Phe	Ser	Tyr	Cys	U
	Phe	Ser	Tyr	Cys	C
	Leu	Ser	Term	Term	A
	Leu	Ser	Term	Trp	G
C	Leu	Pro	His	Arg	U
	Leu	Pro	His	Arg	C
	Leu	Pro	Gln	Arg	A
	Leu	Pro	Gln	Arg	G
A	Ile	Thr	Asn	Ser	U
	Ile	Thr	Asn	Ser	C
	Ile	Thr	Lys	Arg	A
	Met	Thr	Lys	Arg	G
G	Val	Ala	Asp	Gly	U
	Val	Ala	Asp	Gly	C
	Val	Ala	Glu	Gly	A
	Val	Ala	Glu	Gly	G

3. The code is **commaless** (i.e., there are no breaks or markers to distinguish one codon from the next).

4. The code is **nearly universal**. The same codon specifies the same amino acid in almost all species studied; however, some differences have been found in the codons used in mitochondria.

5. The **start codon** (AUG) sets the **reading frame**.

B. **Effect of mutations on proteins**

–Mutations in DNA are transcribed into mRNA and thus may cause changes in the encoded protein.

–The various types of mutations that may occur in DNA have different effects on the encoded protein.

1. **Point mutations** occur when one base in DNA is replaced by another, altering the codon in mRNA.

a. **Silent** mutations do not affect the amino acid sequence of a protein (e.g., CGA to CGG causes no change, since both codons specify arginine).

b. **Missense** mutations result in one amino acid being replaced by another (e.g., CGA to CCA causes arginine to be replaced by proline).

c. **Nonsense** mutations result in premature termination of the growing polypeptide chain (e.g., CGA to UGA causes arginine to be replaced by a stop codon).

2. **Insertions** occur when a base or a number of bases are added to DNA. They may result in a protein with more amino acids than normal.

Amino acid

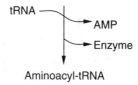

Enzyme [aminoacyl-AMP]

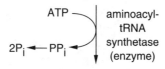

Aminoacyl-tRNA

Figure 3-23. Formation of aminoacyl-tRNA.

3. **Deletions** occur when a base or a number of bases are removed from the DNA. They may result in a protein with fewer amino acids than normal.

4. **Frameshift mutations** occur when the number of bases added or deleted is not a multiple of three. The reading frame is shifted so that completely different sets of codons are read beyond the point where the mutation starts.

C. **Formation of aminoacyl-tRNAs** (Figure 3-23)

  –**Amino acids are activated and attached to** their corresponding **tRNAs** by highly specific enzymes known as aminoacyl-tRNA synthetases.

1. Each **aminoacyl-tRNA synthetase** recognizes a particular amino acid and the tRNAs specific for that amino acid.

2. An **amino acid** first reacts with ATP, forming an enzyme [aminoacyl-AMP] complex and pyrophosphate. Cleavage of pyrophosphate drives this reaction.

3. The **aminoacyl-AMP** then **forms an ester** with the 2' or 3' hydroxyl of a tRNA specific for that amino acid producing an aminoacyl-tRNA and AMP.

4. Once an amino acid is attached to a tRNA, insertion of the amino acid into a growing polypeptide chain depends only on the codon-anticodon interaction (Figure 3-24).

D. **Initiation of translation** (Figure 3-25)

1. **In eukaryotes, methionyl-tRNA$_i^{Met}$** binds to the small **ribosomal subunit**. The **5' cap** of the mRNA binds to the small subunit and the first AUG codon base-pairs with the anticodon on the methionyl-tRNA$_i^{Met}$.

  a. **In bacteria,** the methionine that initiates protein synthesis is **formylated** and is carried by tRNA$_f$.

  b. **Prokaryotes do not contain a 5' cap** on their mRNA. An mRNA sequence upstream from the translation start site (the Shine-Dalgarno sequence) binds to the 3' end of 16S rRNA.

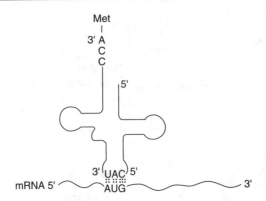

Figure 3-24. Antiparallel binding of aminoacyl-tRNA to mRNA.

2. The **methionine** that initiates protein synthesis is subsequently **removed** from the N-terminus of the polypeptide by a specific aminopeptidase.

3. The **large ribosomal subunit binds,** completing the initiation complex.

  a. The methionyl-tRNA$_i^{Met}$ is bound at the P (peptidyl) site of the complex.

  b. The A (acceptor or aminoacyl) site of the complex is unoccupied.

4. Initiation factors, **ATP,** and **GTP** are required for formation of the initiation complex.

  a. The **initiation factors** are designated IF-1, IF-2, and IF-3 in prokaryotes. In eukaryotes, they are designated eIF-1, eIF-2, and so on. Seven or more may be present.

  b. Release of the initiation factors involves hydrolysis of GTP to GDP and P_i.

E. Elongation of polypeptide chains (see Figure 3-25*B*)

  –The addition of each amino acid to the growing polypeptide chain involves binding of an aminoacyl-tRNA at the A site, formation of a peptide bond, and translocation of the peptidyl-tRNA to the P site.

  1. Binding of aminoacyl-tRNA to the A site

    a. The **mRNA codon** at the A site determines which aminoacyl-tRNA will bind.

      (1) Binding between the codon and the anticodon occurs by base-pairing and is **antiparallel.**

      (2) Internal methionine residues in the polypeptide chain are added in response to AUG codons. They are carried by tRNA$_m^{Met}$, a second tRNA specific for methionine.

    b. An **elongation factor** (EF-Tu in prokaryotes and EF-1 in eukaryotes) and hydrolysis of GTP are required for binding.

  2. Formation of a peptide bond

    a. The **free amino group** in the amino acid of the aminoacyl-tRNA at the A site forms a peptide bond with the carbonyl of the aminoacyl group attached to the tRNA at the **P site.** Formation of the peptide bond is catalyzed by **peptidyl transferase,** which is rRNA.

    b. The tRNA at the P site now does not contain an amino acid. It is "uncharged."

A. Initiation

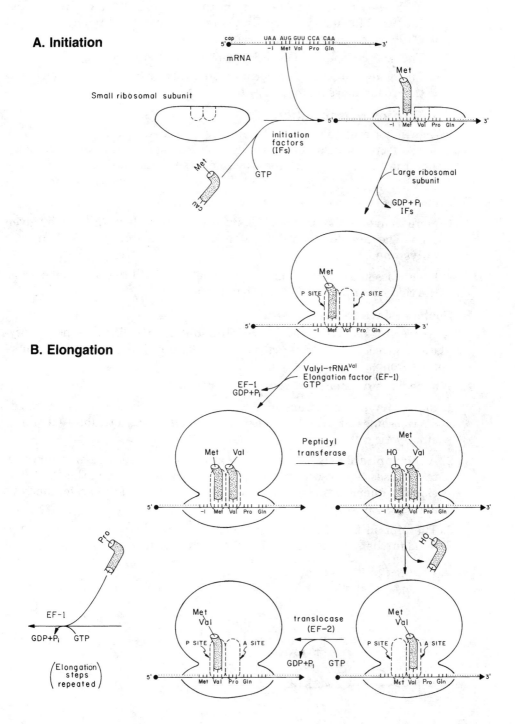

B. Elongation

Figure 3-25. The initiation (*A*) and elongation (*B*) reactions of protein synthesis. EF-1 and EF-2 are eukaryotic elongation factors corresponding to EF-Tu and EF-G in prokaryotes.

c. The growing polypeptide chain is attached to the tRNA in the A site.

3. Translocation of peptidyl-tRNA

a. The peptidyl-tRNA moves from the A site to the P site, and the uncharged tRNA is released from the ribosome. An **elongation factor** (EF-2 in eukaryotes or EF-G in prokaryotes) and the hydrolysis of **GTP** are required for translocation.

b. The next codon in the mRNA is now in the A site.

c. The elongation and translocation steps are repeated until a termination codon moves into the A site.

F. Termination of translation

—When a termination codon (UGA, UAG, or UAA) occupies the A site, release factors cause the newly synthesized polypeptide to be released from the ribosome, and the ribosomal subunits dissociate from the mRNA.

G. Polysomes (Figure 3-26)

1. More than one ribosome may be attached to a single mRNA at any given time. The complex of mRNA with multiple ribosomes is known as a **polysome**.

2. Each ribosome carries a nascent polypeptide chain that grows longer as the ribosome approaches the 3' end of the mRNA.

H. Posttranslational processing

—After synthesis is completed, **proteins may be modified** by phosphorylation, glycosylation, ADP-ribosylation, hydroxylation, and addition of other groups.

I. Synthesis and release of secretory proteins

1. **Secretory proteins,** destined for release from the cell, are synthesized on ribosomes attached to the rough endoplasmic reticulum (**RER**) in eukaryotic cells.

2. A **hydrophobic signal sequence** at the N-terminus of a secretory protein causes the nascent protein to pass into the lumen of the RER. The signal sequence is cleaved from the N-terminus, and the protein may be glycosylated within the RER.

3. The protein travels in vesicles to the **Golgi,** where it may be glycosylated further and is packaged in secretory vesicles.

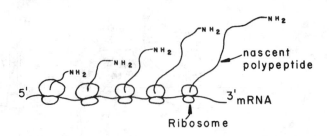

Figure 3-26. A polysome.

4. Secretory vesicles containing the protein travel from the Golgi to the cell membrane. The protein is released from the cell by **exocytosis**.

J. Inhibitors of protein synthesis

1. Compounds used clinically

–The following compounds affect protein synthesis on 70S-type ribosomes of prokaryotes and, therefore, are used as **antibiotics**.

–Because mitochondria contain 70S-type ribosomes that function similarly to those in prokaryotic cells, these compounds also **inhibit mitochondrial protein synthesis**.

a. Streptomycin binds to the 30S ribosomal subunit of prokaryotes and causes misreading of mRNA, thereby preventing formation of the initiation complex.

b. Tetracycline binds to the 30S ribosomal subunit of prokaryotes and inhibits binding of aminoacyl-tRNA to the A site.

c. Chloramphenicol inhibits the peptidyl transferase activity of the 50S ribosomal subunit of prokaryotes.

d. Erythromycin binds to the 50S ribosomal subunit of prokaryotes and prevents translocation.

2. Compounds used for research

a. Puromycin binds at the A site, forms a peptide bond with the growing peptide chain, and prematurely terminates synthesis. It acts in both prokaryotes and eukaryotes.

b. Cycloheximide inhibits peptidyl transferase in eukaryotes.

V. Regulation of Protein Synthesis

- Regulation of protein synthesis in prokaryotes occurs mainly at the transcriptional level and involves genetic units known as operons.
- Operons contain promoter regions where proteins bind and facilitate or inhibit the binding of RNA polymerase.
- When RNA polymerase transcribes the structural genes of an operon, a polycistronic mRNA (i.e., an mRNA that codes for more than one protein) is produced.
- In eukaryotes, regulation of protein synthesis may occur by modification of DNA or at the level of transcription or translation.
  –Genes may be deleted from cells or they may be amplified, rearranged, or methylated.
  –Histones may nonspecifically repress transcription of genes. Inducers cause genes to be activated.
  –Synthesis of some proteins is regulated at the level of translation.

A. Regulation of protein synthesis in prokaryotes

1. Relationship of protein synthesis to nutrient supply

a. Prokaryotes respond to changes in their supply of nutrients in a way that allows them to obtain or conserve energy most efficiently.

   –Prokaryotes, such as *E. coli,* require a source of carbon, which is usually a sugar that is oxidized for energy.

   –A source of nitrogen is also required for the synthesis of amino acids from which structural proteins and enzymes are produced.

b. *E. coli* uses glucose preferentially whenever it is available. The enzymes in the pathways for glucose utilization are made constitutively (i.e., they are constantly being produced).

c. If glucose is not present in the medium but another sugar is available, *E. coli* produces the enzymes and other proteins that allow the cell to derive energy from that sugar. The process by which synthesis of the proteins is regulated is called **induction.**

d. If an amino acid is present in the medium, *E. coli* does not need to synthesize that amino acid and conserves energy by ceasing to produce the enzymes required for its synthesis. The process by which synthesis of these proteins is regulated is called **repression.**

2. Operons
   a. An operon is a **set of genes** that are adjacent to one another in the genome and are coordinately controlled; that is, the genes are either all turned on or all turned off (Figure 3-27).

   b. The **structural genes** of an operon **code for** a series of different **proteins**.

   –A single **polycistronic mRNA** is transcribed from an operon. This single mRNA codes for all the proteins of the operon.

   –A series of start and stop codons on the polycistronic mRNA allows a number of different proteins to be produced from the single mRNA at the translational level.

   c. Transcription begins near a **promoter region,** located at the upstream end of the group of structural genes.

   d. Associated with the promoter is a short sequence, the **operator,** which determines whether the genes are turned on or off.

   e. Binding of a repressor protein to the operator region prevents binding of RNA polymerase to the promoter and **inhibits transcription** of the structural genes of the operon.

   –Repressor proteins are encoded by regulatory genes, which may be located anywhere in the genome.

3. Induction (see Figure 3-27*B*)
   a. Induction is the process where an **inducer** (a small molecule) turns on **transcription** of an operon.

   b. The inducer is frequently a sugar (or a metabolite of the sugar), and the proteins produced from the inducible operon allow the sugar to be metabolized.

   (1) The inducer binds to the **repressor,** inactivating it.
   (2) The **inactive repressor does not bind** to the operator.
   (3) RNA polymerase, therefore, can **bind** to the promoter and **transcribe** the operon.
   (4) The structural **proteins** encoded by the operon are **produced**.

   c. The lactose (*lac*) operon is inducible.

A. In the absence of inducer

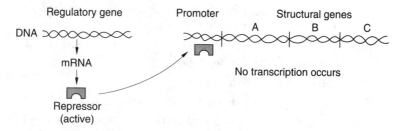

B. In the presence of inducer

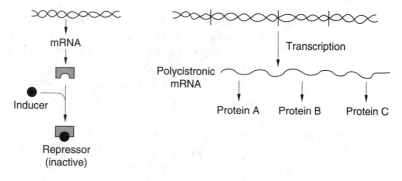

Figure 3-27. An inducible operon (e.g., the *lac* operon).

(1) A metabolite of lactose, **allolactose,** is the inducer.

(2) Proteins produced by the genes of the *lac* operon allow the cell to oxidize lactose as a source of energy.

(3) The *lac* operon is induced only in the absence of glucose.

4. **Repression**

a. Repression is the process where a **corepressor** (a small molecule) **turns off** the **transcription** of an operon.

b. The corepressor is usually an amino acid, and the proteins produced from the repressible operon are involved in the synthesis of the amino acid.

(1) The **corepressor binds to the repressor,** activating it.

(2) The **active repressor binds to the operator**.

(3) **RNA polymerase,** therefore, **cannot bind** to the promoter, and the operon is not transcribed.

(4) The cell stops producing the structural proteins encoded by the operon.

c. The tryptophan (*trp*) operon is repressible.

(1) Tryptophan is the corepressor.

(2) The proteins encoded by the *trp* operon are involved in the synthesis of tryptophan.

(3) The *trp* operon is repressed in the presence of tryptophan, since cells do not need to make the amino acid in this case.

5. **Positive control**

a. Some operons are turned on by mechanisms that activate transcription.

b. When the repressor of the arabinose (*ara*) operon binds arabinose, it changes conformation and becomes an activator that stimulates binding of RNA polymerase to the promoter.

–The operon is then transcribed, and the proteins required for oxidation of arabinose are produced.

6. Catabolite repression (Figure 3-28)

   a. Cells preferentially use **glucose** when it is available.

   b. Some operons (e.g., *lac* and *ara*) are not expressed when glucose is present in the medium. These operons require cAMP for their expression.

   (1) Glucose causes cAMP levels in the cells to decrease.

   (2) When glucose decreases, cAMP levels rise.

   (3) cAMP binds to the catabolite-activator protein (CAP), also called the cAMP-receptor protein (CRP).

   (4) The cAMP-protein complex binds to a site near the promoter of the operon and facilitates binding of RNA polymerase to the promoter.

   c. The *lac* **operon** exhibits catabolite repression.

   (1) In the presence of lactose and the absence of glucose, the *lac* repressor is inactivated and the high levels of cAMP facilitate binding of RNA polymerase to the promoter.

   (2) The operon is transcribed, and the proteins that allow the cells to utilize lactose are produced.

7. Attenuation

   a. In bacterial cells, **transcription and translation** occur simultaneously.

   b. Attenuation occurs by a mechanism by which **rapid translation** of the nascent transcript causes **termination of transcription**.

   c. As the transcript is being produced, if ribosomes attach and rapidly translate the transcript, a secondary structure is generated in the mRNA that is a **termination signal** for RNA polymerase.

   d. If translation is slow, this termination structure does not form, and transcription continues.

   e. The *trp* operon is regulated by attenuation.

8. Factors, such as σ, affect RNA polymerase activity. These factors bind to the core RNA polymerase and increase its ability to bind to specific promoters.

B. Differences between eukaryotes and prokaryotes

1. Eukaryotes contain nuclei. Therefore, transcription is separated from translation. In prokaryotes, transcription and translation occur simultaneously.

2. DNA is complexed with histones in eukaryotes, but not in prokaryotes.

3. The mammalian genome contains about 1000 times more DNA than *E. coli* (10^9 versus 10^6 base pairs).

4. Most mammalian cells are diploid.

5. The major part of the genome of mammalian cells does not code for proteins.

A. In the presence of inducer and glucose

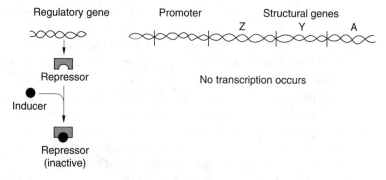

B. In the presence of inducer and absence of glucose

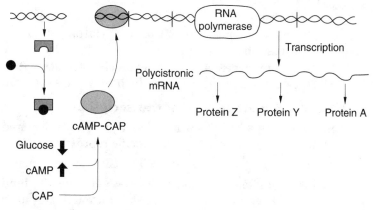

Figure 3-28. Catabolite repression. The cAMP-CAP complex facilitates initiation of transcription by RNA polymerase. Thus, the operon is transcribed only when glucose is low, cAMP is elevated, and the inducer is bound to the repressor, inactivating it. The *lac* operon exhibits catabolite repression.

6. Eukaryotic cells undergo differentiation, and the organisms go through various developmental stages.

7. Many eukaryotic genes, like bacterial genes, are unique (i.e., they exist in one or several copies per genome).

8. Other eukaryotic genes, unlike bacterial genes, have a large number of copies in the genome.

9. There are also relatively **short, repetitive DNA sequences** dispersed throughout the eukaryotic genome that do not code for proteins (e.g., Alu sequences).

10. Eukaryotic genes contain introns. Bacterial genes do not.

C. Regulation of protein synthesis in eukaryotes

   —Regulation may result from changes in genes or from mechanisms that operate at the level of transcription, during processing and transport of RNA, or at the level of translation.

   1. Changes in genes

   a. Genes may be lost (or partially lost) from cells so that functional proteins cannot be produced.

b. Genes may be amplified. For example, the drug methotrexate causes hundreds of copies of the gene for the enzyme dihydrofolate reductase to be produced.

c. Segments of DNA may move from one location to another on the genome, associating with each other in various ways so that different proteins are produced.

   (1) There are a number of different potential sequences (or arrangements) for various portions of an antibody-producing gene.

   (2) During differentiation of lymphocytes, specific sequences are selected and rearranged adjacent to each other on the genome to act as a single transcriptional unit for a specific antibody.

d. Modification of the **bases** in **DNA** affects the **transcriptional activity** of a gene.

   (1) Cytosine may be methylated at its 5 position.

   (2) The greater the extent of methylation, the less readily a gene is transcribed.

   (3) Globin genes are more extensively methylated in nonerythroid cells than in the cells in which they are expressed.

2. Regulation of the level of transcription

   a. Histones, which are small, basic proteins associated with the DNA of eukaryotes, may act as nonspecific repressors.

   b. Specific genes are turned on by positive mechanisms.

   c. Inducers (e.g., steroid hormones) enter cells, bind to proteins, are transported into the nucleus, interact with chromatin, and **activate specific genes** (see Chapter 8).

3. Regulation during processing and transport of mRNA

   –Regulatory mechanisms that occur during capping, polyadenylation, and splicing can alter the nature or quantity of the protein produced from the mRNA.

   a. Alternative polyadenylation or splicing sites can be used to generate different mRNAs.

   (1) Lymphocytes produce a membrane-bound IgM antibody at one stage of development and a soluble form that is secreted at a later stage.

   (2) The gene for this antibody contains two polyadenylation sites, one after the last two exons (which code for a hydrophobic amino acid sequence) and one before these exons.

   (3) When cleavage and poly(A) addition occur after the last two exons, the antibody contains a hydrophobic region that anchors it in the cell membrane.

   (4) When polyadenylation occurs at the first site, the antibody lacks the hydrophobic tail and is secreted from the cell.

   b. mRNAs can be degraded by nucleases after their synthesis in the nucleus and before their translation in the cytoplasm.

   (1) mRNAs have different half-lives. Some are degraded more rapidly than others.

(2) Interferon stimulates synthesis of 2',5'-oligo(A), which activates a nuclease that degrades mRNA.

4. **Protein synthesis** can be regulated at the translational level, during the **initiation or elongation** reactions.

   a. **Heme** stimulates the synthesis of globin by preventing the phosphorylation and consequent inactivation of eIF-2, a factor involved in initiation of protein synthesis.

   b. **Interferon** stimulates the phosphorylation of eIF-2, causing inhibition of initiation.

   c. An excess of **tubulin** arrests elongation and promotes cleavage of the mRNA.

VI. Biotechnology Involving Recombinant DNA

- Newly developed techniques in molecular biology are being used for research and medical diagnosis and provide hope as future therapy for diseases that currently are considered incurable.
- DNA from one organism can be cloned in another organism. Large quantities of the DNA can be produced (and expressed), and large quantities of the protein products of genes can be obtained.
- Genomic DNA, DNA copied from mRNA (cDNA), or chemically synthesized DNA can be cloned and subjected to genetic engineering. DNA libraries can be generated and used to study genes or fragments of genes.
- Restriction enzymes, which cleave within short, specific sequences of DNA, can be used to obtain DNA fragments for study or for insertion into DNA from other sources. The fusion product is known as chimeric DNA.
- The nucleotide sequence of DNA can be determined and used to deduce the amino acid sequence of the protein produced from the DNA.
- Gel electrophoresis can be used to separate DNA fragments.
- Because DNA strands can base-pair with complementary strands of DNA or RNA, a technique known as hybridization has been developed. Labeled DNA can be used as a probe to identify homologous (complementary sequences of) DNA or RNA.

A. Strategies for obtaining copies of genes

1. A **DNA fragment,** which may contain a specific gene, can be isolated from the cellular genome. Genes isolated from eukaryotic cells usually contain **introns**.

2. The **mRNA** for a gene **may be isolated,** and a DNA copy (cDNA) produced. Single-stranded cDNA may be copied to form double-stranded cDNA.

   a. **Reverse transcriptase** is an enzyme that produces cDNA from an RNA template.

   b. cDNA does not contain introns.

3. The DNA that codes for a small protein may be chemically synthesized.

B. Cloning of DNA (Figure 3-29)

  –DNA from one organism ("foreign" DNA, obtained as described above) can be inserted into a DNA vector and used to transform cells from another organism, usually a bacterium, that grows rapidly, replicating the foreign DNA, as well as its own.

  –Large quantities of the foreign DNA can be isolated or, under the appropriate conditions, the DNA can be expressed, and its protein product can be obtained in large quantities.

  1. The DNA to be cloned is cleaved with a **restriction endonuclease**.

   a. Restriction endonucleases recognize short sequences in DNA and cleave both strands within this region (Figure 3-30).

   b. Most of the DNA regions recognized by these enzymes are **palindromes** (i.e., the two strands of DNA have the same base sequence in the 5' to 3' direction).

   (1) *Eco*RI cleaves a region between an A and a G on each strand, generating two products.

   (2) The single-stranded regions of the products allow them to reanneal or to recombine with other DNA that has been cleaved by the same restriction endonuclease.

  2. A **vector** for transferring the foreign DNA into the bacterium, such as a plasmid, is cleaved with the same restriction endonuclease as the foreign DNA. The plasmid should contain genes for antibiotic resistance.

  3. The **plasmid and foreign DNA,** both cleaved by the same restriction enzyme, are mixed together and treated with DNA **ligase**. Some of the interactions produce chimeric plasmids, containing the foreign DNA integrated into the plasmid DNA.

  4. The **plasmids** are introduced into bacterial host cells (**transformation**).

  5. Clones that contain the chimeric plasmid are selected by growing the cells in media containing the antibiotics to which the plasmid has genes for resistance.

C. Isolation of cloned foreign DNA or its protein product

  1. Cells containing an appropriate chimeric plasmid are cultured (see Figure 3-30).

  2. To obtain large quantities of the foreign DNA, the **plasmids** are isolated from cells and treated with the restriction enzyme to release the foreign DNA, which is then isolated.

  3. To obtain large quantities of the **protein product** of the foreign DNA, the cells are grown under conditions that promote synthesis of the protein, and the protein is isolated from the cells.

   a. The DNA used to produce foreign proteins in bacterial cells must not contain introns because bacteria cannot remove them.

   b. Bacterial promoters must be inserted into the DNA so that the gene can be expressed.

D. Polymerase chain reaction (PCR) is an **in vitro technique** used for rapidly producing large amounts of DNA. It is suitable for clinical or forensic

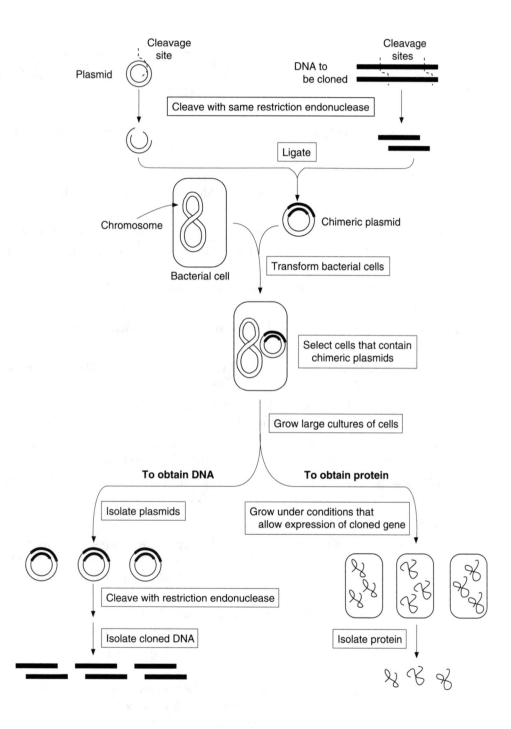

Figure 3-29. A simplified scheme for cloning genes.

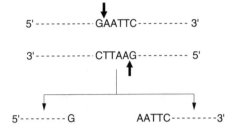

5'----------GAATTC----------3'

3'----------CTTAAG----------5'

5'---------G   AATTC---------3'

Figure 3-30. Action of restriction enzymes. *Eco*RI cleaves this palindrome. Two fragments are produced that contain complementary single-stranded regions (sticky ends).

testing because only a very small sample of DNA is required as the starting material.

1. The DNA sample is denatured by heat.

2. Large quantities of primers are added that bind to each DNA strand when the solution is cooled.

3. DNA polymerase is added, and polymerization is allowed to proceed.

–A heat-stable polymerase (Taq) is usually used, which remains active through many heating and cooling cycles.

4. The heating and cooling cycles are repeated until the DNA is amplified manyfold.

E. DNA sequencing by the Sanger dideoxynucleotide method (Figure 3-31)

1. **Dideoxynucleotides** are added to solutions in which DNA polymerase is catalyzing polymerization of a DNA chain.

2. Because a dideoxynucleotide does not contain a 3'-hydroxyl group, polymerization of the chain is terminated wherever a dideoxynucleotide adds.

3. DNA chains of varying lengths are produced. The shortest chains are nearest the 5' end of the DNA chain (which grows 5' to 3').

4. The sequence of the growing chain can be read (5' to 3') from the bottom to the top of the gel on which the DNA chains are separated.

F. Gel electrophoresis of DNA

1. DNA **chains** of varying length can be **separated** by gel electrophoresis. Agarose gels can be used to distinguish between DNA chains that differ in length by only one nucleotide. Polyacrylamide gels distinguish larger size differences.

2. Because DNA contains negatively charged phosphate groups, it will migrate in an electrical field toward the positive electrode.

3. Shorter chains **migrate** more rapidly through the pores of the gel, so separation depends on length.

4. DNA **bands** in the gel can be **visualized** by various techniques including staining with dyes (e.g., ethidium bromide) and autoradiography (if the gel contains a radioactive compound, which reacts with a photographic film).

5. **Blots** of gels can be made using nitrocellulose paper.

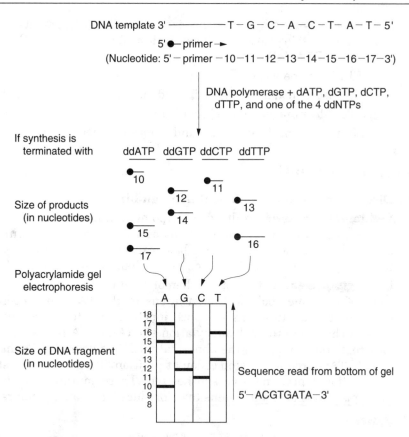

Figure 3-31. DNA sequencing by the dideoxynucleotide method. A DNA template is hybridized with a primer. DNA polymerase is added plus dATP, dGTP, dCTP, and dTTP. Either the primer or the nucleotides must have a radioactive label, so bands can be visualized on the gel by autoradiography. Samples are placed in each of four tubes and one of the four dideoxyribonucleotides (ddNTPs) is added to each tube to cause random termination of synthesis.

G. Use of probes to detect specific DNA or RNA sequences

1. Strands of DNA can hybridize (**base-pair**) with strands of DNA or RNA that contain a complementary sequence.

2. If the DNA contains a label, it can be used as a **probe** to detect complementary DNA or RNA.

   a. Southern blots are produced when radioactive DNA hybridizes with DNA on a nitrocellulose blot of a gel.

   b. Northern blots are produced when radioactive DNA hybridizes with RNA on a nitrocellulose blot of a gel.

   c. Western blot is a related technique in which proteins are separated by gel electrophoresis and probed with antibodies.

H. Use of variations in DNA sequences

1. **Polymorphisms** (variations in DNA sequences) occur frequently in the genome in coding or noncoding regions.

2. Variations may occur in restriction sites so that, for example, a sequence within a gene may be cleaved by a specific restriction enzyme but its allele may not.

3. The **variations** may cause the size of restriction fragments to differ for homologous chromosomes (**restriction fragment length polymorphisms [RFLPs]**).

–RFLPs may be used

   a. To distinguish homozygous individuals from heterozygous individuals

   b. To determine genetic relationships

   c. In the diagnosis of disease and in forensic medicine

VII. Clinical Correlations

A. Diseases related to abnormal hemoglobin

Sickle cell anemia results from a point mutation (GAG to GTG) that causes valine to replace glutamate at position 6 in the β-globin chain. In hemoglobin Wayne, deletion of a base causes a frameshift that produces the wrong sequence of amino acids in the chain beyond position 127.

In the **thalassemias** (a group of hemolytic anemias), mutations affect all steps of RNA metabolism. Substitutions in the TATA box decrease promoter function. Mutations in splice junctions create alternative splicing sites. A change in the polyadenylation site (AATAAA to AATAGA) causes incorrect processing, and the abnormal mRNA is degraded. A change from CAG to TAG produces a stop codon at position 39 that causes a shortened, nonfunctional protein to be synthesized. These mutations cause insufficient quantities of globin chains to be produced, and an anemia results.

B. Cancer

Cancer is a group of diseases in which cells are not responsive to the normal restraints on growth. The major causes of cancer are radiation, chemicals, and viruses. Chemicals and radiation cause damage to DNA, which, if not repaired rapidly, produces mutations that can result in cancer.

Burning organic material (e.g., cigarettes) produces chemicals such as benzo(a)pyrene that covalently bind to the bases in DNA, producing mutations that lead to **lung cancer**. Ultraviolet light, including that from the sun, produces pyrimidine dimers in DNA that lead to **skin cancer**. This condition is particularly pronounced in people with xeroderma pigmentosum because their DNA repair system does not function normally.

Oncogenes are genes that cause cancer. Their counterparts in normal cells, proto-oncogenes, are involved in growth and development. If oncogenes enter cells as a consequence of viral infection or if the normal proto-oncogenes are expressed abnormally, cancer may result. Many oncogenes encode proteins that are related to growth factors, to receptors for growth factors, or to proteins produced by growth factors. Some oncogene products enter the nucleus and activate genes. According to the oncogene theory, viruses may cause cancer by inserting additional or abnormal copies of proto-oncogenes into cells or by inserting promoters into regions that regulate the expression of these genes. Proto-oncogenes may be amplified. The gene for the proto-oncogene or its control region may undergo mutations due to radiation or chemicals. Alteration of the product or of the level of expression of a proto-oncogene produces changes in the growth characteristics of cells that may result in cancer.

The **treatment** of cancer frequently involves drugs that interfere with DNA synthesis. For example, 5-fluorouracil (5-FU) prevents the conversion of dUMP to dTMP, reducing the level of thymine nucleotides required for DNA synthesis. Methotrexate prevents formation of tetrahydrofolate from its more oxidized precursors. As a result, the formation of both thymine for DNA synthesis and the purines for DNA and RNA synthesis is inhibited.

C. Treatment of viral infections

When viruses infect cells, they convert the cells' DNA-, RNA-, and protein-producing machinery to the generation of viral genes and proteins (i.e., to the production of new viruses). Few drugs are currently effective against viral infections.

Azidothymidine (**AZT**), an analogue of thymidine, is phosphorylated in the cell and inhibits retroviral reverse transcriptase (which is used to make DNA copies of viral RNA) and serves as a DNA chain terminator. It is being used to treat HIV infections associated with AIDS.

D. Treatment of bacterial infections

Antibiotics that selectively affect bacterial function and have minimal side effects in humans are usually selected to treat bacterial infections. Rifampicin, which inhibits the initiation of prokaryotic RNA synthesis, is used to treat tuberculosis. Streptomycin, tetracycline, chloramphenicol, and erythromycin inhibit protein synthesis on prokaryotic ribosomes and are used for many infections. Chloramphenicol affects mitochondrial ribosomes and must be used with caution.

E. Diagnosis of disease

Biotechnology is being used in the diagnosis of disease. For example, samples of DNA are treated with restriction endonucleases and then subjected to electrophoresis on gels. Using the Southern blot technique, the restriction fragments on the gel are hybridized with radioactive cDNA for the gene of interest. If a mutation occurs in a cleavage site for the restriction enzyme, a pattern of restriction fragments that differs from normal will be found. Sickle cell genes, for instance, produce a fragment of 1350 base pairs when treated with *Mst*III rather than the normal fragment of 1150 base pairs. Techniques using recombinant DNA can be used to produce the probes (e.g., cDNA) for screening human samples and to generate large quantities of proteins, such as polypeptide hormones, for therapeutic purposes.

Review Test

Directions: Each of the numbered items or incomplete statements in this section is followed by answers or by completions of the statement. Select the **one** lettered answer or completion that is **best** in each case.

1. In DNA, on a molar basis

(A) adenine equals thymine
(B) adenine equals uracil
(C) guanine equals adenine
(D) cytosine equals thymine
(E) cytosine equals uracil

2. Which of the following sequences is complementary to the DNA sequence 5'–AAGTC-CGA–3'?

(A) 5'–AAGUCCGA–3'
(B) 3'–TTCAGGCT–5'
(C) 5'–TTCAGGCT–3'
(D) 3'–TCGGACTT–5'

3. DNA contains which one of the following components?

(A) Nitrogenous bases joined by phosphodiester bonds
(B) Negatively charged phosphate groups in the interior of the molecule
(C) Base pairs stacked along the central axis of the molecule
(D) Two strands that run in the same direction
(E) The sugar ribose

4. Which RNA contains 7-methylguanine at the 5' end?

(A) 5S RNA
(B) rRNA
(C) hnRNA
(D) tRNA

5. Thymine is present in which type of RNA?

(A) mRNA
(B) rRNA
(C) hnRNA
(D) tRNA

6. The action of DNA polymerases requires

(A) a 5'-hydroxyl group
(B) dUTP
(C) NAD$^+$ as a cofactor
(D) a 3'-hydroxyl group
(E) CTP

7. Which of the following statements concerning replication of DNA is TRUE?

(A) It progresses in both directions away from each point of origin on the chromosome
(B) It requires a DNA template that is copied in its 5' to 3' direction
(C) It occurs during the M phase of the cell cycle
(D) It produces one newly synthesized double helix and one composed of the two parental strands

8. When base-pairing occurs in loops of RNA, adenine is hydrogen-bonded to

(A) guanine
(B) thymine
(C) cytosine
(D) uracil

Questions 9–13

Use the figure below to answer questions 9–13.

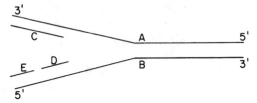

9. When synthesis of segment C begins, which other segment is also being synthesized?

(A) A
(B) B
(C) C
(D) D
(E) E

10. Segment C is synthesized

(A) from the middle toward both ends simultaneously
(B) toward the replication fork
(C) away from the replication fork
(D) in a 3' to 5' direction

11. Segment E is synthesized

(A) from the middle toward both ends simultaneously
(B) toward the replication fork
(C) away from the replication fork
(D) after segment D

12. The enzyme that joins segments D and E is

(A) the replicative DNA polymerase
(B) RNA polymerase
(C) an endonuclease
(D) polynucleotide ligase
(E) a repair DNA polymerase

13. If adenine is the first base on the template strand corresponding to the initiation point for segment E, which precursor molecule would serve as the substrate that formed the first nucleotide in segment E?

(A) dUTP
(B) UTP
(C) dTTP
(D) TTP
(E) dTMP

14. Which of the following statements concerning Okazaki fragments is TRUE?

(A) They are produced by restriction enzymes
(B) They are synthesized on the leading strand during replication
(C) They are regions of DNA that do not code for the amino acids in a protein
(D) They are relatively short polydeoxyribonucleotides with a few ribonucleotide residues at the 5' end
(E) They are products of the action of RNase on hnRNA

15. A bacterial mutant grows normally at 32°C but at 42°C accumulates short segments of newly synthesized DNA. Which of the following enzymes is most likely to be defective in this mutant?

(A) An endonuclease
(B) DNA polymerase
(C) An exonuclease
(D) An unwinding enzyme (helicase)
(E) Polynucleotide ligase

16. Which of the following phrases describes nucleosomes?

(A) Single ribosomes attached to mRNA
(B) Complexes of DNA and all the histones except H4
(C) Subunits of chromatin
(D) Structures that contain DNA in the core with histones wrapped around the surface
(E) Complexes of protein and the 45S rRNA precursors found in the nucleolus

17. The base pair shown is normally found in

(A) DNA
(B) RNA
(C) both DNA and RNA
(D) neither DNA nor RNA

18. In an embryo that lacked nucleoli, the synthesis of which type of RNA would be most directly affected?

(A) tRNA
(B) rRNA
(C) mRNA
(D) 5S RNA
(E) hnRNA

19. Eukaryotic genes that produce mRNA

(A) contain a TATA box downstream from the start site of transcription
(B) may contain a CAAT box in the 5' flanking region
(C) are transcribed by RNA polymerase III
(D) contain long stretches of thymine nucleotides that produce the poly(A) tail of mRNA
(E) do not contain intervening sequences or introns

20. If a fragment of DNA containing the sequence 5'–AGCCAATT–3' serves as the template for transcription, the RNA that is produced will have the sequence

(A) 5'–AGCCAAUU–3'
(B) 5'–UCGGUUAA–3'
(C) 5'–UUAACCGA–3'
(D) 5'–AAUUGGCU–3'

21. A person ate mushrooms picked in a wooded area. Shortly thereafter, he was rushed to the hospital, where he died. He had no previous medical problems. The cause of his death was most likely the RNA polymerase inhibitor

(A) rifampicin
(B) α-amanitin
(C) streptolydigin
(D) actinomycin D

22. When benzo(a)pyrene (a carcinogen in cigarette smoke) binds to DNA, it forms a bulky covalent adduct on guanine residues. The consequence is that

(A) cells are rapidly transformed into cancer cells
(B) glycosylases remove the benzpyrene residues
(C) a repair process usually removes and replaces the damaged region of DNA
(D) UV light cleaves the benzpyrene from the guanine residue

23. Patients with xeroderma pigmentosum suffer DNA damage when they are exposed to UV light because UV light causes the formation of

(A) purine dimers in DNA
(B) pyrimidine dimers in DNA
(C) deoxyribose dimers in DNA
(D) anhydride bonds between phosphate groups in DNA

24. Patients with xeroderma pigmentosum develop skin cancer when they are exposed to sunlight because they have a deficiency in

(A) primase
(B) recombinase
(C) glycosylase
(D) an enzyme that acts early in the excision repair pathway
(E) an enzyme essential to repair mismatched bases

25. A common mutagenic event is the deamination of cytosine in DNA to form uracil. If the damaged strand is replicated, a CG base pair in DNA will be converted to a

(A) TA base pair
(B) GC base pair
(C) GG base pair
(D) UG base pair

26. If cytosine in DNA is deaminated, the uracil residue that results may be removed by

(A) an endonuclease
(B) an exonuclease
(C) a glycosylase
(D) polynucleotide ligase
(E) a repair DNA polymerase

27. An aminoacyl-tRNA exhibits which one of the following characteristics?

(A) It is produced by a synthetase that is specific for the amino acid, but not the tRNA
(B) It is composed of an amino acid esterified to the 5' end of the tRNA
(C) It requires GTP for its synthesis from an amino acid and a tRNA
(D) It contains an anticodon that is complementary to the codon for the amino acid

28. Which one of the following point mutations would NOT produce a change in the protein translated from an mRNA?

(A) UCA → UAA
(B) UCA → CCA
(C) UCA → UCU
(D) UCA → ACA
(E) UCA → GCA

29. The structure shown is

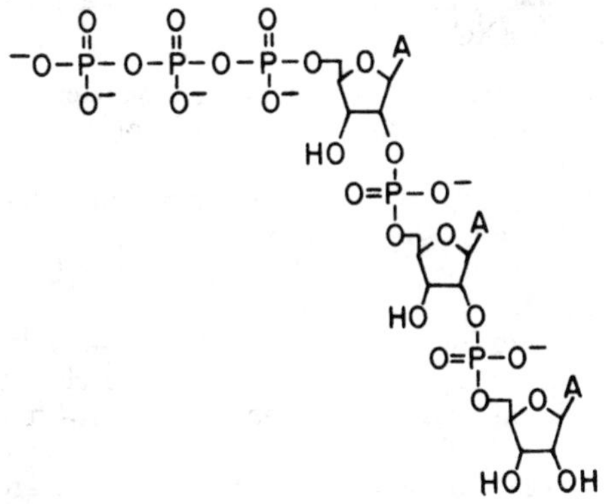

(A) a segment of the poly(A) tail of mRNA
(B) a region of DNA that, when transcribed to mRNA, codes for phenylalanine
(C) a sequence found at the 3' end of all tRNAs
(D) not found in DNA or RNA

30. Which of the following statements about methionine is TRUE?

(A) It is the amino acid used for initiation of the synthesis of proteins
(B) It is generally found at the N-terminus of proteins isolated from cells
(C) It requires a codon other than AUG to be added to growing polypeptide chains
(D) It is formylated when it is bound to tRNA in eukaryotic cells

31. Which of the following statements about bacteria is correct?

(A) They contain 80S ribosomes
(B) They initiate protein synthesis with methionyl-tRNA
(C) They are insensitive to chloramphenicol
(D) They synthesize proteins on mRNA that is in the process of being transcribed

32. Which of the following is NOT required for initiation of protein synthesis in the cytoplasm of eukaryotic cells?

(A) A 40S ribosomal subunit
(B) eIF-2
(C) Methionyl-tRNA$_i^{Met}$
(D) GTP
(E) EF-2

33. Which of the following is NOT required for the elongation reactions of protein synthesis in eukaryotes?

(A) Peptidyl transferase
(B) GTP
(C) Formylmethionyl-tRNA
(D) Elongation factor 2 (EF-2)
(E) mRNA

34. The mechanism for termination of protein synthesis in eukaryotes requires

(A) a peptidyl-tRNA that cannot bind at the P site
(B) the codon UGA, UAG, or AUG in the A site
(C) nuclease cleavage of mRNA
(D) release factors

35. Proteins that are secreted from cells

(A) contain methionine as the N-terminal amino acid
(B) are produced from translation products that have a signal sequence at the C-terminal end
(C) are synthesized on ribosomes that bind to proteins on the endoplasmic reticulum
(D) contain a hydrophobic sequence at the C-terminal end that is embedded in the membrane of secretory vesicles
(E) contain carbohydrate residues that bind to receptors on the interior of lysosomal membranes

36. Tetracycline, streptomycin, and erythromycin are effective antibiotics because they inhibit

(A) RNA synthesis in prokaryotes
(B) RNA synthesis in eukaryotes
(C) protein synthesis in prokaryotes
(D) protein synthesis on cytoplasmic ribosomes of eukaryotes
(E) protein synthesis on mitochondrial ribosomes of eukaryotes

Questions 37 and 38

In the year 2020, a new antibiotic will be available. Studies performed using an in vitro protein-synthesizing system produced the expected polypeptide in response to a synthetic mRNA with the following sequence:

AUGUUCUUCUUCUUCUUCUUCUAA

37. The polypeptide produced in the in vitro system was

(A) Met-Phe-Phe-Phe-Phe-Phe-Phe
(B) Met-Ser-Ser-Ser-Ser-Ser-Ser
(C) Met-Leu-Leu-Leu-Leu-Leu-Leu
(D) Met-Phe-Ser-Leu-Phe-Ser-Leu

38. When this in vitro protein-synthesizing system was treated with the new antibiotic, the only product was methionyl-phenylalanyl-tRNA. What process in protein synthesis is inhibited by the antibiotic?

(A) Binding of an aminoacyl-tRNA to the A site of the ribosome
(B) Initiation
(C) Translocation
(D) Peptidyl transferase catalyzed formation of a peptide bond

39. In bacterial operons that are inducible

(A) the inducer binds to the repressor and activates it
(B) the inducer stimulates binding of RNA polymerase to the promoter
(C) a regulatory gene produces an inactive repressor
(D) structural genes that are adjacent on the DNA are coordinately expressed in response to the inducer
(E) each of the structural genes produces a separate mRNA

40. When cAMP levels are relatively high in *E. coli*

(A) lactose is not required for transcription of the *lac* operon
(B) glucose levels in the medium are low
(C) the repressor is bound to the *lac* operon if lactose is present
(D) the enzymes for the metabolism of lactose are not induced

41. Each of the following statements about regulation of protein synthesis in eukaryotes is correct EXCEPT

(A) a gene that is methylated is transcribed more readily than one that is not methylated
(B) genes may undergo rearrangements that allow cells to produce new proteins
(C) red blood cells do not produce hemoglobin because they lack the appropriate gene
(D) recognition of alternative polyadenylation sites allows cells to produce closely related proteins that have different functions
(E) steroid hormones activate genes that are nonspecifically repressed by histones

42. Gene transcription rates and mRNA levels were determined for an enzyme that is induced by glucocorticoids. Compared to untreated levels, glucocorticoid treatment caused a 10-fold increase in the gene transcription rate and a 20-fold increase in both mRNA levels and enzyme activity. These data indicate that a primary effect of glucocorticoid treatment is to

(A) increase the rate of mRNA translation
(B) decrease the rate of mRNA translation
(C) increase mRNA stability
(D) decrease mRNA stability

43. Use the following figure to answer question 43.

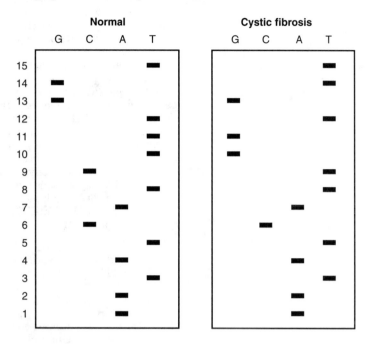

The gene for cystic fibrosis (CF) has been isolated and sequenced. The gel pattern for the DNA sequence of the region that differs from the normal gene is shown above. The positions of the bases, starting at the 5' end of this sequence, are indicated on the left. In this region

(A) there is no homology between the normal and CF genes
(B) the first nine bases of the CF gene are the same as the normal gene
(C) the four bases at positions 12–15 of the normal gene are the same as those at positions 9–12 of the CF gene
(D) the CF gene has a 3-base insertion
(E) the CF gene has a 4-base deletion

44. Use the following figure to answer question 44.

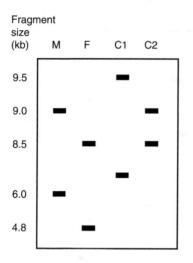

Fragment size (kb)   M   F   C1   C2

Two male infants were born on the same day in the same hospital. Because of concern that the infants had been switched in the hospital nursery, genetic tests based on a DNA restriction fragment that exhibits polymorphism (RFLP, restriction fragment length polymorphism) were performed. Blood was drawn from the parents and the infants, the DNA extracted, and polymerase chain reaction (PCR) performed. The DNA was then treated with the restriction enzyme *Ban*I, and the fragments were separated by gel electrophoresis. The results of a Southern blot test are shown above. A radioactive probe was used that bound to a sequence within the *Ban*I fragments that exhibited polymorphism. Which of the two infants, C1 or C2, is the genetic offspring of this mother (M) and father (F)?

(A) C1 could be the offspring of these parents
(B) C2 could be the offspring of these parents
(C) Both infants could be the offspring of these parents (i.e., this test cannot discriminate)
(D) Either of these infants could be related to the mother, but neither could be related to the father
(E) Neither infant could be related to this mother or this father

45. Which region (A to D) of the DNA strands shown could serve as the template for transcription of the region of an mRNA that contains the initial codon for translation of a protein 300 amino acids in length?

```
             ┌─── A ───┐              ┌─── B ───┐
5' . . . . . . A G A T G C C C T A A G G T C A T T G T T . . . . . 3'
3' . . . . . . T C T A C G G G A T T C C A G T A A C A A . . . . . 5'
             └─── D ───┘              └─── C ───┘
```

(A) A
(B) B
(C) C
(D) D

Directions: Each group of items in this section consists of lettered options followed by a set of numbered items. For each item, select the **one** lettered option that is most closely associated with it. Each lettered option may be selected once, more than once, or not at all.

Questions 46–49

Match each characteristic below with the most appropriate enzyme.

(A) DNA polymerase
(B) RNA polymerase
(C) Polynucleotide ligase
(D) Reverse transcriptase

46. Used by retroviruses to copy their RNA genome

47. Uses a nucleotide containing uracil as a precursor

48. Initiates the synthesis of polynucleotide strands

49. Does not produce inorganic pyrophosphate

Questions 50–53

For each characteristic below, select the most appropriate RNA of eukaryotes.

(A) mRNA
(B) hnRNA
(C) rRNA
(D) tRNA

50. Contains introns and a cap at the 5' end

51. Contains no introns but has a poly(A) tail

52. Does not travel from the nucleus to the cytoplasm

53. Is produced by RNA polymerase I

Questions 54–56

Match each description with the most appropriate term.

(A) Introns
(B) Exons
(C) TGA
(D) TAA

54. Generally absent from bacterial genes

55. Removed from transcripts by a splicing process

56. Codes for amino acids in proteins

Questions 57–61

Match each of the effects below with the most appropriate drug.

(A) Rifampicin
(B) 5-Fluorouracil
(C) Erythromycin
(D) Streptomycin
(E) Tetracycline

57. Prevents formation of the translation initiation complex

58. Binds to RNA polymerase and prevents transcription

59. Inhibits translocation

60. Prevents synthesis of DNA

61. Prevents binding of aminoacyl-tRNAs to the A site

Questions 62–65

In the diagram shown below, the letters indicate the phases of the cell cycle. For each statement, select the letter corresponding to the most appropriate phase.

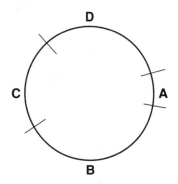

62. Anticancer drugs inhibit replication

63. Colchicine prevents formation of the mitotic spindle

64. Screening tests for chromosomal abnormalities can be performed

65. Liver cells enter this phase when they have been stimulated to divide

Answers and Explanations

1–A. On a molar basis, DNA contains equal amounts of adenine and thymine and of guanine and cytosine. Uracil is not found in DNA.

2–B. Complementary sequences base-pair with each other. The strands run in opposite directions. 5'–AAGTCCGA–3' base-pairs with 3'–TTCAGGCT–5' (or with the RNA sequence 3'–UUCAGGCU–5').

3–C. DNA chains are composed of nucleosides joined by 3',5'-phosphodiester bonds. Each nucleoside consists of a nitrogenous base linked to deoxyribose. Two DNA chains, oriented in opposite directions, base-pair with each other and are twisted to form a double helix. The bases are stacked on top of each other, forming a "spiral staircase" in the center of the molecule, and the sugar-phosphate backbone, in which the phosphates are negatively charged, is wrapped around the outside.

4–C. hnRNA is capped at its 5' end during transcription. The cap, which contains 7-methylguanine, is retained when hnRNA is converted to mRNA.

5–D. Each tRNA contains one thymine residue in the ribothymidine that is found in the TψC loop.

6–D. DNA polymerase requires dATP, dTTP, dGTP, and dCTP as precursors that add to the 3'-hydroxyl group at the 3' end of the growing chain.

7–A. Replication occurs during the S phase of the cell cycle. "Bubbles" on the parental DNA serve as points of origin from which replication proceeds in both directions simultaneously. (Replication is bidirectional.) DNA template strands are always copied in a 3' to 5' direction. New strands are synthesized 5' to 3'. The daughter molecules each contain one parental strand and one newly synthesized strand. (Replication is semiconservative.)

8–D. RNA contains uracil instead of thymine. Uracil base-pairs with adenine.

9–E. Segment C is synthesized in a 5' to 3' direction. When synthesis of C begins, E is also being synthesized.

10–B. Segment C is synthesized in a 5' to 3' direction, toward the replication fork.

11–C. Segment E is synthesized in a 5' to 3' direction (following the template 3' to 5') away from the replication fork. Segment E is synthesized before segment D. Therefore, the overall direction of synthesis of the lagging strand is toward the replication fork; this occurs as short fragments are synthesized in the opposite direction and then are joined together.

12–D. Polynucleotide ligase joins DNA chains together. Polymerases can only add one nucleotide at a time to the 3' end of a growing chain. An endonuclease cleaves DNA strands.

13–B. The first nucleotide to be added must base-pair with adenine. However, DNA polymerase cannot initiate synthesis of strands. RNA polymerase (the primase) produces a short primer that DNA polymerase can extend. Therefore, the first nucleoside must be a ribonucleotide that base-pairs with adenine, that is, UTP. (Nucleoside triphosphates serve as precursors for the polymerases.)

14–D. Okazaki fragments are synthesized on the lagging strand. They consist of a few ribonucleotides (added by RNA polymerase) to which deoxyribonucleotides were added by DNA polymerase.

15–E. The short segments of newly synthesized DNA that accumulate at 42°C are Okazaki fragments. They are usually joined together by polynucleotide ligase, which is probably temperature-sensitive in this mutant. If the ligase is not functioning, Okazaki fragments would not be joined during replication, so the cells would contain short fragments of DNA. Endonucleases and exonucleases cleave DNA strands in the middle and at the ends, respectively. They do not join fragments together, nor does DNA polymerase. Unwinding enzymes "unzip" the parental strands.

16–C. Nucleosomes are subunits of chromatin (the complex of DNA and proteins in the nucleus). They consist of a core composed of two molecules of each of the four histones—H2A, H2B, H3, and H4—with 140 base pairs of DNA wrapped around the surface. Histone H1 binds to the linker DNA that joins one nucleosome core to the next.

17–B. The figure shows adenine on the right base-paired to uracil on the left. Uracil is present in RNA but not in DNA.

18-B. In nucleoli, rRNA genes are transcribed to produce the 45S rRNA precursor, which is trimmed, modified, and complexed with proteins to form ribosomal subunits. Therefore, the synthesis of rRNA would be most directly affected. The embryo probably would not survive.

19–B. Eukaryotic genes contain TATA and CAAT boxes in the 5' flanking region, upstream from the start site for transcription. RNA polymerase II transcribes these genes, producing hnRNA, which is modified and processed to form mRNA. A cap is added at the 5' end and poly(A) is added to the 3' end posttranscriptionally; they are not encoded in the DNA. These genes contain introns, which are removed during processing of hnRNA.

20–D. The RNA will run in the opposite direction and will be complementary: A pairing with T, U with A, G with C, and C with G. Thus, the RNA will be 3'–UCGGUUAA –5' (which is 5'–AAUUGGCU–3').

21-B. The poison in poisonous mushrooms is α-amanitin, an inhibitor of eukaryotic RNA polymerases.

22–C. If the damage is repaired rapidly enough, transformation does not occur. Bulky lesions in DNA are repaired by the process of nucleotide excision. A short segment of nucleotides is removed and replaced, with the undamaged strand serving as the template.

23–B. UV light causes the formation of pyrimidine dimers in DNA.

24–D. The damage to DNA caused by ultraviolet light (pyrimidine dimers) can be repaired by the excision repair pathway. In some cases, the missing enzyme is a repair endonuclease.

25–A. When the DNA is replicated, U in the template strand will pair with A in the daughter strand. Subsequent rounds of replication will cause the original CG base pair to become a TA base pair.

26–C. Although other enzymes are involved in the repair of DNA, only glycosylases remove bases by cleaving *N*-glycosidic bonds.

27–D. Aminoacyl-tRNA synthetases react with ATP and an amino acid to form an enzyme [aminoacyl-AMP] complex. The amino acid is then transferred to the 3' end of tRNA. These enzymes are very specific for both the amino acid and the tRNA. The codon on mRNA for the amino acid is complementary to the anticodon on the tRNA.

28–C. UCA is a codon for serine. It is converted to a termination codon by mutation A, to a proline codon by mutation B, to a threonine codon by mutation D, and to an alanine codon by mutation E. Only mutation C would produce no change in the protein, since UCU is also a codon for serine.

29–D. The structure contains 2',5'-phosphodiester bonds, so it is not part of DNA or RNA, both of which have 3',5'-phosphodiester bonds. This compound, 2',5'-oligo(A), is produced in response to interferon and activates a nuclease that degrades mRNA.

30–A. Methionine, the amino acid that initiates the synthesis of proteins, is subsequently cleaved from the protein. The only codon for methionine is AUG, which serves as the codon for methionine residues within a protein as well as the initiating residue. The methionyl-tRNA for initiation is formylated in bacterial cells and in mitochondria.

31–D. Bacteria contain 70S ribosomes that are sensitive to chloramphenicol. They initiate protein synthesis with methionine that is formylated. Because bacteria do not have nuclei, ribosomes bind to mRNA as it is being synthesized, so that translation begins before transcription is completed.

32–E. Methionyl-tRNA, initiation factor eIF-2, and GTP bind to the 40S ribosomal subunit. The cap of mRNA binds, secondary structures unwind, and the ribosomal subunit moves along the mRNA until the first AUG codon pairs with the anticodon on the tRNA. The 60S subunit is added, and protein synthesis is initiated. EF-2 is used during the elongation process.

33–C. mRNA supplies the codons, aminoacyl-tRNA and GTP provide energy, peptidyl transferase catalyzes the formation of peptide bonds, and elongation factor 2 translocates the peptidyl-tRNA. Formylmethionyl-tRNA is involved in initiation of protein synthesis in prokaryotes.

34–D. Protein synthesis is normally terminated by release factors, not by nuclease cleavage of mRNA or peptidyl-tRNA that cannot bind at the P site. UGA and UAG are stop codons, but AUG is the initiation codon.

35–C. Proteins destined for secretion contain a signal sequence at the N-terminal end that causes the ribosomes on which they are being synthesized to bind to the RER. As they are being produced, they enter the cisternae of the RER, where the signal sequence, including the initial methionine, is removed. Carbohydrate groups may be attached in the RER or the Golgi. Secretory vesicles bud from the Golgi, and the proteins are secreted from the cell by the process of exocytosis. If the proteins have a hydrophobic sequence that embeds in the membrane, they remain attached and are not secreted.

36–C. These antibiotics inhibit protein synthesis in prokaryotes; thus, they can be used to treat bacterial infections. One of their undesirable side effects, however, is that they also inhibit protein synthesis on mitochondrial ribosomes (which are of the 70S prokaryotic class).

37–A. AUG is the initiation codon for methionine. The subsequent bases, read sequentially in sets of three, would all produce phenylalanine, except the termination codon UAA.

38–C. Initiation and formation of the first peptide bond occurred; therefore, A, B, and D were not affected. The antibiotic most likely affects the translocation step.

39–D. In induction, a regulatory gene produces an active repressor, which is inactivated by binding to the inducer. The inducer removes the repressor from the operator rather than stimulating binding of RNA polymerase. The structural genes are coordinately expressed. Transcription yields a single, polycistronic mRNA, which is translated to produce a number of different proteins.

40–B. cAMP is involved in catabolite repression. Bacterial cells preferentially use glucose. When glucose is low, cAMP rises. cAMP binds to a protein and complexes near the *lac* promoter region, facilitating binding of RNA polymerase. Lactose must be present to inactivate the repressor, so that the operon may be expressed.

41–A. A gene that is methylated is less readily transcribed.

42–C. If the rate of degradation of the mRNA is not altered by glucocorticoids, the increase in mRNA levels should reflect the increase in transcription rate. Because the increase in mRNA level is greater than the increase in transcription rate, the glucocorticoids must also be increasing mRNA stability (i.e., decreasing the rate of degradation).

43–C. Sequences are read from the bottom to the top of the gel. In this region, the sequences of the CF and normal genes are identical for the first eight bases. Positions 12–15 of the normal gene and 9–12 of the CF gene are identical. Therefore, there is a 3-base deletion in the CF gene corresponding to bases 9–11 of the normal gene.

44–B. Every chromosome has a homologue. Therefore, there will be two copies of every DNA sequence in the genome. Child C2 could have obtained the 9.0-kb restriction fragment from this mother, and the 8.5-kb fragment from this father. According to this test, child C1 is not genetically related to either this mother or this father.

45–B. Although D contains the sequence 3'–TAC–5', which produces a start codon (5'–AUG–3') in the mRNA, there is a stop codon (3'–ATT–5' in the DNA) in frame with this start codon. Sequence B, read 3' to 5' (from right to left), would produce a start codon in the mRNA transcribed from it. There are no stop codons in this sequence, so it *could* produce a protein 300 amino acids in length. Sequences A and C do not contain triplets corresponding to the start codon in mRNA.

46–D. Reverse transcriptase produces a DNA copy of an RNA template.

47–B. UTP is a substrate for RNA polymerase.

48–B. Only RNA polymerase can initiate the synthesis of strands.

49–C. When nucleoside triphosphates add to growing DNA or RNA strands, pyrophosphate is released. Ligase joins two polynucleotide strands together.

50–B. Both hnRNA and mRNA are capped, but only hnRNA contains introns.

51–A. Both hnRNA and mRNA contain poly(A) tails, but hnRNA contains introns that are removed as it is processed to form mRNA.

52–B. Only hnRNA does not travel from the nucleus to the cytoplasm.

53–C. RNA polymerase I produces the large rRNA precursor. Polymerase II produces hnRNA (and consequently mRNA). Polymerase III produces tRNA and 5S rRNA.

54–A. Bacterial genes generally do not contain introns.

55–A. Introns are removed from RNA transcripts by splicing mechanisms.

56–B. Exons code for amino acids in proteins. Introns do not. TGA and TAA produce stop codons in mRNA.

57–D. Streptomycin causes misreading of mRNA codons and, thereby, prevents formation of the initiation complex in prokaryotes.

58–A. Rifampicin binds to RNA polymerase and prevents initiation of transcription in prokaryotes.

59–C. Erythromycin prevents translocation during protein synthesis in prokaryotes.

60–B. 5-Fluorouracil prevents conversion of dUMP to dTMP. DNA synthesis is inhibited because of a lack of thymine nucleotides, so this compound is used to treat cancer.

61–E. Tetracycline prevents the binding of aminoacyl-tRNAs to the A site on ribosomes.

62–C. Replication occurs during the S phase.

63–A. The mitotic spindle functions during mitosis (M phase).

64–A. Condensed chromosomes can be observed under the microscope during the M phase (mitosis).

65–B. Liver cells in a nondividing state (G_0) can be stimulated to reenter the cell cycle at the G_1 phase.

4

Generation of ATP from Metabolic Fuels

Overview

- ATP transfers energy from the processes that produce it to those that use it.
- Most of the carbons of the glucose, fatty acids, glycerol, and amino acids derived from food are converted, sometimes by circuitous routes, to acetyl CoA. ATP is produced by some of these reactions (Figure 4-1).
- Acetyl CoA is oxidized in the TCA cycle. CO_2 is released, and electrons are passed to NAD^+ and FAD, producing NADH and $FADH_2$.
- NADH and $FADH_2$ transfer the electrons to O_2 via the electron transport chain. Energy from this transfer of electrons is used to produce ATP by the process of oxidative phosphorylation.

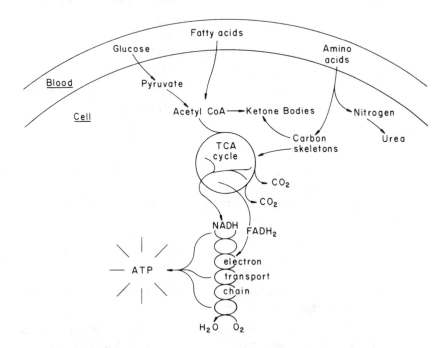

Figure 4-1. The generation of ATP from fuels in the blood.

● Cofactors, many of which are minerals or compounds produced from vitamins, aid the enzymes that catalyze these metabolic reactions.

I. Bioenergetics

● For a biochemical reaction
$$aA + bB \rightleftharpoons cC + dD$$
the change in free energy (ΔG) is related to the concentrations of the substrates and products and to the change in the standard free energy of the reaction at pH 7 ($\Delta G^{\circ\prime}$). The change in standard free energy is determined by the chemical bonds that are being broken and formed.

● Reactions with a negative ΔG proceed spontaneously. Those with a positive ΔG do not. If $\Delta G = 0$, the reaction is at equilibrium, and
$$K_{eq} = \frac{[C]^c [D]^d}{[A]^a [B]^b}$$
(where the substrates and products are at their equilibrium concentrations). Therefore,
$$\Delta G^{\circ\prime} = -RT \ln K_{eq}$$

● The change in free energy can be used to predict the direction in which a reaction will proceed. It gives no indication of the speed with which the reaction will occur.

A. The change in free energy in biologic systems

1. In biologic systems, the change in free energy (the energy available to do useful **work** at constant pressure and temperature) is defined by the equation
$$\Delta G = \Delta H - T\Delta S$$
where ΔG is the change in free energy, ΔH is the change in enthalpy, ΔS is the change in entropy, and T is the absolute temperature in $^\circ K$.

2. For a biochemical reaction, the change in free energy can be used to predict the **direction** in which the reaction will proceed.

3. For the reaction in which
$$aA + bB \rightleftharpoons cC + dD$$
(where the upper-case letters symbolize the molecule; the lower-case letters indicate the number of molecules), the free energy change depends on the concentrations of substrates and products and on the value of the constant $\Delta G^{\circ\prime}$.
$$\Delta G = \Delta G^{\circ\prime} + RT \ln \frac{[C]^c [D]^d}{[A]^a [B]^b}$$
where $\Delta G^{\circ\prime}$ is the standard free energy change at pH 7, R is the gas constant, and [ ] means concentration.

4. If **ΔG is negative,** the reaction will proceed spontaneously with the release of energy. If **ΔG is positive,** the reaction will not proceed spontaneously. If

ΔG is 0, the reaction is at **equilibrium,** and, although substrates react to form products and products react to form substrates, there is no net change in the concentrations.

B. The equilibrium constant and the change in free energy

1. At **equilibrium, ΔG = 0** and

$$\Delta G^{\circ\prime} = -RT \ln \frac{[C]^c \, [D]^d}{[A]^a \, [B]^b}$$

2. The **equilibrium constant** (K_{eq}) is related to the **concentrations** of substrates and products **at equilibrium,**

$$K_{eq} = \frac{[C]^c \, [D]^d}{[A]^a \, [B]^b}$$

(where the substrates and products are at their equilibrium concentrations). Therefore,

$$\Delta G^{\circ\prime} = -RT \ln K_{eq}$$

–If $K_{eq} = 1$, $\Delta G^{\circ\prime} = 0$.
–If $K_{eq} > 1$, $\Delta G^{\circ\prime}$ is negative.
–If $K_{eq} < 1$, $\Delta G^{\circ\prime}$ is positive.

3. The larger and more negative the $\Delta G^{\circ\prime}$, the less substrate relative to product is required to produce a negative ΔG (i.e., the more likely the reaction is to proceed spontaneously).

4. For a sequence of reactions that have common intermediates, the standard free energy changes are **additive** (Table 4-1).

C. The relevance of free energy changes to biologic systems

1. The **rate** of a reaction is not related to its free energy change.

   a. A reaction with a large negative free energy change does not necessarily proceed rapidly.

   b. The **speed** with which a reaction proceeds depends on the properties of the **enzyme** that catalyzes the reaction.

   –An enzyme increases the rate at which a reaction reaches equilibrium. It does not affect **K_{eq}** (the relative concentrations of substrates and products at equilibrium).

2. Most biochemical reactions exist in pathways; therefore, other reactions are constantly adding substrate and removing product.

   –The relative activities of the enzymes that catalyze the individual reactions of a pathway differ.

   –Some reactions are near equilibrium ($\Delta G = 0$). Their direction can be

Table 4-1. The Additive Nature of Free Energy Changes

glucose + P_i → glucose-6-P + H_2O	$\Delta G^{\circ\prime}$ = +3.3 kcal/mole
ATP + H_2O → ADP + P_i	$\Delta G^{\circ\prime}$ = –7.3 kcal/mole
Sum: glucose + ATP → glucose-6-P + ADP	$\Delta G^{\circ\prime}$ = –4.0 kcal/mole

readily altered by small changes in the concentrations of their substrates or products.

–Other reactions are far from equilibrium. Allosteric factors that alter the activity of these enzymes can change the overall flux through the pathway.

II. Properties of Adenosine Triphosphate

- ATP (adenosine triphosphate) contains the base adenine, the sugar ribose, and three phosphate groups joined to each other by two anhydride bonds.
- ATP is synthesized mainly by the process of oxidative phosphorylation.
- The free energy released when ATP is hydrolyzed is used to drive reactions that require energy.
- ATP can transfer phosphate groups to other compounds such as glucose, forming ADP.
- ADP can accept phosphate groups from compounds such as phosphocreatine, forming ATP.

A. The structure of ATP

–ATP is composed of the base **adenine,** the sugar **ribose,** and **three phosphate groups** (Figure 4-2).

1. **Adenosine** (a nucleoside) consists of the base adenine linked by an *N*-glycosidic bond to the sugar ribose.

2. **AMP** (adenosine monophosphate) is a nucleotide that contains the nucleoside adenosine in which the 5'-hydroxyl of the sugar is esterified to an acidic group of inorganic phosphate.

3. **ADP** (adenosine diphosphate) contains a second phosphate group attached to the phosphate of AMP by an anhydride bond.

4. **ATP** contains a third phosphate group linked to ADP by another anhydride bond.

B. The functions of ATP

–ATP plays a central role in **energy exchanges** in the body.

1. ATP is constantly being **consumed and regenerated**.

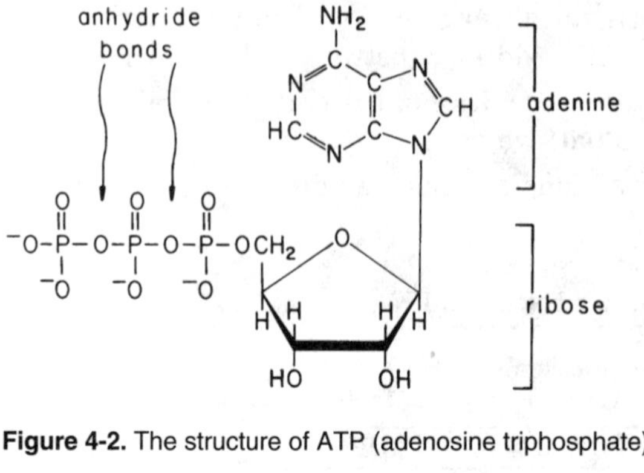

Figure 4-2. The structure of ATP (adenosine triphosphate).

    a. It is consumed by processes such as muscular contraction, active transport, and biosynthetic reactions.

    b. It is regenerated by the oxidation of foodstuffs.

2. The **free energy** that is released when ATP is hydrolyzed is used to **drive reactions** that require energy.

    a. ATP may be hydrolyzed to **ADP** and inorganic phosphate (**P$_i$**) or to **AMP** and pyrophosphate (**PP$_i$**). ATP, ADP, and AMP may be interconverted by the adenylate kinase reaction

$$ATP + AMP \rightleftharpoons 2\,ADP$$

    b. Other nucleoside triphosphates (GTP, UTP, and CTP) are sometimes used to drive biochemical reactions. They may be derived from ATP.

3. When ATP is hydrolyzed, **$\Delta G^{\circ\prime}$ = –7.3 kcal/mole**.

    a. The anhydride bonds of ATP are often called "high-energy bonds."

    b. **$\Delta G^{\circ\prime}$ is large,** however, not because a single bond is broken, but because the products of hydrolysis are more stable than ATP.

4. ATP can transfer phosphate groups to compounds such as glucose, forming ADP.

5. ADP can accept phosphate groups from compounds such as phosphoenolpyruvate, phosphocreatine, or 1,3-bisphosphoglycerate, forming ATP.

III. Electron Carriers and Vitamins

- Certain cofactors of enzymes transfer electrons from foodstuffs to O_2, a process that generates energy for the synthesis of ATP.
- NAD$^+$ (derived from the vitamin niacin) and FAD (derived from the vitamin riboflavin) pass electrons to the electron transport chain. FMN and coenzyme Q (ubiquinone) pass the electrons to heme-containing cytochromes, which transfer the electrons to O_2. As a consequence of these processes, ATP is produced.
- Other cofactors involved in deriving energy from food include coenzyme A (derived from the vitamin pantothenate), thiamine pyrophosphate, and lipoic acid.
- Additional cofactors derived from water-soluble vitamins are involved in a variety of metabolic reactions. These include NADPH (derived from the vitamin niacin), biotin, pyridoxal phosphate (derived from vitamin B$_6$), tetrahydrofolate (derived from the vitamin folate), vitamin B$_{12}$, and vitamin C.
- The fat-soluble vitamins (A, D, E, and K) are also involved in metabolism.

A. Major cofactors in generation of ATP from foodstuffs

1. NAD$^+$ (nicotinamide adenine dinucleotide) and **FAD** (flavin adenine dinucleotide)

    —As food is oxidized to CO_2 and H_2O, electrons are transferred mainly to **NAD$^+$ and FAD.**

    a. NAD$^+$ accepts a hydride ion, which reacts with its nicotinamide ring (Figure 4-3). NAD$^+$ is reduced, the substrate (RH$_2$) is oxidized, and a proton is released.

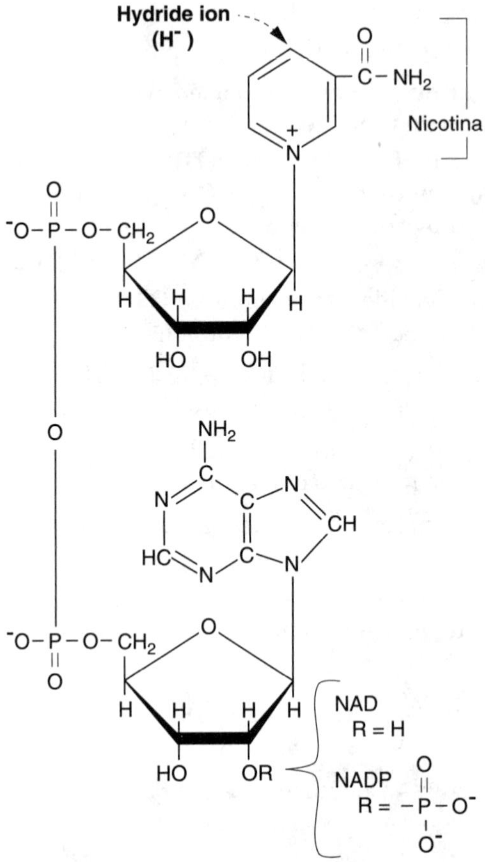

Figure 4-3. The structure of NAD$^+$ and NADP$^+$. R differs for NAD$^+$ and NADP$^+$ as indicated. The *arrow* indicates the position where a hydride ion (H$^-$) covalently binds when NAD$^+$ or NADP$^+$ is reduced.

$$NAD^+ + RH_2 \rightleftharpoons NADH + H^+ + R$$

(1) NAD$^+$ is frequently involved in oxidizing a hydroxyl group to a ketone.

$$\underset{\displaystyle R-CH-R_1}{\overset{\displaystyle OH}{|}} + NAD^+ \rightleftharpoons \underset{\displaystyle R-C-R_1}{\overset{\displaystyle O}{\|}} + NADH + H^+$$

(2) The **nicotinamide ring** of NAD$^+$ is derived from the vitamin **niacin** (nicotinic acid). It is also produced to a limited extent from the amino acid **tryptophan**.

b. FAD accepts two hydrogen atoms (with their electrons) that react with its isoalloxazine ring (Figure 4-4). FAD is reduced and the substrate is oxidized.

$$FAD + RH_2 \rightleftharpoons FADH_2 + R$$

(1) FAD is frequently involved in reactions that produce a double bond.

$$R-CH_2-CH_2-R_1 + FAD \rightleftharpoons R-CH=CH-R_1 + FADH_2$$

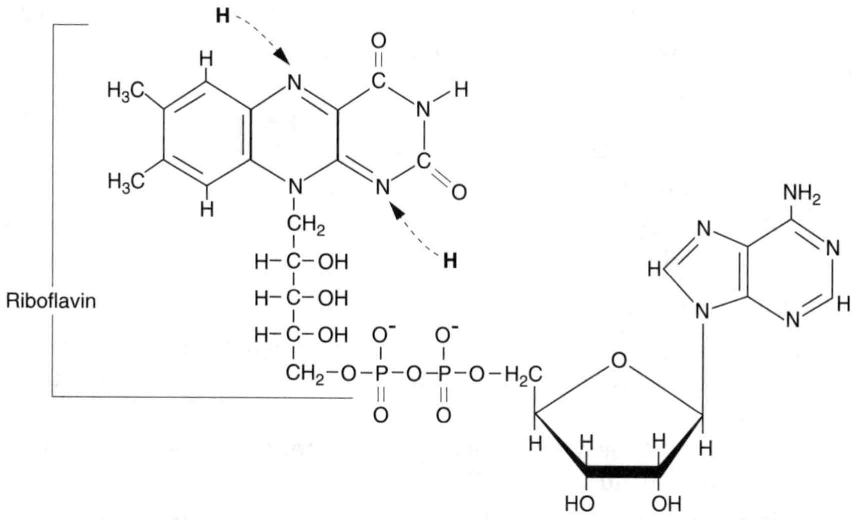

Figure 4-4. The structure of FAD. *Arrows* indicate positions where hydrogens covalently bind when FAD is reduced to FADH$_2$. FMN consists only of the riboflavin moiety plus one phosphate.

(2) FAD is derived from the vitamin **riboflavin** by the addition of a phosphate group that forms an anhydride bond with the phosphate of AMP.

2. The electron transport chain

–The reduced cofactors, NADH and FADH$_2$, transfer electrons to the electron transport chain.

   a. NADH transfers electrons to FMN (flavin mononucleotide), which transfers the electrons through Fe-S centers to **coenzyme Q** (Figure 4-5).

   –FMN is derived from **riboflavin** by the addition of a phosphate group.

   b. FADH$_2$ transfers electrons through Fe-S centers directly to **coenzyme Q** (CoQ).

   –FADH$_2$ is not free in solution like NAD$^+$ and NADH; it is tightly bound to enzymes. Thus, it is a **prosthetic group**.

   –Coenzyme Q may be synthesized in the body. It is not derived from a vitamin.

   c. The reduced form of coenzyme Q transfers electrons to the **cytochromes** of the electron transport chain.

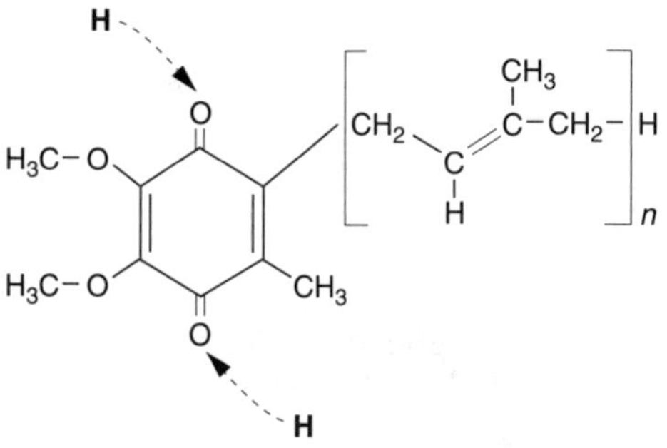

Figure 4-5. The structure of coenzyme Q (CoQ), or ubiquinone. Hydrogen atoms can bind, one at a time, as indicated by the *arrows*.

–Each cytochrome consists of a heme group associated with a protein.

–The **iron** of the heme group is reduced when the cytochrome accepts an electron.

$$Fe^{3+} \rightleftharpoons Fe^{2+}$$

–Heme is synthesized from glycine and succinyl CoA in humans (Figure 4-6). It is not derived from a vitamin.

  d. At the end of the electron transport chain, the **electrons are transferred to O_2,** which is reduced to H_2O.

B. Other cofactors involved in the oxidation of food

  1. **Coenzyme A (CoASH)** contains a sulfhydryl group that reacts with carboxylic acids to form **thioesters** such as acetyl CoA, succinyl CoA, and palmitoyl CoA (Figure 4-7).

   a. The $\Delta G^{\circ\prime}$ for hydrolysis of the thioester bond is -7.5 kcal/mole (a high-energy bond).

   b. Coenzyme A contains AMP linked by a phosphate group to **pantothenic acid,** which is joined to β-mercaptoethylamine.

    –Pantothenic acid (a vitamin) is also present in the fatty acid synthase complex.

  2. **Thiamine pyrophosphate** (Figure 4-8A) is involved in decarboxylation of α-keto acids and is the cofactor for the **transketolase** of the pentose phosphate pathway.

   a. The α-carbon of the α-keto acid becomes covalently attached to thiamine pyrophosphate, and the carboxyl group is released as CO_2.

   b. **Thiamine pyrophosphate** is formed from ATP and the vitamin thiamine.

  3. **Lipoic acid** is also involved in the oxidative decarboxylation of α-keto acids (Figure 4-9).

   a. After an α-keto acid is decarboxylated, the remainder of the compound is oxidized as it is transferred from thiamine pyrophosphate to **lipoic acid,** which is reduced in the reaction.

   b. The oxidized compound, which forms a thioester with lipoate, is then transferred to the sulfur of coenzyme A.

   c. Because there is a limited amount of lipoate in the cell, reduced lipoate must be reoxidized so that it can be reutilized in these types

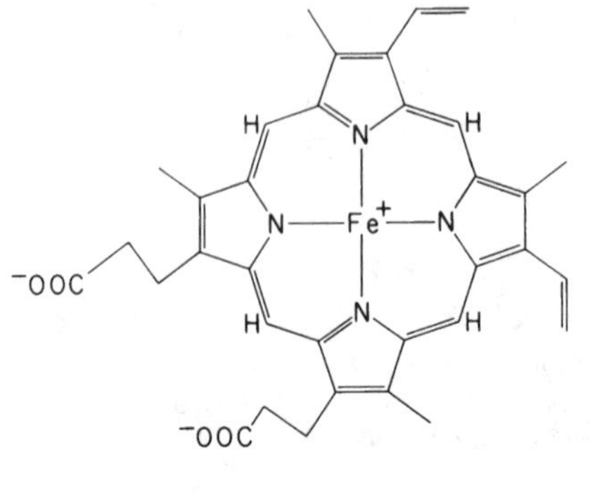

Figure 4-6. The general structure of the heme group, which is present in hemoglobin and the cytochromes b, c, and c_1.

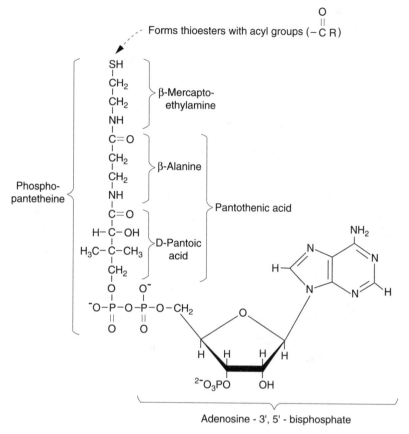

Figure 4-7. The structure of coenzyme A. The *arrow* indicates where acyl (e.g., acetyl, succinyl, and fatty acyl) groups bind to form thioesters.

of reactions. It is reoxidized by FAD, which becomes reduced ($FADH_2$) and is subsequently reoxidized by NAD^+.

d. Lipoic acid is not derived from a vitamin.

4. Action of α-keto acid dehydrogenases

a. The sequence of reactions involving thiamine pyrophosphate, lipoic acid, coenzyme A, FAD, and NAD^+ is utilized by pyruvate dehydrogenase, the enzyme complex that oxidatively decarboxylates pyruvate, forming **acetyl CoA**.

b. Analogous reactions are utilized by **α-ketoglutarate dehydrogenase,** which catalyzes the conversion of α-ketoglutarate to succinyl CoA, and the **α-keto acid dehydrogenase** complex that is involved in the oxidation of the branched-chain amino acids.

C. Other cofactors derived from water-soluble vitamins

1. NADPH (the reduced form of $NADP^+$) provides reducing equivalents for synthesis of **fatty acids** and other compounds.

–$NADP^+$ is identical to NAD^+ except that it contains an additional phosphate group esterified to the 2'-hydroxyl of the AMP moiety (see Figure 4-3).

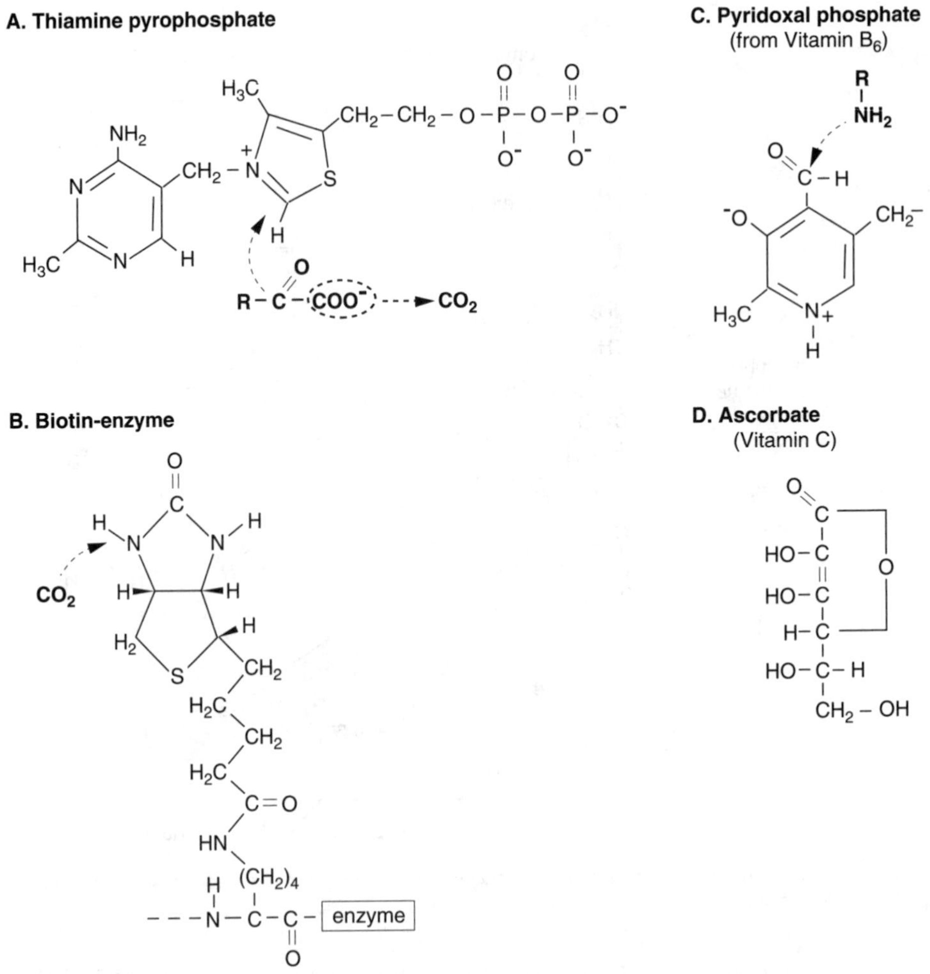

A. Thiamine pyrophosphate

C. Pyridoxal phosphate
(from Vitamin B$_6$)

B. Biotin-enzyme

D. Ascorbate
(Vitamin C)

Figure 4-8. The structures of thiamine pyrophosphate (*A*), biotin (*B*), pyridoxal phosphate (*C*), and ascorbate (*D*). *Arrows* indicate the reactive sites. When an α-keto acid binds to thiamine pyrophosphate, the keto group attaches and the carboxyl group is released as CO_2.

2. **Biotin** is involved in the carboxylation of **pyruvate** (which forms oxaloacetate), **acetyl CoA** (which forms malonyl CoA), and other compounds.

   –The vitamin biotin is covalently linked to a lysyl residue of the enzyme (see Figure 4-8*B*).

3. **Pyridoxal phosphate,** in its aldehyde form, interacts with an amino acid to form a Schiff's base (see Figure 4-8*C*). Various products may be generated, depending on the enzyme.

   a. Many amino acids may be **transaminated**.

   b. Some amino acids may be **decarboxylated;** others may be **deaminated**.

   c. **Pyridoxal phosphate** is derived from vitamin B$_6$ (pyridoxine).

4. **Tetrahydrofolate** (see Figure 7-14) transfers one-carbon units from compounds such as serine to compounds such as dUMP (to form dTMP).

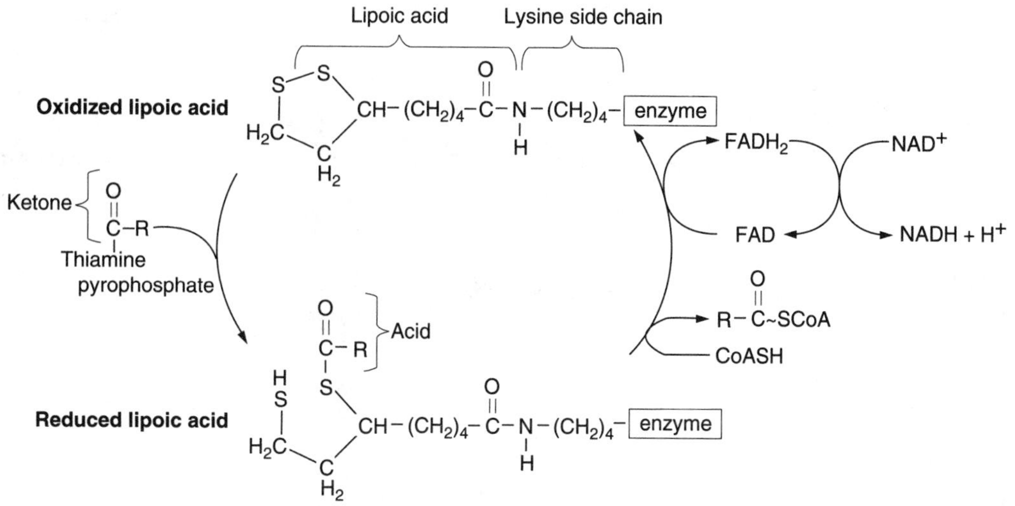

Figure 4-9. Role of lipoic acid in oxidative decarboxylation of α-keto acids.

    a. The transferred carbon is more reduced than CO_2.

    b. Tetrahydrofolate is synthesized from the vitamin folate.

  5. Vitamin B_{12} (see Figure 7-17), which contains cobalt, is involved in two reactions in the body.

    a. It **transfers methyl groups** from tetrahydrofolate to homocysteine (to form methionine).

    b. It is involved in the **conversion of methylmalonyl CoA** to succinyl CoA.

  6. Vitamin C (ascorbic acid) has at least three functions in the body (see Figure 4-8*D*).

    a. It is involved in **hydroxylation reactions,** such as the hydroxylation of prolyl residues in the precursor of collagen.

    b. It functions in the **absorption of iron**.

    c. It is an **antioxidant**.

D. Fat-soluble vitamins (Figure 4-10)

  1. Vitamin K is involved in the activation of precursors of prothrombin and other clotting factors by carboxylation of glutamate residues.

  2. Vitamin A is necessary for the light reactions of vision, for normal growth and reproduction, and for differentiation and maintenance of epithelial tissues. Vitamin A is an antioxidant and has anticancer activity.

    a. Δ^{11}-*cis*-Retinal binds to the protein opsin, forming rhodopsin.

    b. Light causes Δ^{11}-*cis*-retinal in rhodopsin to be converted to all *trans*-retinal, which dissociates from opsin, causing changes that allow light to be perceived by the brain.

  3. Vitamin E serves as an antioxidant.

    a. It prevents free radicals from oxidizing polyunsaturated fatty acids.

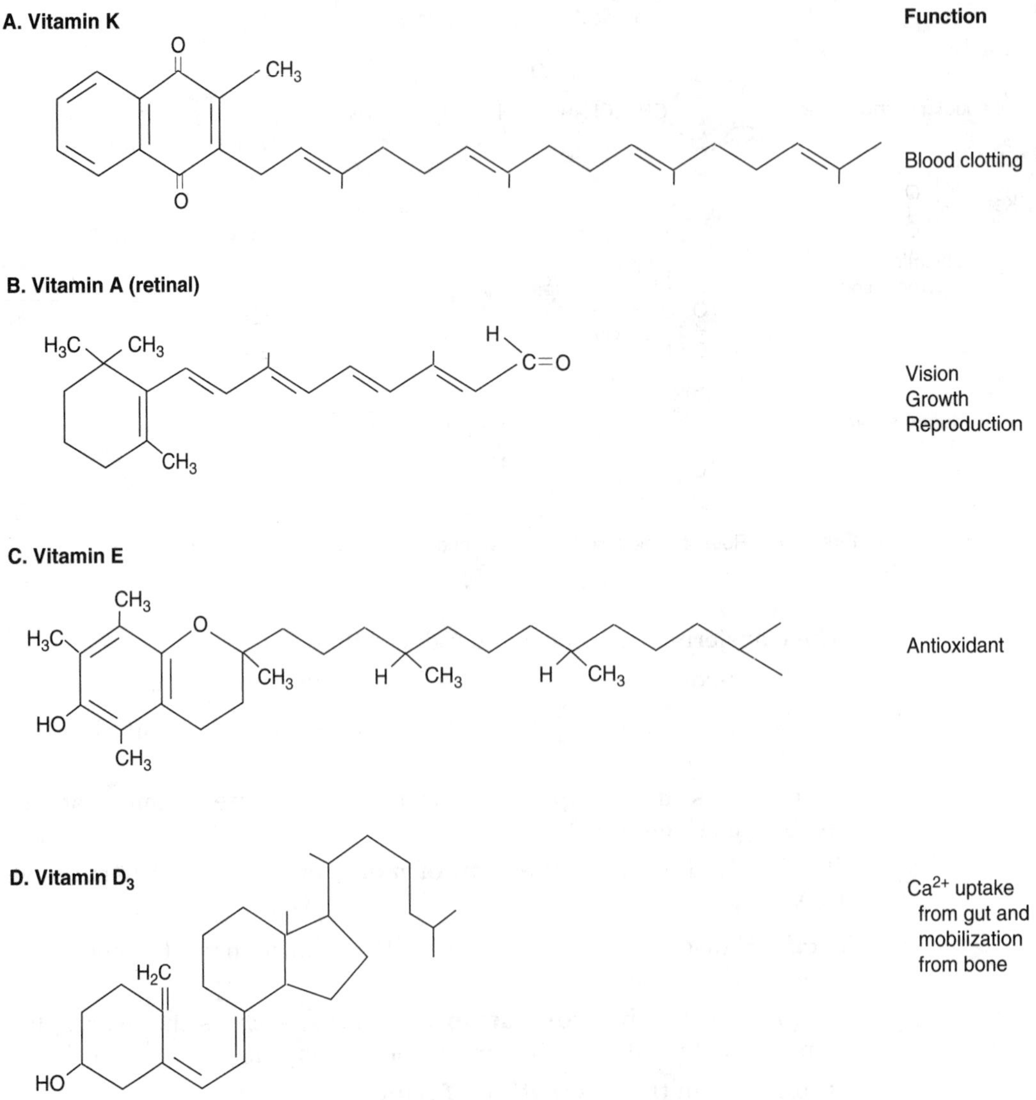

A. Vitamin K

Function

Blood clotting

B. Vitamin A (retinal)

Vision
Growth
Reproduction

C. Vitamin E

Antioxidant

D. Vitamin D$_3$

Ca^{2+} uptake
from gut and
mobilization
from bone

Figure 4-10. The fat-soluble vitamins and their major functions.

  b. A major consequence is that the integrity of membranes, which contain fatty acid residues esterified to phospholipids, is maintained.

  4. Vitamin D (as 1,25-dihydroxycholecalciferol) is involved in calcium metabolism (see Chapter 8).

IV. TCA Cycle

- The TCA (tricarboxylic acid cycle, also known as the citric acid cycle or the Krebs cycle) is the major energy-producing pathway in the body. The cycle occurs in mitochondria.
- Foodstuffs feed into the cycle as acetyl CoA and are oxidized for energy.
- The cycle also serves in the synthesis of fatty acids, amino acids, and glucose.

- The cycle starts with the 4-carbon compound oxaloacetate, adds 2 carbons from acetyl CoA, loses 2 carbons as CO_2, and regenerates the 4-carbon compound oxaloacetate.
- Electrons are transferred by the cycle to NAD^+ and FAD.
- As the electrons subsequently are passed to O_2 by the electron transport chain, ATP is generated by the process of oxidative phosphorylation.
- ATP is also generated from GTP, produced in one reaction of the cycle by a substrate-level phosphorylation.

A. The reactions of the TCA cycle (Figure 4-11)

–All the enzymes of the TCA cycle are in the **mitochondrial matrix** except succinate dehydrogenase, which is on the inner mitochondrial membrane.

1. **Acetyl CoA** condenses with **oxaloacetate,** forming citrate.

   a. Enzyme: **citrate synthase**.

   b. Cleavage of the high-energy thioester bond in acetyl CoA provides the energy for this condensation.

   c. **Citrate** (the product) is an inhibitor of this reaction.

2. **Citrate** is isomerized to **isocitrate** by a rearrangement of the molecule. Aconitate serves as an enzyme-bound intermediate.

   a. Enzyme: **aconitase**.

   b. The equilibrium of this reaction favors citrate.

3. **Isocitrate** is oxidized to **α-ketoglutarate** in the first oxidative decarboxylation reaction. CO_2 is produced, and the electrons are passed to NAD^+ to form $NADH + H^+$.

   a. Enzyme: **isocitrate dehydrogenase**.

   b. This enzyme, a key regulatory enzyme of the TCA cycle, is allosterically activated by ADP and inhibited by NADH.

4. **α-Ketoglutarate** is converted to **succinyl CoA** in a second oxidative decarboxylation reaction. CO_2 is released, and the keto group of α-ketoglutarate is oxidized to an acid, which combines with CoASH to form succinyl CoA. $NADH + H^+$ are produced.

   a. Enzyme: **α-ketoglutarate dehydrogenase**.

   b. This enzyme requires five cofactors (see III B).

5. **Succinyl CoA** is cleaved to succinate. Cleavage of the high-energy thioester bond of succinyl CoA provides energy for the substrate-level phosphorylation of GDP to GTP. Since this does not involve the electron transport chain, it is not an oxidative phosphorylation.

   a. Enzyme: **succinyl CoA synthetase**.

   b. This enzyme was named for the reverse reaction.

6. **Succinate** is oxidized to fumarate. Two hydrogens are removed together with their electrons from succinate and transferred to FAD, forming $FADH_2$.

   a. Enzyme: **succinate dehydrogenase**.

   b. This enzyme is present on the inner **mitochondrial membrane**. The other enzymes of the cycle are in the matrix.

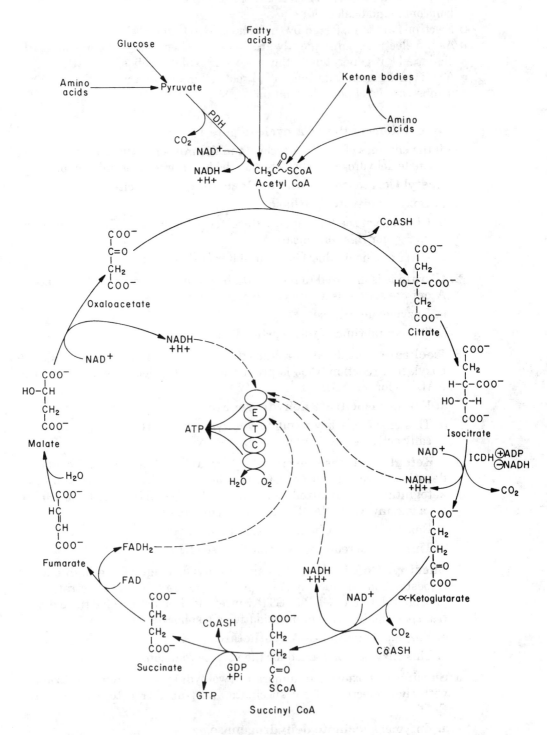

Figure 4-11. The TCA cycle and the precursors of acetyl CoA. *ICDH* = isocitrate dehydrogenase; *PDH* = pyruvate dehydrogenase; *ETC* = electron transport chain; ⊕ = activator; ⊖ = inhibitor. The (~) indicates that the thioesters acetyl CoA and succinyl CoA are high-energy compounds.

7. Water adds across the double bond of fumarate, generating malate. The enzyme is **fumarase**.

8. Malate is oxidized regenerating oxaloacetate and thus completing the cycle. Two hydrogens along with their electrons are removed from the α-carbon and hydroxyl group of malate. They are passed to NAD^+, producing $NADH + H^+$.

    a. Enzyme: **malate dehydrogenase**.

    b. The equilibrium of this reaction favors malate.

B. Energy production by the TCA cycle

1. The NADH and $FADH_2$ (produced by the cycle) donate electrons to the electron transport chain. For each **NADH,** approximately 3 **ATP** are generated, and from each $FADH_2$ approximately 2 ATP are generated by the passage of these electrons to O_2 (oxidative phosphorylation). In addition, GTP is produced when succinyl CoA is cleaved. GTP produces ATP (GTP $+$ ADP $\rightleftharpoons$ GDP $+$ ATP).

2. The **total energy** generated by one round of the cycle, starting with 1 acetyl CoA, is approximately **12 ATP**.

C. Regulation of the TCA cycle

–The TCA cycle is regulated by the **cell's need for energy** in the form of ATP. An ATP synthase on the inner mitochondrial membrane, the electron transport chain, and the TCA cycle act in concert to produce ATP.

1. There are limited amounts of the adenine nucleotides (ATP, ADP, and AMP) in the cell.

2. When ATP is utilized, ADP and inorganic phosphate (P_i) are produced.

3. When **ADP levels become high** relative to ATP—that is, when the cell needs energy—the reactions of the electron transport chain are accelerated. NADH is rapidly oxidized; consequently, the TCA cycle speeds up.

–One aspect of this process is that ADP allosterically activates isocitrate dehydrogenase.

4. When the **concentration of ATP is high**—that is, when the cell has an adequate energy supply—the electron transport chain slows down, NADH builds up, and consequently the TCA cycle is inhibited.

    a. NADH allosterically inhibits isocitrate dehydrogenase.

    b. Isocitrate accumulates and, because the aconitase equilibrium favors citrate, the concentration of citrate rises.

    c. Citrate inhibits citrate synthase, the first enzyme of the cycle.

    d. High NADH (and low NAD^+) levels also affect the reactions of the cycle that generates NADH, resulting in a slowing of the cycle by mass action.

    e. Oxaloacetate is converted to malate when NADH is high and, therefore, less substrate (OAA) is available for the citrate synthase reaction.

D. Vitamins required for reactions of the TCA cycle

1. Niacin is utilized in the synthesis of the nicotinamide portion of **NAD,** which is used in the isocitrate dehydrogenase, α-ketoglutarate dehydrogenase, and malate dehydrogenase reactions.

2. **Riboflavin** is utilized in the synthesis of **FAD,** which is used in the succinate dehydrogenase reaction. FAD is also required by α-ketoglutarate dehydrogenase and pyruvate dehydrogenase.

3. Four vitamins and lipoic acid are needed for coenzymes of **α-ketoglutarate dehydrogenase,** a multienzyme complex (see III B 2–4).

   a. **Thiamine pyrophosphate** (which contains the vitamin thiamine)

   b. **Lipoic acid** (a cofactor but not a vitamin)

   c. **CoASH** (which contains the vitamin pantothenate)

   d. **FAD** (which contains riboflavin)

   e. **NAD⁺** (which requires niacin for its synthesis)

   f. The **net reaction** is

   $$\alpha\text{-ketoglutarate} + NAD^+ + CoASH \rightarrow \text{succinyl CoA} + NADH + H^+ + CO_2$$

E. Pyruvate dehydrogenase complex

1. **Reaction sequence**

   a. Pyruvate dehydrogenase, a multienzyme complex located exclusively in the mitochondrial matrix, catalyzes the oxidative decarboxylation of pyruvate, forming acetyl CoA.

   b. The reactions catalyzed by the pyruvate dehydrogenase complex are analogous to those catalyzed by the α-ketoglutarate dehydrogenase complex. These enzyme complexes require the same five coenzymes, four of which contain vitamins.

2. **Regulation of pyruvate dehydrogenase**

   a. In contrast to α-ketoglutarate dehydrogenase, pyruvate dehydrogenase exists in a phosphorylated (inactive) form and a dephosphorylated (active) form.

   b. A kinase associated with the multienzyme complex phosphorylates the pyruvate decarboxylase subunit, inactivating the pyruvate dehydrogenase complex.

   c. A phosphatase dephosphorylates the subunit and activates the complex.

   d. The products of the pyruvate dehydrogenase reaction, **acetyl CoA and NADH,** activate the kinase, and the substrates **CoASH and NAD⁺** inactivate the kinase. The kinase is also inactivated by ADP.

   e. When the concentration of substrates is high, the dehydrogenase is active and pyruvate is converted to acetyl CoA. When the concentration of products is high, the dehydrogenase is relatively inactive.

F. Synthetic function of the TCA cycle (Figure 4-12)

–**Intermediates** of the TCA cycle are utilized in the fasting state for the production of **glucose** and in the fed state for the synthesis of **fatty acids**. Intermediates of the TCA cycle are also used to synthesize **amino acids** or to convert one amino acid to another.

1. **Anaplerotic reactions** replenish intermediates of the TCA cycle as they are removed for the synthesis of glucose, fatty acids, amino acids, or other compounds.

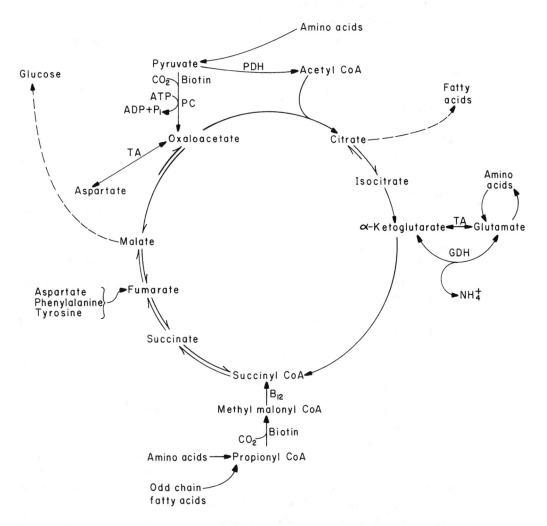

Figure 4-12. Anaplerotic and biosynthetic reactions involving TCA cycle intermediates. Synthetic reactions that form fatty acids and glucose are indicated by *broken lines*. *PC* = pyruvate carboxylase; *GDH* = glutamate dehydrogenase.

a. A key anaplerotic reaction is catalyzed by **pyruvate carboxylase,** which carboxylates pyruvate, forming oxaloacetate.

   (1) Pyruvate carboxylase requires **biotin,** a cofactor that is commonly involved in CO_2 fixation reactions.

   (2) Pyruvate carboxylase, found in the liver, brain, and adipose tissue (but not in muscle), is activated by acetyl CoA.

b. Amino acids may produce intermediates of the TCA cycle through anaplerotic reactions.

   (1) Glutamate may be converted to α-ketoglutarate.

   (2) Aspartate may be transaminated to form **oxaloacetate**.

   (3) Valine, isoleucine, methionine, and threonine may produce propionyl CoA, which is converted to methylmalonyl CoA and, subsequently, to succinyl CoA, an intermediate of the TCA cycle.

(4) Phenylalanine, tyrosine, and aspartate may form fumarate.

2. Synthesis of glucose

    a. The synthesis of glucose occurs by the pathway of **gluconeogenesis,** which involves intermediates of the TCA cycle.

    b. As glucose is synthesized, **malate or oxaloacetate** is removed from the TCA cycle. These intermediates of the cycle are replenished by anaplerotic reactions.

        (1) Pyruvate, produced from lactate or alanine, is converted by pyruvate carboxylase to oxaloacetate, which forms malate.

        (2) Various amino acids that supply carbon for gluconeogenesis may also be converted to intermediates of the TCA cycle.

3. Synthesis of fatty acids

    a. The pathway for fatty acid synthesis from glucose includes reactions of the TCA cycle.

    b. From glucose, pyruvate is produced and converted to oxaloacetate (by pyruvate carboxylase) and to acetyl CoA (by pyruvate dehydrogenase).

    c. Oxaloacetate and acetyl CoA condense to form **citrate,** which is used for fatty acid synthesis.

    d. Pyruvate carboxylase catalyzes the anaplerotic reaction that replenishes the TCA cycle intermediates.

4. Synthesis of amino acids

    a. Synthesis of amino acids from glucose involves intermediates of the TCA cycle.

        (1) Glucose may be converted to pyruvate, which forms oxaloacetate, which by transamination forms **aspartate** and, subsequently, **asparagine**.

        (2) Glucose may be converted to pyruvate, which forms both oxaloacetate and acetyl CoA, which condense, forming citrate. Citrate forms isocitrate and then α-ketoglutarate, from which **glutamate, glutamine, proline, and arginine** can be produced.

    b. Interconversion of amino acids involves intermediates of the TCA cycle. For example, the carbons of **glutamate** can feed into the TCA cycle at the α-ketoglutarate level and traverse the cycle, forming oxaloacetate, which may be transaminated to **aspartate**.

V. Electron Transport Chain and Oxidative Phosphorylation

- ATP is generated as a result of the energy produced when electrons from NADH and $FADH_2$ are passed to molecular oxygen by a series of electron carriers, collectively known as the electron transport chain. The components of the chain includes FMN, Fe-S centers, coenzyme Q, and a series of cytochromes (b, c_1, c, and aa_3).
- According to the chemiosmotic theory, the energy derived from the transfer of electrons through the electron transport chain is used to pump protons across the inner mitochondrial membrane from the matrix to the cytosolic side. An electrochemical gradient is generated, consisting of a proton gradient and a membrane potential.

- Protons move back into the matrix through the ATP synthase complex, causing ATP to be produced from ADP and inorganic phosphate.
- ATP is transported from the mitochondrial matrix to the cytosol in exchange for ADP (the ATP-ADP antiport system).
- The oxidation of one NADH generates approximately 3 ATP, while the oxidation of one $FADH_2$ generates approximately 2 ATP.
- Because energy generated by the transfer of electrons through the electron transport chain to O_2 is used in the synthesis of ATP, the overall process is known as oxidative phosphorylation.
- Electron transport and ATP synthesis occur simultaneously and are tightly coupled.
- NADH and $FADH_2$ are oxidized only if ADP is available for conversion to ATP (i.e., if ATP is being utilized and converted to ADP).

A. Overview of the electron transport chain

–NADH or $FADH_2$, or both, are produced by glycolysis, β-oxidation of fatty acids, the TCA cycle, and other oxidative reactions.

1. **Electrons** from NADH and $FADH_2$ are passed to the components of the electron transport chain, which are located in the inner mitochondrial membrane.

   a. **NADH** freely diffuses from the matrix to the membrane, while **$FADH_2$** is tightly bound to enzymes that produce it within the inner mitochondrial membrane.

   b. Mitochondria are separated from the cytoplasm by two membranes. The inner membrane contains infoldings known as **cristae**.

   c. The soluble interior of mitochondria is called the **matrix**.

2. The **transfer of electrons** from NADH to O_2 occurs in three stages, each of which involves a large protein complex in the inner mitochondrial membrane.

3. Each complex uses the energy from electron transfer to **pump protons** to the cytosolic side of the membrane.

4. An **electrochemical potential** or proton-motive force is generated.

   a. The electrochemical potential is composed of both a membrane potential and a pH gradient.

   b. The cytosolic side of the membrane is more acid than the matrix.

5. The inner mitochondrial membrane is impermeable to protons. The **protons** can reenter the matrix only through the ATP synthase complex (the F_0-F_1/ATPase), causing ATP to be generated.

   –The ATP synthase complex contains proteins (F_0) that form a channel in the inner mitochondrial membrane, through which protons can flow, and a stalk that is attached to an ATP-synthesizing head (F_1) that projects into the matrix.

6. During the transfer of electrons through the electron transport chain, some of the **energy is lost as heat**.

7. The electron transport chain has a large negative $\Delta G^{\circ\prime}$, so electrons flow from NADH (or $FADH_2$) toward O_2.

B. The three major stages of electron transport (Figure 4-13)

1. Transfer of electrons from NADH to coenzyme Q

a. Electrons are passed by the NADH dehydrogenase complex from NADH to FMN (flavin mononucleotide).

(1) NADH is produced by the α-ketoglutarate dehydrogenase, isocitrate dehydrogenase, and malate dehydrogenase reactions of the TCA cycle, by the pyruvate dehydrogenase reaction that converts pyruvate to acetyl CoA, by β-oxidation of fatty acids, and by other oxidation reactions.

(2) NADH produced in the mitochondrial matrix diffuses to the inner mitochondrial membrane where it passes electrons to FMN, which is tightly bound to a protein.

b. FMN passes the electrons through a series of iron-sulfur (Fe-S) complexes to coenzyme Q (CoQ), which may accept electrons one at a time, forming first the semiquinone and then ubiquinol.

c. The energy produced by these electron transfers is used to pump protons to the cytosolic side of the inner mitochondrial membrane.

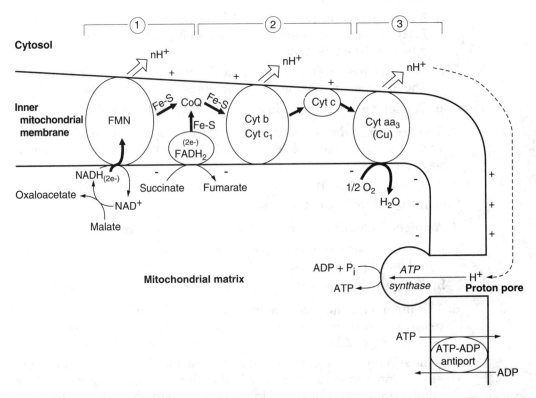

Figure 4-13. The electron transport chain and oxidative phosphorylation. *Heavy arrows* indicate the flow of electrons. *Fe-S* = iron-sulfur centers; FMN = flavin mononucleotide; CoQ = coenzyme Q (ubiquinone); Cyt = cytochrome. nH^+ indicates that an undetermined number of protons are pumped from the matrix to the cytosolic side. The *numbers* at the top of the figure correspond to the three major stages of electron transfer described in the text in V B.

d. As the protons flow back into the **matrix** through pores in the ATP synthase complex, approximately one ATP is generated for each NADH that transfers two electrons to CoQ.

2. Transfer of electrons from coenzyme Q to cytochrome c

  a. Electrons are transferred from coenzyme Q through Fe-S centers to cytochromes b and c_1, which transfer the electrons to cytochrome c. The protein complex involved in these transfers is called **cytochrome reductase**.

  (1) These cytochromes each contain heme as a prosthetic group but have different apoproteins.

  (2) In the **ferric (+3)** state, the heme iron can accept one electron and be reduced to the **ferrous (+2)** state.

  (3) Because the cytochromes can only carry one electron at a time, two molecules in each cytochrome complex must be reduced for every molecule of NADH that is oxidized.

  b. The energy produced by the transfer of electrons from coenzyme Q to cytochrome c is used to pump protons across the inner mitochondrial membrane.

  c. As the protons flow back into the matrix through pores in the ATP synthase complex, approximately one ATP is generated for every $CoQH_2$ that transfers two electrons to cytochrome c.

3. Transfer of electrons from cytochrome c to oxygen

  a. Cytochrome c transfers electrons to the cytochrome aa_3 complex, which transfers the electrons to molecular oxygen, reducing it to water. Cytochrome oxidase catalyzes this transfer of electrons.

  (1) Cytochromes a and a_3 each contain heme a and two different proteins that each contain copper.

  (2) Two electrons are required to reduce one atom of oxygen; therefore, for each NADH that is oxidized, one-half O_2 is converted to H_2O.

  b. The energy produced by the transfer of electrons from cytochrome c to O_2 is used to pump protons across the inner mitochondrial membrane.

  c. As the protons flow back into the matrix, approximately one ATP is generated for every two electrons that are transferred from cytochrome c to oxygen—that is, for every one-half O_2 that is reduced to H_2O.

C. ATP production

  –The production of ATP is coupled to the transfer of electrons through the electron transport chain to O_2. The overall process is known as **oxidative phosphorylation**.

  –The exact amount of ATP that is generated by this process has not been unequivocally established.

  1. For every NADH that is oxidized, one-half O_2 is reduced to H_2O and approximately 3 ATP are produced.

  2. For every $FADH_2$ that is oxidized, approximately 2 ATP are generated because the electrons from $FADH_2$ enter the chain via coenzyme Q, bypassing the NADH dehydrogenase step.

D. The ATP-ADP antiport

–ATP produced within mitochondria is transferred to the cytosol in exchange for ADP by a transport protein in the inner mitochondrial membrane known as the **ATP-ADP antiport** (see Figure 4-13).

E. Inhibitors of electron transport and oxidative phosphorylation

1. Agents that act on components of the electron transport chain

–If there is a block at any point in the electron transport chain, all carriers before the block will accumulate in their reduced states, while those after the block will accumulate in their oxidized states. As a result, O_2 will not be consumed, ATP will not be generated, and the TCA cycle will slow down.

a. Rotenone, a fish poison, complexes with NADH dehydrogenase, causing NADH to accumulate. It does not block the transfer of electrons to the chain from $FADH_2$.

b. Antimycins (antibiotics) block the passage of electrons through the cytochrome b-c_1 complex.

c. Cyanide and carbon monoxide, poisons commonly used for suicide, combine with cytochrome oxidase and block the transfer of electrons to O_2.

2. Inhibitors of ATP synthesis

–Because the synthesis of ATP and electron transport are coupled, if the ATP synthase complex is inhibited or if an adequate supply of ADP is not available, ATP synthesis will be inhibited, O_2 will not be consumed, the carriers of the electron transport chain will accumulate in their reduced states, and the TCA cycle will slow down.

a. Oligomycin binds to the ATP synthase complex and prevents the conversion of ADP to ATP.

b. Other compounds, such as atractyloside and bongkrekate (poisons from certain plants), **block the ATP-ADP antiport** (the exchange of ATP for ADP across the inner mitochondrial membrane).

3. Uncoupling agents

a. Agents such as **dinitrophenol** and ionophores allow protons from the cytosol to reenter the matrix without going through the pore in the ATP synthase complex, thus uncoupling electron transport and ATP production.

b. Uncouplers cause the rate of O_2 consumption and the rate of electron transport to increase. The TCA cycle is stimulated, causing the rate of CO_2 production to increase.

c. ATP production does not occur because the proton gradient across the inner mitochondrial membrane is dissipated.

d. The **energy** generated by the increased rate of respiration (electron transport and O_2 consumption) is **lost as heat**.

VI. Clinical Correlations

A. Vitamin deficiencies and excesses

Vitamin deficiencies usually occur because of an insufficient dietary content or decreased conversion of the vitamin to its coenzyme derivatives caused by drugs, diseases, or other factors. Decreased vitamin absorption from the gut, plasma transport, tissue storage, binding to proteins, or increased excretion may also play a role. Often multiple deficiencies occur, usually of the water-soluble vitamins, which are stored in limited amounts that may be depleted within weeks. Stores of the fat-soluble vitamins are larger and, therefore, are depleted more slowly. Toxicity due to excessive intake of the fat-soluble vitamins can develop.

1. **Thiamine deficiency** resulting in **beriberi** usually occurs because of excessive alcohol intake, which interferes with thiamine absorption. In the early stages, the person tires easily and the limbs feel heavy and weak. In an advanced stage of "wet" beriberi, symptoms include neuromuscular problems, enlarged heart (with tachycardia), peripheral edema, and weakness and malaise. These symptoms usually respond to thiamine administration.

2. **Riboflavin deficiency** is characterized by a sore mouth and burning and itching eyes. Fissures may appear at the corners of the mouth, and the tongue may become magenta-colored.(glossitis)

3. Severe **niacin deficiency** results in **pellagra,** which is characterized by the four D's: dermatitis, diarrhea, dementia, and death.

4. Early symptoms of **vitamin A deficiency** are night blindness and dry eyes. **Excessive doses of vitamin A** cause headaches, dizziness, a sore mouth, skin desquamation, and hair loss.

5. **Vitamin D deficiencies** in young children cause **rickets.**

6. **Vitamin K deficiency** results in a tendency to **hemorrhage.** Vitamin K is synthesized by the normal intestinal flora. Deficiencies occur in infants because fetal stores are low and the fetal gut is sterile. In adults, deficiencies result from long-term administration of antibiotics that destroy the intestinal flora. Anticoagulant drugs, such as warfarin or dicoumarol, prevent vitamin K from being metabolically activated.

B. Cyanide poisoning

Cyanide binds to Fe^{3+} in cytochrome aa_3. As a result, O_2 cannot receive electrons, respiration is inhibited, energy production is halted, and death occurs rapidly.

C. Malignant hyperthermia

The major inhalation anesthetics (halothane, ether, and methoxyflurane) trigger a reaction in susceptible people, which results in the uncoupling of oxidative phosphorylation from electron transport. ATP production decreases, heat is generated and the temperature rises markedly, the TCA cycle is stimulated, and excessive CO_2 production leads to a respiratory acidosis.

D. Acute myocardial infarction

Coronary arteries frequently narrow because of atherosclerotic plaques. **Coronary occlusions** may occur and regions of heart muscle may be deprived of blood and, therefore, of oxygen for prolonged periods of time. **Lack of oxygen** causes inhibition of the processes of electron transport and oxidative phosphorylation, which results in a decreased production of ATP. **Heart muscle,** suffering from a lack of energy required for contraction and maintenance of membrane integrity, becomes **damaged**. Enzymes from the damaged cells leak into the blood. If the damage is relatively mild, recovery is possible. If heart function is severely compromised, death may result.

Review Test

Directions: Each of the numbered items or incomplete statements in this section is followed by answers or by completions of the statement. Select the **one** lettered answer or completion that is **best** in each case.

1. The structure shown is part of each of the following molecules EXCEPT

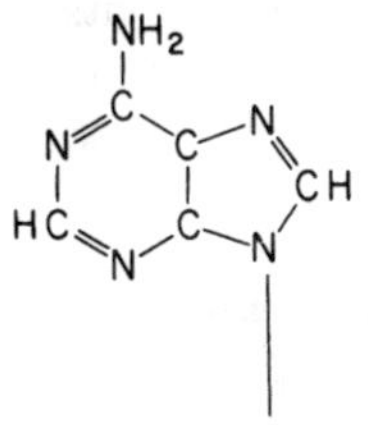

(A) NAD$^+$
(B) coenzyme A
(C) FAD
(D) thiamine pyrophosphate
(E) ATP

2. The reaction $A + B \rightleftharpoons P + Q$ occurs in a liter of solution, and the amounts (in mmoles) of each of the components at equilibrium are A, 10; B, 5; P, 5; and Q, 5. The equilibrium constant (K_{eq}) for the reaction is

(A) 0.50
(B) 0.67
(C) 1.00
(D) 1.50
(E) 2.00

Questions 3 and 4

Refer to the following equation when answering questions 3 and 4.

$$fumarate + H_2O \rightleftharpoons malate$$

3. When measured in the absence of fumarase, the $\Delta G^{\circ'}$ for this reaction is 0 kcal/mole (neglecting any terms associated with H_2O). The equilibrium constant for this reaction is

(A) 0
(B) 0.5
(C) 1.0
(D) 10.0
(E) 50.0

4. Fumarase was added to a solution that initially contained 20 µM fumarate. After the establishment of equilibrium, the concentration of malate was

(A) 2 µM
(B) 5 µM
(C) 10 µM
(D) 20 µM
(E) 50 µM

Questions 5 and 6

Refer to the following reactions and associated values when answering questions 5 and 6.

Reaction	Approximate $\Delta G^{o'}$ (kcal/mole)
Acetate + 2 O_2 → 2 CO_2 + 2 H_2O	-243
NADH + H^+ + ½ O_2 → NAD^+ + H_2O	-53
$FADH_2$ + ½ O_2 → FAD + H_2O	-41
GTP → GDP + P_i	-8
ATP → ADP + P_i	-8

5. Of the total energy available from the oxidation of acetate, what percentage is transferred via the TCA cycle to NADH, $FADH_2$, and GTP?

(A) 38%
(B) 42%
(C) 82%
(D) 86%
(E) 100%

6. What percentage of the energy available from the oxidation of acetate is converted to ATP?

(A) 3%
(B) 30%
(C) 40%
(D) 85%
(E) 100%

7. If the enzyme concentration for a biochemical reaction is increased 100-fold, the equilibrium constant for the reaction will

(A) decrease twofold
(B) remain the same
(C) increase in proportion to the enzyme concentration
(D) change inversely with the enzyme concentration

8. All of the following are electron carriers in the electron transport chain EXCEPT

(A) cytochromes
(B) coenzyme Q
(C) Fe-S centers
(D) hemoglobin
(E) riboflavin

9. In the tricarboxylic acid cycle, thiamine pyrophosphate

(A) accepts electrons from the oxidation of pyruvate and α-ketoglutarate
(B) accepts electrons from the oxidation of isocitrate
(C) forms a covalent intermediate with the α-carbon of α-ketoglutarate
(D) forms a thioester with the sulfhydryl group of CoASH
(E) forms a thioester with the sulfhydryl group of lipoic acid

10. Each of the following vitamins is required for reactions in the oxidation of pyruvate to CO_2 and H_2O EXCEPT

(A) pantothenate
(B) niacin
(C) thiamine
(D) biotin
(E) riboflavin

Question 11

Refer to the following compounds to answer question 11.

$$\begin{array}{c} OH \\ | \\ {}^-OOC - CH - CH - CH_2 - COO^- \\ | \\ COO^- \end{array}$$

Compound A

$${}^-OOC - CH = CH - COO^-$$
Compound B

11. The segment of the TCA cycle in which Compound A is converted to Compound B

(A) yields 5 moles of high-energy phosphate bonds per mole of A
(B) requires a coenzyme synthesized in the human from niacin (nicotinamide)
(C) is catalyzed by enzymes located solely in the mitochondrial membrane
(D) produces 1 mole of CO_2 for every mole of Compound A oxidized
(E) requires GTP to drive one of the reactions

12. The reactions of the TCA cycle oxidizing succinate to oxaloacetate

(A) require coenzyme A
(B) include an isomerization reaction
(C) produce one high-energy phosphate bond
(D) require both NAD^+ and FAD
(E) produce one GTP from GDP + P_i

13. Each of the following statements concerning pyruvate dehydrogenase is true EXCEPT

(A) it is an example of a multienzyme complex
(B) it requires thiamine pyrophosphate as a cofactor
(C) it produces oxaloacetate from pyruvate
(D) it is converted to an inactive form by phosphorylation
(E) it is inhibited when NADH levels increase

14. The principl function of the TCA cycle is to

(A) generate CO_2
(B) transfer electrons from the acetyl portion of acetyl CoA to NAD^+ and FAD
(C) oxidize the acetyl portion of acetyl CoA to oxaloacetate
(D) generate heat from the oxidation of the acetyl portion of acetyl CoA
(E) dispose of excess pyruvate and fatty acids

15. During exercise, stimulation of the TCA cycle results principally from

(A) allosteric activation of isocitrate dehydrogenase by increased NADH
(B) allosteric activation of fumarase by increased ADP
(C) a rapid decrease in the concentration of four-carbon intermediates
(D) product inhibition of citrate synthase
(E) stimulation of the flux through a number of enzymes by a decreased $NADH/NAD^+$ ratio

16. CO_2 production by the TCA cycle would be increased to the greatest extent by a genetic abnormality that resulted in

(A) a 50% increase in the concentration of ADP in the mitochondrial matrix
(B) a 50% increase in the oxygen content of the cell
(C) a 50% decrease in the V_m of α-ketoglutarate dehydrogenase
(D) a 50% increase in the K_m of isocitrate dehydrogenase

17. A man presents to the emergency department after ingesting an insecticide. His respiration rate is very low. Information from the Poison Control Center indicates that this particular insecticide binds to and completely inhibits cytochrome c. Therefore, in this man's mitochondria

(A) coenzyme Q would be in the oxidized state
(B) cytochromes a and a_3 would be in the reduced state
(C) the rate of ATP synthesis would be approximately zero
(D) the rate of CO_2 production would be increased

18. Which one of the following statements best describes the consequence of ingesting a compound that stimulates ATP hydrolysis by plasma membrane Na^+-K^+ ATPase?

(A) The pH gradient across the mitochondrial membranes would increase
(B) The rate of conversion of NADH to NAD^+ in the mitochondria would decrease
(C) Heat production would decrease
(D) The transfer of electrons to O_2 would increase

19. A chemist wanting to lose weight obtained dinitrophenol (DNP). Before using the DNP, the chemist consulted her physician and was informed that DNP was an uncoupling agent and was dangerous to use for weight loss. Which of the following changes would have occurred in her mitochondria if she had ingested enough DNP?

(A) O_2 consumption would decrease
(B) CO_2 production would decrease
(C) The proton gradient would increase
(D) NADH would be oxidized more rapidly
(E) Temperature would decrease

Directions: Each group of items in this section consists of lettered options followed by a set of numbered items. For each item, select the **one** lettered option that is most closely associated with it. Each lettered heading may be selected once, more than once, or not at all.

Questions 20–24

Match each cofactor with the vitamin that is required for its synthesis.

(A) Riboflavin
(B) Pantothenic acid
(C) Niacin
(D) Vitamin B_6

20. NAD^+
21. FAD
22. Coenzyme A
23. FMN
24. Pyridoxal phosphate

Questions 25–28

Match each process with the appropriate vitamin.

(A) Vitamin A
(B) Vitamin C
(C) Vitamin D
(D) Vitamin K

25. Blood clotting
26. Calcium metabolism
27. Collagen synthesis
28. Vision

Questions 29–33

Match each description with the appropriate compound.

(A) $$^-OOC - CH_2 - \overset{\displaystyle OH}{\overset{|}{CH}} - COO^-$$

(B) $$^-OOC - CH_2 - CH_2 - \overset{\displaystyle O}{\overset{\|}{C}} - COO^-$$

(C) $$^-OOC - CH_2 - \overset{\displaystyle OH}{\overset{|}{\underset{\underset{\displaystyle COO^-}{|}}{C}}} - CH_2 - COO^-$$

(D) $$^-OOC - CH_2 - CH_2 - COO^-$$

29. An intermediate in the conversion of citrate to succinyl CoA in the TCA cycle
30. Converted to isocitrate by the enzyme aconitase
31. Formed by the addition of water across the double bond of fumarate
32. Oxidized to oxaloacetate by malate dehydrogenase
33. Generated in a reaction that produces GTP

Questions 34–38

Match each enzyme below with the vitamin or vitamins required for its activity.

(A) Thiamine
(B) Niacin
(C) Thiamine and niacin
(D) Neither thiamine nor niacin

34. Pyruvate dehydrogenase
35. Malate dehydrogenase
36. Pyruvate carboxylase
37. α-Ketoglutarate dehydrogenase
38. Succinate dehydrogenase

Questions 39–42

Match each item below with the appropriate enzyme or enzymes.

(A) Isocitrate dehydrogenase
(B) Malate dehydrogenase
(C) Both isocitrate and malate dehydrogenase
(D) Neither dehydrogenase

39. Regulated allosterically by ADP
40. Liberates CO_2
41. Reduces a cofactor that transfers electrons to the electron transport chain
42. Utilizes FAD as a cofactor

Answers and Explanations

1–D. The structure shown is an adenine moiety. It is not present in thiamine pyrophosphate.

2–A. $K_{eq} = [P][Q]/[A][B] = (5 \times 5)/(10 \times 5) = 0.5$.

3–C. $\Delta G^{\circ\prime} = 0 = -1400 \log K_{eq}$. Therefore, $K_{eq} = 1$. $\log K_{eq} = \log 1 = 0$.

4–C. $K_{eq} = 1 = [\text{Malate}]/[\text{Fumarate}] = X/(20-X)$. Therefore, $(20-X) = X$, $20 = 2X$, and $X = 10$.

5–D. In the TCA cycle, 3 NADHs are produced ($3 \times 53 = 159$ kcal), 1 $FADH_2$ (41 kcal), and 1 GTP (8 kcal). The percentage of the total energy available from oxidation of acetate that is transferred to these compounds is, therefore, 208/243 kcal or 86%.

6–C. About 12 ATP are produced by the TCA cycle (12×8 kcal = 96 kcal). The percentage of the total energy available from oxidation of acetate that is converted to ATP is 96/243, or 40%.

7–B. An enzyme increases the rate at which the reaction reaches equilibrium but does not change the concentration of reactants and products at equilibrium; that is, the K_{eq} is not affected by an enzyme.

8–D. Although heme is contained in the cytochromes of the electron transport chain, the protein globin is not present.

9–C. Thiamine pyrophosphate forms a covalent intermediate with the α-carbon of α-ketoglutarate.

10–D. Pyruvate dehydrogenase and α-ketoglutarate dehydrogenase require four vitamins for synthesis of their coenzymes (thiamine, pantothenate, niacin, and riboflavin). Niacin is also required for the NAD^+ utilized by isocitrate dehydrogenase and malate dehydrogenase. Riboflavin is required for the FAD utilized by succinate dehydrogenase and the FMN of the electron transport chain. Biotin is not required.

11–B. In the conversion of isocitrate (Compound A) to fumarate (Compound B), 2 CO_2, NADH (which contains niacin), 1 GTP, and 1 $FADH_2$ are produced. A total of approximately 9 ATP are generated. The enzymes for these reactions are all located in the mitochondrial matrix except succinate dehydrogenase, which is in the inner mitochondrial membrane. GTP does not drive any of the reactions.

12–D. FAD is required for conversion of succinate to fumarate, and NAD^+ is required for conversion of malate to oxaloacetate. Five ATP are generated. Coenzyme A is not required, and no isomerization reactions occur. GTP is produced when succinyl CoA is converted to succinate.

13–C. Pyruvate dehydrogenase converts pyruvate to acetyl CoA. The enzyme contains a dehydrogenase component that oxidatively decarboxylates pyruvate, a dihydrolipoyl transacetylase that transfers the acetyl group to coenzyme A, and a dihydrolipoyl dehydrogenase that reoxidizes lipoic acid. Thiamine pyrophosphate, lipoic acid, coenzyme A, NAD^+, and FAD serve as cofactors for these reactions. In addition, a kinase is present that phosphorylates and inactivates the decarboxylase component. Acetyl CoA and NADH activate this kinase, thereby inactivating pyruvate dehydrogenase. A phosphatase dephosphorylates the kinase, thereby reactivating pyruvate dehydrogenase.

14–B. Although the TCA cycle produces CO_2 and oxaloacetate and generates heat, these are not its major functions. It does not "dispose" of excess pyruvate and fatty acids, it oxidizes them in a controlled manner to generate energy. The principal function of the cycle is to pass electrons to NAD^+ and FAD, which transfer them to the electron transport chain. The net result is the production of ATP.

15–E. NADH decreases during exercise (if it increased, it would slow the cycle). Fumarase is not activated by ADP. Four-carbon intermediates of the cycle are recycled. Their concentration does not decrease. Product inhibition of citrate synthase would slow the cycle. During exercise, the TCA cycle is stimulated because the NADH/NAD^+ ratio decreases and stimulates flux through isocitrate dehydrogenase, α-ketoglutarate dehydrogenase, and malate dehydrogenase.

16–A. If the V_m of α-ketoglutarate dehydrogenase decreased, flux through the TCA cycle would decreases; therefore, CO_2 production would decrease. If the K_m of isocitrate dehydrogenase increased, higher concentrations of isocitrate would be required for the cycle to operate at its normal rate. O_2 is normally present in excess and is not rate-limiting. The only change that would increase the rate of CO_2 production by the cycle would be an increase of ADP, which would allosterically activate isocitrate dehydrogenase.

17–C. If cytochrome c cannot function, all components of the electron transport chain between it and O_2 remain in the oxidized state, and the components of the chain before cytochrome c are reduced. The electron transport chain will not function, O_2 will not be consumed, a proton gradient will not be generated, and ATP will not be produced. NADH will not be oxidized, so the TCA cycle will slow down and, therefore, CO_2 production will decrease.

18–D. If ATP were rapidly hydrolyzed, ADP levels would increase. Therefore, the rate of ATP synthesis would increase. The pH gradient across the mitochondrial membrane would decrease, the rate of electron transport would increase, O_2 consumption would increase, NADH would be oxidized more rapidly, and heat production would increase.

19–D. An "uncoupler" dissipates the proton gradient across the inner mitochondrial membrane. Therefore, ATP is not produced, and energy is liberated as heat. The low proton gradient causes the electron transport chain to speed up, O_2 consumption increases, NADH is rapidly oxidized, the TCA cycle speeds up, and CO_2 production increases.

20–C. NAD^+ contains niacin.

21–A. FAD contains riboflavin.

22–B. Coenzyme A contains pantothenic acid.

23–A. FMN contains riboflavin.

24–D. Pyridoxal phosphate contains vitamin B_6.

25–D. Vitamin K is involved in blood clotting.

26–C. Vitamin D is involved in calcium metabolism.

27–B. Vitamin C is required for the hydroxylation of proline and lysine residues in the precursor of collagen.

28–A. Vitamin A is required for formation of the visual pigments.

29–B. α-Ketoglutarate is an intermediate in the conversion of citrate to succinyl CoA.

30–C. Citrate is converted to isocitrate by aconitase.

31–A. Malate is formed by the addition of water across the double bond of fumarate.

32–A. Malate is oxidized to oxaloacetate by malate dehydrogenase.

33–D. Succinate is produced from succinyl CoA in a reaction that generates GTP.

34–C. Pyruvate dehydrogenase requires five coenzymes: thiamine pyrophosphate, lipoic acid, coenzyme A, FAD, and NAD^+, which is synthesized from niacin.

35–B. Malate dehydrogenase requires NAD^+, which is synthesized from niacin.

36–D. Pyruvate carboxylase requires biotin.

37–C. α-Ketoglutarate dehydrogenase requires the same five coenzymes as pyruvate dehydrogenase.

38–D. Succinate dehydrogenase requires FAD.

39–A. Isocitrate dehydrogenase is allosterically activated by ADP.

40–A. Isocitrate dehydrogenase removes CO_2 from isocitrate to form α-ketoglutarate.

41–C. Isocitrate dehydrogenase and malate dehydrogenase produce NADH, which interacts with the electron transport chain.

42–D. Neither isocitrate dehydrogenase nor malate dehydrogenase utilizes FAD.

5

Carbohydrate Metabolism

Overview

- Dietary carbohydrates include starch, sucrose, lactose, and indigestible fiber.
- The major product of digestion of carbohydrates is glucose, and some galactose and fructose are also produced.
- Glucose is a major fuel source that is oxidized by cells for energy. After a meal, it is converted to glycogen or to triacylglycerols and stored.
- Glucose also may be converted to compounds such as proteoglycans, glycoproteins, and glycolipids.
- When glucose enters cells, it is converted to glucose 6-phosphate, which is a pivotal compound in several metabolic pathways.
  - The major fate of glucose 6-phosphate is to enter the pathway of glycolysis, which produces pyruvate and generates NADH and ATP.
  - Glucose 6-phosphate can be converted to glucose 1-phosphate and then to UDP-glucose, which is used for the synthesis of glycogen or compounds such as the proteoglycans.
  - Glucose 6-phosphate can also enter the pentose phosphate pathway, which produces NADPH for reactions such as the biosynthesis of fatty acids and ribose for nucleotide production.
- Fructose and galactose are converted to intermediates in the pathways by which glucose is metabolized.
- Glycogen is the major storage form of carbohydrate in animals. The largest stores are in muscle and liver. Liver glycogen is used to maintain blood glucose during fasting or exercise, and muscle glycogen is used to generate ATP for muscle contraction.
- The maintenance of blood glucose is a major function of the liver, which produces glucose by glycogenolysis and gluconeogenesis.

I. Carbohydrate Structure

- Carbohydrates are compounds that contain at least three carbon atoms, a number of hydroxyl groups, and usually an aldehyde or ketone group. They may contain phosphate, amino, or sulfate groups.
- Monosaccharides, the simplest carbohydrates, may be of the D- or L-series.
- Monosaccharides form rings that usually contain five or six members and are called furanoses and pyranoses, respectively. The hydroxyl

group on the anomeric carbon, the carbonyl carbon, may be in either the α or the β configuration.

● Monosaccharides may be joined by *O*-glycosidic bonds to form disaccharides, oligosaccharides, and polysaccharides.
● Nucleotides contain *N*-glycosidic bonds.
● Monosaccharides may be oxidized to the corresponding acids or reduced to the corresponding polyols.

A. Monosaccharides

1. Nomenclature

a. The simplest monosaccharides have the formula $(CH_2O)_n$. Those with three carbons are called **trioses;** four, **tetroses;** five, **pentoses;** and six, **hexoses**.

b. They are called **aldoses** or **ketoses,** depending on whether their most oxidized functional group is an aldehyde or a ketone (Figure 5-1).

2. D and L sugars

a. The configuration of the asymmetric carbon atom farthest from the aldehyde or ketone group determines whether a monosaccharide belongs to the D or L series. In the D form, the hydroxyl group is on the right; in the L form, it is on the left (see Figure 5-1).

b. An asymmetric carbon atom has four different chemical groups attached to it.

c. **Sugars of the D series,** which are related to D-glyceraldehyde, are the most common in nature (Figure 5-2).

3. Stereoisomers, enantiomers, and epimers

a. **Stereoisomers** have the same chemical formula but differ in the position of the hydroxyl groups on **one or more** of their asymmetric carbons.

—A sugar with n asymmetric centers usually has 2n stereoisomers unless it has a plane of symmetry.

b. **Enantiomers** are stereoisomers that are mirror images of each other.

c. **Epimers** are stereoisomers that differ in the position of the hydroxyl group at only one asymmetric carbon. For example, D-glucose and D-galactose are epimers that differ at carbon 4 (see Figure 5-2).

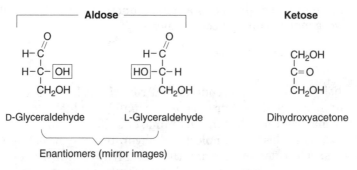

Figure 5-1. Examples of trioses, the smallest monosaccharides.

$$
\begin{array}{ccc}
\text{D-Glucose} & \text{D-Galactose} & \text{D-Fructose}
\end{array}
$$

Epimers

Figure 5-2. Some common hexoses of the D configuration.

4. Ring structures of carbohydrates

a. Although **monosaccharides** are often drawn as straight chains (Fischer projections), they exist mainly as ring structures in which the aldehyde or ketone group has reacted with a hydroxyl group in the same molecule (Figure 5-3).

–The **anomeric carbon** (which originally was part of the aldehyde or ketone group) forms a **hemiacetal** or **hemiketal,** generating a new asymmetric center.

–The ring structures are often drawn as **Haworth projections** with the ring in the plane of the paper and the substituents extending above or below the ring.

b. The ring forms are **furanoses** and **pyranoses** and contain five and six members, respectively.

c. The **hydroxyl group** on the anomeric carbon may be in the α or β configuration. In the α **configuration,** the hydroxyl group on the anomeric carbon is on the right in the Fischer projection and below the plane of the ring in the Haworth projection. In the β **configuration,** it is on the left in the Fischer projection and above the plane in the Haworth projection (Figure 5-4).

d. In solution, **mutarotation occurs**. The α and β forms equilibrate via the straight-chain aldehyde form.

B. Glycosides

1. Formation of glycosides

a. Glycosidic bonds form when the **hydroxyl** group on the anomeric carbon of a monosaccharide reacts with an –OH or –NH group of another compound.

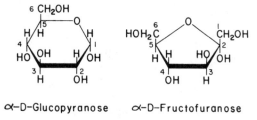

α–D–Glucopyranose α–D–Fructofuranose

Figure 5-3. Furanose and pyranose rings.

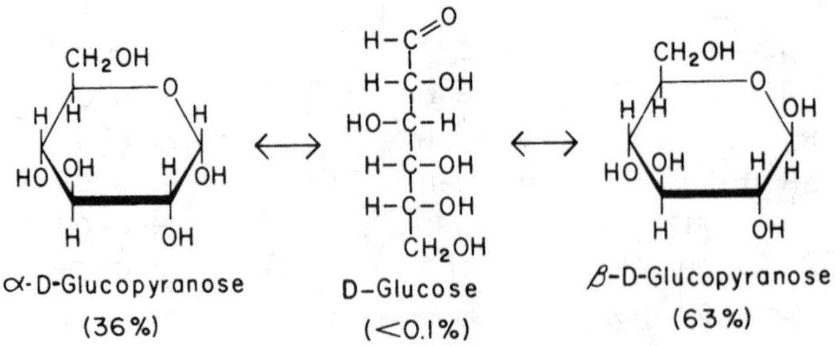

Figure 5-4. Mutarotation of glucose in solution. The percentage of each form is indicated.

    b. α-Glycosides or **β-glycosides** are produced depending on the position of the atom attached to the anomeric carbon of the sugar.

2. O-Glycosides

    a. Monosaccharides may be linked via *O*-glycosidic bonds to another monosaccharide, forming *O*-glycosides.

    b. Disaccharides contain two monosaccharides. Sucrose, lactose, and maltose are common disaccharides (Figure 5-5).

    c. Oligosaccharides contain up to 12 monosaccharides (see II B and C).

    d. Polysaccharides contain more than 12 monosaccharides; for example, glycogen, starch, and glycosaminoglycans (see Figures 5-8 and 5-12).

3. N-Glycosides

    a. Monosaccharides may be linked via *N*-glycosidic bonds to an organic base, forming *N*-glycosides.

    b. Nucleotides contain *N*-glycosidic bonds.

C. Derivatives of carbohydrates

    1. Phosphate groups may be attached to carbohydrates.

        a. Glucose and fructose may be phosphorylated on carbons 1 and 6.

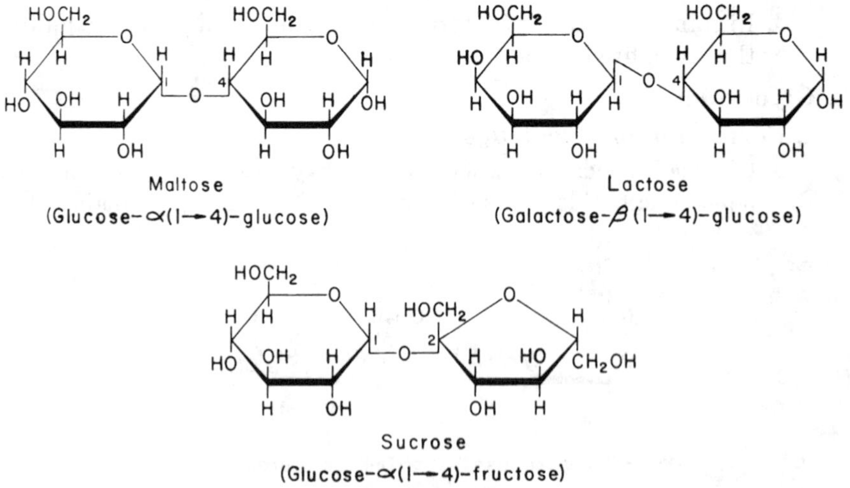

Figure 5-5. The most common disaccharides.

b. Phosphate groups may link sugars to nucleotides, as in UDP-glucose (see Figure 5-13).

2. Amino groups are often acetylated and may be linked to sugars (e.g., glucosamine and galactosamine).

3. Sulfate groups are often found on sugars (e.g., chondroitin sulfate and other glycosaminoglycans) [Figure 5-6].

D. Oxidation of carbohydrates

1. Oxidized forms

  a. The anomeric carbon of an aldose (C1) may be oxidized to an acid.

  (1) Glucose may be oxidized to form **gluconic acid**.

  (2) 6-Phosphogluconate is an important intermediate in the pentose phosphate pathway.

  b. Carbon 6 of a hexose may be oxidized to a uronic acid.

  (1) Glucose may form **glucuronic acid**. A major function of glucuronic acid is to make lipid compounds more water soluble (e.g., bilirubin diglucuronide).

  (2) Uronic acids are found in glycosaminoglycans of proteoglycans (see Figure 5-6).

2. Test for reducing sugars

  –Reducing sugars contain a free anomeric carbon that can be oxidized.

  a. When the anomeric carbon is oxidized, another compound is reduced. If the reduced product of this reaction is colored, the intensity of the color can be used to determine the amount of the reducing sugar that has been oxidized.

  b. This reaction is the basis of the reducing-sugar test, which is used by many clinical laboratories to determine **blood glucose levels**.

  (1) If other reducing sugars (e.g., fructose or galactose) are present in the blood, they may interfere with this test for blood glucose.

  (2) The reducing-sugar test is done in an alkaline solution in which fructose forms glucose and mannose as the result of endiol formation. Therefore, **fructose** gives a **positive reaction** and is a reducing sugar.

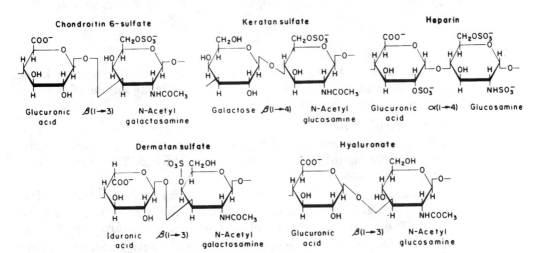

Figure 5-6. Repeating disaccharides of some glycosaminoglycans.

    c. Maltose and lactose are reducing sugars; **sucrose is not** because the anomeric carbons of both the glucose and fructose residues are joined in the disaccharide (see Figure 5-5).

E. Reduction of carbohydrates

    1. The aldehyde or ketone group of a sugar can be reduced to a hydroxyl group, forming a **polyol** (polyalcohol).

    2. Glucose may be reduced to **sorbitol** and galactose to **galactitol**.

II. Proteoglycans, Glycoproteins, and Glycolipids

- Proteoglycans consist of long linear chains of glycosaminoglycans attached to a core protein. Each chain is composed of a repeating disaccharide that usually is negatively charged and contains a hexosamine and a uronic acid. Sulfate groups are often present.
  –The glycosaminoglycans are synthesized from UDP-sugars.
- Glycoproteins contain smaller polysaccharide chains that are usually branched.
  –In addition to glucose and galactose and their amino derivatives, glycoproteins contain mannose, L-fucose, and *N*-acetylneuraminic acid (NANA).
  –The polysaccharide chain grows by the sequential addition of monosaccharide units to serine or threonine residues in a protein.
  –Some branched carbohydrate chains may be synthesized on dolichol phosphate and transferred to the amide nitrogen of an asparagine residue in a protein.
- Glycolipids are members of the class of sphingolipids.
  –The carbohydrate portion is synthesized from UDP-sugars that add to the hydroxymethyl group of ceramide (derived from sphingosine) and then sequentially to the nonreducing end of the chain.
  –*N*-Acetylneuraminic acid (derived from CMP-NANA) often forms branches from the main chain.
- Proteoglycans, glycoproteins, and glycolipids are synthesized in the endoplasmic reticulum and Golgi complex; they are degraded by the action of lysosomes.

    A. Proteoglycans are found in the extracellular matrix or ground substance of connective tissue, synovial fluid of joints, vitreous humor of the eye, secretions of mucus-producing cells, and in cartilage.

      1. Structure of proteoglycans

        a. Proteoglycans consist of a core protein with long unbranched polysaccharide chains (**glycosaminoglycans**) attached. The overall structure resembles a bottle brush (Figure 5-7).

        b. These chains are **composed of repeating disaccharide units,** which usually contain a uronic acid and a hexosamine (see Figure 5-6). The uronic acid is generally D-glucuronic or L-iduronic acid.

        c. The amino group of the hexosamine is usually acetylated, and sulfate groups are often present on carbons 4 and 6.

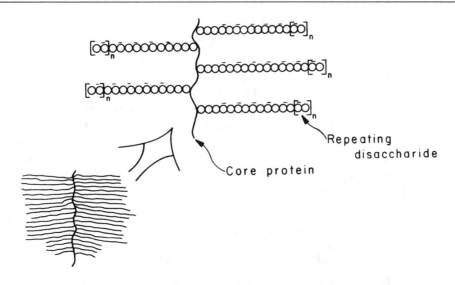

Figure 5-7. "Bottle brush" structure of a proteoglycan with a magnified segment.

    d. A sequence consisting of two galactose residues and a xylose connects the chain of repeating disaccharides to the core protein.

2. Synthesis of proteoglycans

    a. The **protein** is synthesized on the endoplasmic reticulum (ER).

    b. Sugars are added to serine or threonine residues of the protein. UDP-sugars serve as the precursors of the glycosaminoglycans.

    c. In the ER and the Golgi, the chains grow by sequential addition of sugars to the nonreducing end.

        (1) Sulfate groups, donated by 3'-phosphoadenosine-5'-phosphosulfate (PAPS), are added after the hexosamine is incorporated into the chain.

        (2) Because of the uronic acid and sulfate groups, the glycosaminoglycans are negatively charged, causing the chains to be heavily hydrated.

    d. Proteoglycans are secreted from the cell.

    e. Proteoglycans may associate with hyaluronic acid, a glycosaminoglycan, forming large aggregates, which act as molecular sieves, which can be penetrated by small but not by large molecules. A link protein noncovalently joins each proteoglycan to a hyaluronic acid chain.

3. Degradation of proteoglycans by lysosomal enzymes

    a. Because proteoglycans are located outside the cell, they are taken up by endocytosis. The endocytic vesicles fuse with lysosomes.

    b. Lysosomal enzymes specific for each monosaccharide remove the sugars, one at a time, from the nonreducing end of the chain.

    c. Sulfatases remove the sulfate groups before the sugar residue is hydrolyzed.

B. Glycoproteins serve as enzymes, hormones, antibodies, and structural proteins. They are found in extracellular fluids and in lysosomes and are attached to the cell membrane. They are involved in cell–cell interactions.

1. **Structure of glycoproteins**

   a. The carbohydrate portion of glycoproteins differs from that of proteo-glycans in that it is shorter and often branched.

      (1) Glycoproteins contain mannose, L-fucose, and *N*-acetylneuraminic acid (NANA) in addition to glucose, galactose, and their amino derivatives. NANA is a member of the class of sialic acids.

      (2) The antigenic determinants of the ABO and Lewis blood group substances are sugars at the ends of these carbohydrate branches.

   b. The carbohydrates may be attached to the protein via the hydroxyl groups of **serine and threonine** residues or the amide N of **asparagine**.

2. **Synthesis of glycoproteins**

   a. The protein is synthesized on the ER. In the ER and the Golgi, the **carbohydrate chain** is produced by the sequential addition of monosaccharide units to the nonreducing end. UDP-sugars, GDP-mannose, GDP-L-fucose, and CMP-NANA act as precursors.

   b. For some glycoproteins, the initial sugar is added to serine or threonine residues in the protein and the carbohydrate chain is then elongated.

   c. **Dolichol phosphate** is involved in the synthesis of glycoproteins in which the carbohydrate moiety is attached to the amide N of asparagine.

      (1) Dolichol phosphate, a long-chain alcohol containing about 20 five-carbon isoprenoid units, can be synthesized from acetyl CoA.

      (2) Sugars are added sequentially to **dolichol phosphate,** which is associated with the membrane of the ER.

      (3) The **branched polysaccharide chain** is transferred to an amide N of an **asparagine** residue in the protein.

      (4) In the ER and the Golgi, sugars may be removed from the chain and other sugars may be added.

   d. **Glycoproteins** may be **segregated** into lysosomes within the cell, they may be **attached** to the cell membrane, or they may be **secreted** by the cell. When a glycoprotein is attached to the cell membrane, the carbohydrate portion extends into the extracellular space and a hydrophobic segment of the protein is anchored in the membrane.

3. **Degradation of glycoproteins**

   –Lysosomal enzymes specific for each monosaccharide remove sugars sequentially from the nonreducing ends of the chains.

C. **Glycolipids**

   1. Glycolipids (**or sphingolipids**) are derived from the lipid **sphingosine** (see Figure 6-15). This class of compounds includes cerebrosides and gangliosides. Some bacterial toxins and viruses use them as receptors.

      a. **Cerebrosides** are synthesized from ceramide and UDP-sugars.

      b. In **gangliosides,** CMP-NANA adds *N*-acetylneuraminic acid residues as branches from the linear oligosaccharide chain.

   2. Glycolipids are found in the cell membrane with the carbohydrate portion extending into the extracellular space.

   3. They are **degraded by** lysosomal enzymes.

III. Digestion of Carbohydrates

- The major dietary carbohydrates are starch, sucrose, and lactose.
- In the mouth, salivary α-amylase acts on starch, cleaving α-1,4 linkages between glucose residues.
- In the intestine, pancreatic α-amylase continues the digestion of starch.
- Enzymes associated with the brush border of intestinal epithelial cells digest sucrose, lactose, and the products generated from starch by α-amylase.
- The final products of carbohydrate digestion—glucose, fructose, and galactose—are absorbed by intestinal epithelial cells and enter the blood.

A. Dietary carbohydrates (mainly starch, sucrose, and lactose) constitute about 50% of the calories in the average diet in the United States.

    1. Starch, sucrose, and lactose

        a. Starch, the storage form of carbohydrate in plants, is similar in structure to glycogen (Figure 5-8).

          –Starch contains amylose (long unbranched chains with glucose units linked α-1,4) and amylopectin (α-1,4–linked chains with α-1,6–linked branches). Amylopectin has fewer branches than glycogen.

        b. Sucrose (table sugar) contains glucose and fructose residues linked via their anomeric carbons (see Figure 5-5).

        c. Lactose (milk sugar) contains galactose linked β-1,4 to glucose (see Figure 5-5).

    2. Small amounts of free **glucose and fructose** are also present in the diet.

B. Digestion of dietary carbohydrates in the mouth (Figure 5-9)

    –In the mouth, **salivary α-amylase** cleaves starch by breaking α-1,4 linkages between glucose residues within the chains (see Figure 5-8). Dextrins (linear and branched oligosaccharides) are the major products that enter the stomach.

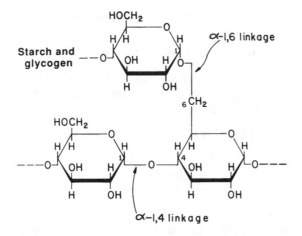

Figure 5-8. α-1,4 and α-1,6 linkages between glucose residues in starch and glycogen.

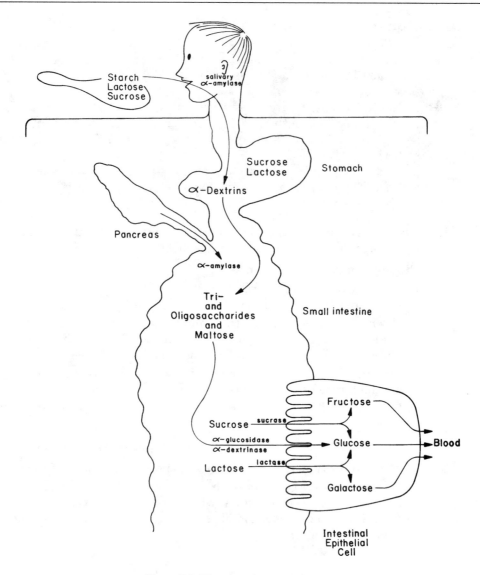

Figure 5-9. Digestion of carbohydrates.

C. Digestion of carbohydrates in the intestine (see Figure 5-9)

—The stomach contents pass into the intestine where **bicarbonate** secreted by the pancreas neutralizes the stomach acid, raising the pH into the optimal range for the action of the intestinal enzymes.

1. Digestion by pancreatic enzymes (see Figure 5-9)

a. The pancreas secretes an **α-amylase** that acts in the lumen of the small intestine and, like the salivary enzyme, cleaves α-1,4 linkages between glucose residues.

b. The products of pancreatic α-amylase are the disaccharide maltose, the trisaccharide maltotriose, and small oligosaccharides containing α-1,4 and α-1,6 linkages.

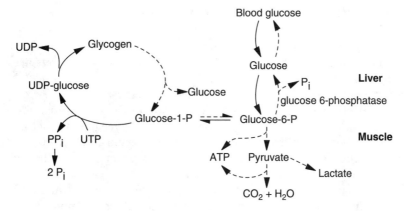

Figure 5-10. Overview of glycogen synthesis and degradation. *Solid arrows* = glycogen synthesis; *broken arrows* = glycogen degradation.

2. Digestion by enzymes of intestinal cells

a. Enzymes, produced by intestinal epithelial cells and **located in** their **brush borders,** complete the conversion of starch to glucose (see Figure 5-9).

(1) An **α-glucosidase** cleaves glucose residues from the nonreducing ends of oligosaccharides and also cleaves the α-1,4 bond of maltose, releasing the two glucose residues.

(2) An **α-dextrinase** cleaves α-1,6 linkages, releasing glucose residues from branched oligosaccharides.

b. Dietary disaccharides are also digested by enzymes in the brush borders of intestinal epithelial cells.

(1) Sucrase converts sucrose to glucose and **fructose**.

(2) Lactase (a β-galactosidase) converts lactose to glucose and galactose.

3. Absorption of glucose, fructose, and galactose

–Glucose, fructose, and galactose, the final products generated by digestion of dietary carbohydrates, are absorbed by intestinal epithelial cells.

a. They are transported into the cells on **carriers,** moving down a concentration gradient.

b. Glucose also moves into cells on a **transport protein** that carries sodium ions in addition to the monosaccharide—a secondary active transport process. A Na^+–K^+ ATPase pumps Na^+ into the blood, and glucose moves down a concentration gradient from the cell into the blood.

D. Carbohydrates that cannot be digested

–**Indigestible polysaccharides** are part of the dietary fiber that passes through the intestine into the feces. Because enzymes produced by human cells cannot cleave the β-1,4 bonds of cellulose, this polysaccharide is indigestible.

IV. Glycogen Structure and Metabolism

- Glycogen, the major storage form of carbohydrate in animals, consists of chains of α-1,4–linked glucose residues with branches that are attached by α-1,6 linkages.
- Glycogen is synthesized from glucose.
  - –UDP-glucose supplies the glucose moieties, which are added to the nonreducing ends of a glycogen primer by glycogen synthetase.
  - –Branches are produced by the branching enzyme glucosyl 4:6 transferase.
- Glycogen degradation produces glucose 1-phosphate as the major product, but free glucose is also formed.
  - –Glucose units are removed from the nonreducing ends of glycogen chains by phosphorylase, which produces glucose 1-phosphate.
  - –Three of the four glucose units at a branch point are moved by a glucosyl 4:4 transferase to the nonreducing end of another chain.
  - –The remaining glucose unit that is linked α-1,6 at the branch point is released as free glucose by an α-1,6-glucosidase.
- Liver glycogen is used to maintain blood glucose during fasting or exercise.
  - –Its breakdown is stimulated by glucagon and by epinephrine via a mechanism that involves cAMP.
- Muscle glycogen is utilized to generate ATP for muscle contraction.
  - –Epinephrine, via cAMP, stimulates muscle glycogen breakdown.

A. Glycogen structure

–**Glycogen** is a large, **branched polymer** consisting of D-glucose residues (Figure 5-11).

1. The **linkage** between glucose residues is α-1,4 except at branch points where the **linkage** is α-1,6. Branching is more frequent in the interior of the molecule and less frequent at the periphery, the average being an α-1,6 branch every 8–10 residues.

2. One glucose unit with a free anomeric carbon is located at the reducing end of each glycogen molecule. This glucose is attached to the protein glycogenin.

3. The **glycogen molecule** branches like a tree and has many nonreducing ends at which addition and release of glucose residues occur during synthesis and degradation, respectively.

B. Glycogen synthesis

–**UDP-glucose is the precursor** for glycogen synthesis.

1. **Synthesis of UDP-glucose** (Figure 5-12)

   a. Glucose enters the cell and is phosphorylated to glucose 6-phosphate by hexokinase (or by glucokinase in the liver). ATP provides the phosphate group.

   b. **Phosphoglucomutase** converts glucose 6-phosphate to glucose 1-phosphate.

   c. **Glucose 1-phosphate** reacts with UTP, forming **UDP-glucose** in a reaction catalyzed by UDP-glucose pyrophosphorylase. Inorganic pyrophosphate (PP_i) is released in this reaction.

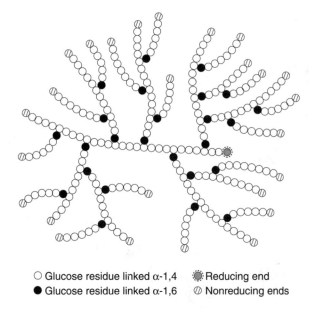

○ Glucose residue linked α-1,4   ◉ Reducing end
● Glucose residue linked α-1,6   ⊘ Nonreducing ends

Figure 5-11. The structure of glycogen.

–**PP**$_i$ is cleaved by a pyrophosphatase to 2 P$_i$, producing energy that drives the process in the direction of glycogen synthesis.

2. **Action of glycogen synthetase** (Figure 5-13)
   a. **Glycogen synthetase is** the key regulatory enzyme for glycogen synthesis.
   b. **Glucose residues** from UDP-glucose are transferred to the nonreducing ends of a glycogen primer by the enzyme glycogen synthetase.
      –UDP is released and may be reconverted to UTP by reaction with ATP.

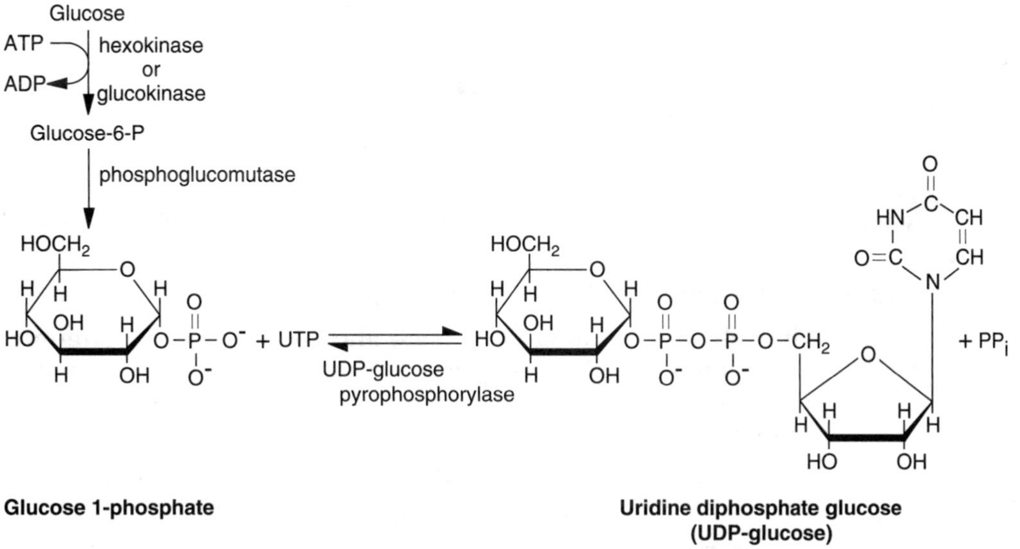

Glucose 1-phosphate

Uridine diphosphate glucose (UDP-glucose)

Figure 5-12. Formation of UDP-glucose.

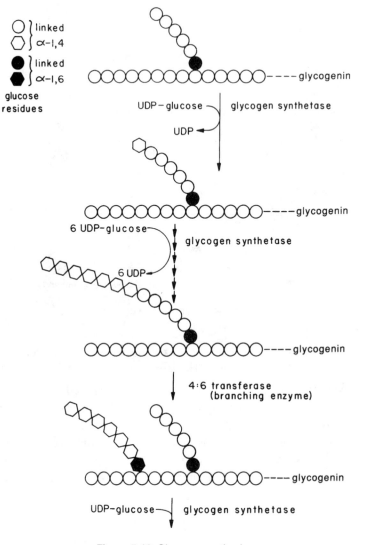

glucose
residues

○ } linked
⬡ } α-1,4

● } linked
⬣ } α-1,6

Figure 5-13. Glycogen synthesis.

   c. The primers, which are attached to glycogenin, are glycogen molecules that were partially degraded in liver during fasting or in muscle and liver during exercise.

3. Formation of branches (see Figure 5-13)

   a. When a chain contains 11 or more glucose residues, an **oligomer,** 6–8 residues in length, is removed from the nonreducing end of the chain. It is **reattached** via an α-1,6 linkage to a glucose residue within an α-1,4-linked chain.

   b. These branches are formed by the branching enzyme, a **glucosyl 4:6 transferase** that breaks an α-1,4 bond and forms an α-1,6 bond.

   c. The new branch points are at least 4 residues away from, and an average of 7–11 residues away from, previously existing branch points.

4. Growth of glycogen chains

a. Glycogen synthetase continues to add glucose residues to the nonreducing ends of newly formed branches as well as to the ends of the original chains.

b. As the chains continue to grow, additional branches are produced by the branching enzyme.

C. Glycogen degradation (Figure 5-14)

1. Action of phosphorylase

a. Phosphorylase, the key regulatory enzyme for glycogen degradation, removes glucose residues, one at a time, from the nonreducing ends of glycogen molecules.

b. Phosphorylase uses inorganic phosphate (P_i) to cleave α-1,4 bonds, producing **glucose 1-phosphate**.

c. Phosphorylase can act only until it is four glucose units from a branch point.

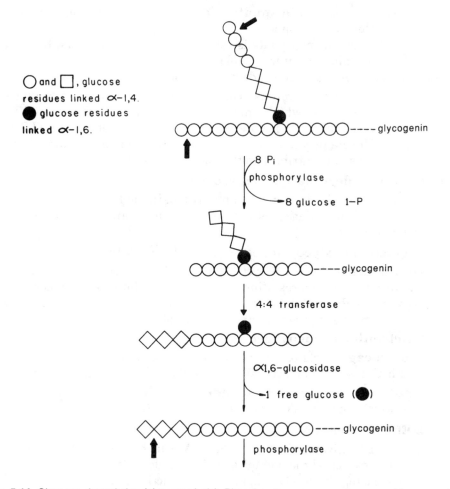

Figure 5-14. Glycogen degradation (glycogenolysis). Phosphorylase removes glucose residues one at a time, beginning at the *heavy arrows*. The three residues (⦵) adjacent to the branch point are transferred as a trisaccharide by the 4:4 transferase.

2. Removal of branches

–The four units remaining at a branch are removed by the debranching enzyme, which has both glucosyl 4:4 transferase and α-1,6-glucosidase activity.

a. Three of the four **glucose residues** that remain at the branch point are removed as a trisaccharide and attached to the nonreducing end of another chain by a 4:4 transferase, which cleaves an α-1,4 bond and forms a new α-1,4 bond.

b. The last glucose unit, which is linked α-1,6 at the branch point, is hydrolyzed by α-1,6-glucosidase, forming free glucose.

3. Degradation of glycogen chains

–The **phosphorylase/debranching process is repeated,** generating glucose 1-phosphate and free glucose in about a 10:1 ratio that reflects the length of the chains in the outer region of the glycogen molecule.

4. Fate of glucosyl units released from glycogen (see Figure 5-10)

a. In the **liver,** glycogen is degraded to maintain blood glucose.

(1) **Glucose 1-phosphate** is converted by phosphoglucomutase to glucose 6-phosphate.

(2) Inorganic phosphate is released by **glucose 6-phosphatase,** and free glucose enters the blood.

b. In **muscle,** glycogen is degraded to provide energy for contraction.

(1) Phosphoglucomutase converts glucose 1-phosphate to glucose 6-phosphate, which enters the pathway of glycolysis and is converted either to lactate or to CO_2 and H_2O, generating ATP.

(2) Muscle does not contain glucose 6-phosphatase and, therefore, does not contribute to the maintenance of blood glucose.

D. Lysosomal degradation of glycogen

–Glycogen may be degraded by an **α-glucosidase** located in lysosomes.
–Lysosomal degradation is not necessary for maintaining normal blood glucose levels.

E. Regulation of glycogen degradation (Figure 5-15)

–**Hormones** that use 3',5'-cyclic AMP (cAMP) as a second messenger stimulate a mechanism, resulting in the phosphorylation of enzymes.
–**Glycogen degradation** is stimulated and synthesis is inhibited when the enzymes of glycogen metabolism are phosphorylated.

1. Activation of adenyl cyclase

a. **Glucagon** acts on liver cells and **epinephrine** (adrenaline) acts on both liver and muscle cells to stimulate glycogen degradation.

b. These hormones activate **adenyl cyclase** in the cell membrane, which converts ATP to cAMP (Figure 5-16).

2. Effects of cAMP

a. cAMP activates **protein kinase A,** which consists of two regulatory and two catalytic subunits. cAMP binds to the regulatory subunits, releasing the catalytic subunits in an active form.

b. Protein kinase A phosphorylates glycogen synthetase, causing it to be less active, thereby decreasing glycogen synthesis.

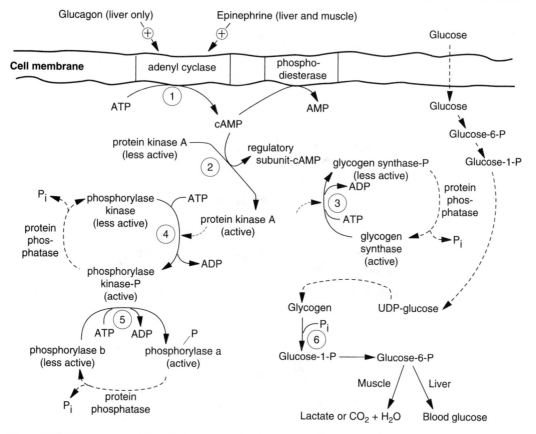

Figure 5-15. Hormonal regulation of glycogen synthesis and degradation. *Solid lines* indicate reactions that predominate when glucagon or epinephrine is elevated (steps 1 through 6 indicated by *circled numbers*); *dashed lines* indicate those that predominate when insulin is elevated.

    c. Protein kinase A causes phosphorylase to be activated, thereby increasing glycogen degradation. The process occurs in two steps.
      (1) Protein kinase A phosphorylates phosphorylase kinase.
      (2) Phosphorylase kinase adds a phosphate to phosphorylase b, converting it to its active form, phosphorylase a, which cleaves glucose residues from the nonreducing ends of glycogen chains, producing glucose 1-phosphate (see Figure 5-14).

3. The cAMP cascade
   –The cAMP-activated process is a cascade in which the initial hormonal signal is amplified manyfold.
    a. One **hormone molecule,** by activating the enzyme adenyl cyclase, produces many cAMPs, which activate the enzyme protein kinase A.
    b. One active **protein kinase A molecule** phosphorylates many phosphorylase kinase molecules, which convert many molecules of phosphorylase b to phosphorylase a.
    c. One **molecule of phosphorylase a** produces many molecules of glucose 1-phosphate from glycogen.
    d. The net result is that one hormone molecule can generate tens of thousands of molecules of glucose 1-phosphate.

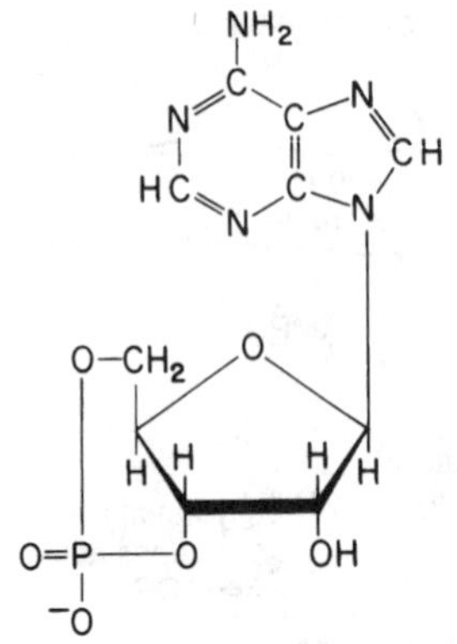

Figure 5-16. Cyclic AMP (cAMP).

4. Additional **regulatory mechanisms** in muscle

–In addition to cAMP-mediated regulation, AMP and Ca^{2+} stimulate glycogen breakdown in muscle.

a. Phosphorylase is activated by the rise in **AMP,** which occurs during muscle contraction by the following reactions:

$$2\ \text{ATP} \xrightarrow{\text{contraction}} 2\ \text{ADP} + 2\ P_i$$

$$2\ \text{ADP} \xrightarrow{\substack{\text{adenylate kinase} \\ \text{(myokinase)}}} \text{AMP} + \text{ATP}$$

$$\text{Sum: ATP} \longrightarrow \text{AMP} + 2\ P_i$$

b. Phosphorylase kinase is activated by Ca^{2+}, which is released from the sarcoplasmic reticulum during muscle contraction.

–Ca^{2+} binds to calmodulin, which serves as a subunit of phosphorylase kinase.

F. Regulation of glycogen synthesis (see Figure 5-15)

–**Insulin,** which is elevated after a meal, stimulates the synthesis of glycogen in liver and muscle.

1. Factors that promote glycogen synthesis in the liver

a. In the fed state, **glucagon is low,** and the cAMP cascade is not activated.

(1) cAMP is converted to AMP by a phosphodiesterase.

(2) As cAMP decreases, the regulatory subunits rejoin the catalytic subunits of protein kinase A, and the enzyme is inactivated.

(3) Dephosphorylation of phosphorylase kinase and phosphorylase a causes these enzymes to be inactivated. **Insulin** causes activation of the **phosphatases** that dephosphorylate these enzymes.

b. Glycogen synthesis is promoted by activation of glycogen synthetase and by the increased concentration of glucose, which enters liver cells from the hepatic portal vein.

–The less active phosphorylated form of glycogen synthetase is dephosphorylated, causing the enzyme to become active. Insulin causes activation of the phosphatase that catalyzes this reaction.

Table 5–1. Effect of Insulin on Glucose Transport Systems of Various Tissues

Tissues	Insulin Effect on Glucose Transport
Liver	0
Brain	0
Red blood cell	0
Adipose	+
Muscle	+

0 indicates no effect; + indicates stimulation.

2. Factors that promote glycogen synthesis in muscle

    a. After a meal, muscle will have low levels of cAMP, AMP, and Ca^{2+} if it is not contracting and epinephrine is low. Consequently, muscle glycogen degradation will not occur.

    b. Insulin stimulates glycogen synthesis by mechanisms similar to those in the liver.

    c. In addition, insulin stimulates the transport of glucose into muscle cells, providing increased substrate for glycogen synthesis (Table 5-1).

V. Glycolysis

- Glycolysis is the pathway by which glucose is oxidized to pyruvate. It occurs in the cytosol of all cells of the body.
- In the initial reactions, a hexose is twice phosphorylated by ATP and then cleaved to yield two triose phosphates.
  - Glucose is phosphorylated to glucose 6-phosphate, which is isomerized to fructose 6-phosphate.
  - Fructose 6-phosphate is phosphorylated by the key regulatory enzyme, phosphofructokinase. The product is fructose 1,6-bisphosphate, which is cleaved, forming two triose phosphates.
- In the second sequence of reactions, the triose phosphates produce ATP.
- Glycolysis produces ATP, NADH, and pyruvate.
  - ATP is produced directly by reactions catalyzed by phosphoglycerate kinase and pyruvate kinase.
  - Although NADH produced in the cytosol cannot directly enter mitochondria, reducing equivalents may be shuttled into this organelle where they generate ATP.
  - Pyruvate may enter mitochondria and be converted to acetyl CoA, which is oxidized by the TCA cycle, generating additional ATP.
  - Pyruvate may be converted to oxaloacetate by a reaction that replenishes intermediates of the TCA cycle.
  - Pyruvate may also be reduced to lactate or transaminated to alanine.

A. Transport of glucose into cells

    1. Glucose usually travels across the cell membrane on a transport protein.

    2. Insulin stimulates glucose transport into muscle and adipose cells by causing glucose transport proteins within cells to move to the cell membrane (see Table 5-1).

    3. Insulin is not required for the transport of glucose into tissues such as liver, brain, and red blood cells.

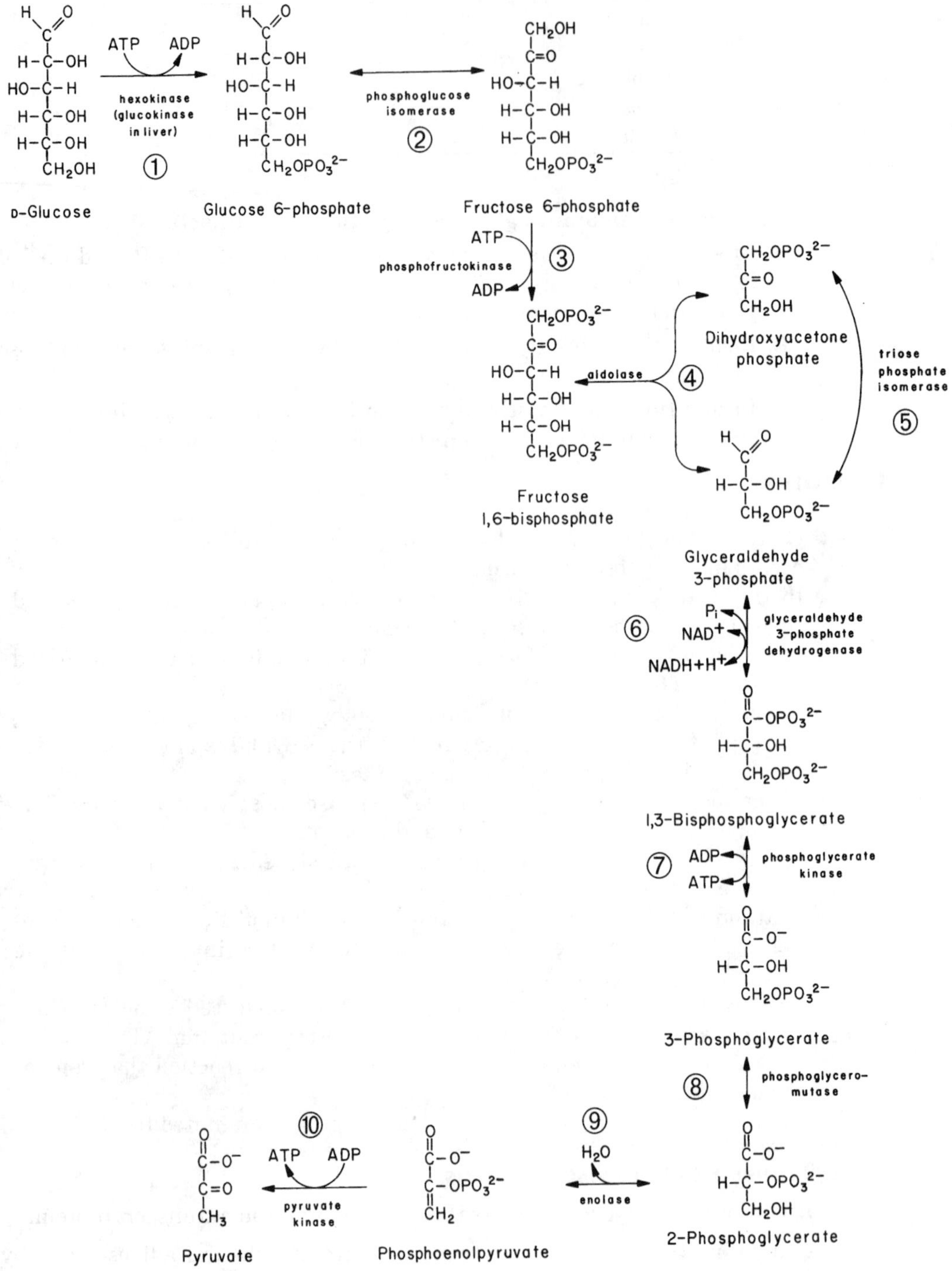

Figure 5-17. The reactions of glycolysis. The numbers correspond to those in V B in the text. These reactions occur in the cytolsol.

B. Reactions of glycolysis (Figure 5-17)

1. Glucose is converted to **glucose 6-phosphate** in a reaction that utilizes ATP and produces ADP.

    –Enzymes: **hexokinase** in all tissues and also, in the liver, **glucokinase**. Both of these enzymes are subject to regulatory mechanisms.

2. Glucose 6-phosphate is isomerized to fructose 6-phosphate.

    –Enzyme: **phosphoglucose isomerase**

3. **Fructose 6-phosphate** is phosphorylated by ATP, forming fructose 1,6-bisphosphate and ADP. This reaction is the first committed step in glycolysis.

    –Enzyme: **phosphofructokinase 1** (PFK1)

    –PFK1 is regulated by a number of effectors.

4. **Fructose 1,6-bisphosphate** is cleaved to form the triose phosphates, glyceraldehyde 3-phosphate and dihydroxyacetone phosphate.

    –Enzyme: **aldolase**

5. **Dihydroxyacetone phosphate** is isomerized to glyceraldehyde 3-phosphate.

    –Enzyme: **triose phosphate isomerase**

    Note: *The net result of reactions 1 through 5 is that two moles of glyceraldehyde 3-phosphate are formed from one mole of glucose.*

6. **Glyceraldehyde 3-phosphate** is oxidized by NAD^+ and reacts with inorganic phosphate (P_i) to form 1,3-bisphosphoglycerate and $NADH + H^+$.

    –Enzyme: **glyceraldehyde 3-phosphate dehydrogenase**

    –The aldehyde group of glyceraldehyde 3-phosphate is oxidized to a carboxylic acid, which forms a high-energy anhydride with inorganic phosphate.

7. **1,3-Bisphosphoglycerate** reacts with ADP to produce 3-phosphoglycerate and ATP.

    –Enzyme: **phosphoglycerate kinase**

8. **3-Phosphoglycerate** is converted to 2-phosphoglycerate by transfer of the phosphate group from carbon 3 to carbon 2.

    –Enzyme: **phosphoglyceromutase**

9. **2-Phosphoglycerate** is dehydrated to phosphoenolpyruvate (PEP), which contains a high-energy enol phosphate.

    –Enzyme: **enolase**

10. **Phosphoenolpyruvate** reacts with ADP to form pyruvate and ATP in the last reaction of glycolysis.

    –Enzyme: **pyruvate kinase**. Pyruvate kinase is more active in the fed state than in the fasting state.

C. Special reactions in red blood cells

1. In red blood cells, significant quantities of 1,3-bisphosphoglycerate are converted to 2,3-bisphosphoglycerate (BPG), a compound that decreases the affinity of hemoglobin for oxygen.

2. 2,3-Bisphosphoglycerate is dephosphorylated to form inorganic phosphate and 3-phosphoglycerate, an intermediate that reenters the glycolytic pathway.

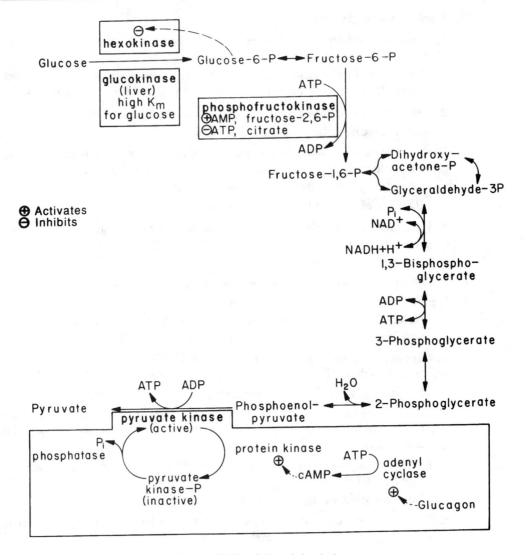

Figure 5-18. Regulation of glycolysis.

D. Regulatory enzymes of glycolysis (Figure 5-18)

    1. Hexokinase is found in most tissues and is geared to provide glucose 6-phosphate for ATP production even when blood glucose is low.

        a. Hexokinase has a **low K_m** for glucose (about 0.1 mM). Therefore, it is working near its maximum rate even at fasting blood glucose levels (about 5 mM).

        b. Hexokinase is inhibited by its product, glucose 6-phosphate. Therefore, it is most active when glucose 6-phosphate is being rapidly utilized.

    2. Glucokinase is found in the **liver** and functions at a significant rate only after a meal.

        a. Glucokinase has a **high K_m for glucose** (about 10 mM). Therefore, it is very active after a meal when glucose levels in the hepatic portal vein are high, and it is relatively inactive during fasting when glucose levels are low.

b. Glucokinase is induced when insulin levels are high.

c. Glucokinase is not inhibited by glucose 6-phosphate at physiologic concentrations.

3. Phosphofructokinase 1 (**PFK1**) is regulated by several factors. It functions at a rapid rate in the liver when blood glucose is high, or in cells such as muscle when there is a need for ATP.

   a. PFK1 is activated by **fructose 2,6-bisphosphate,** an important regulatory mechanism in the liver (Figure 5-19).

   (1) Fructose 2,6-bisphosphate is formed from fructose 6-phosphate by phosphofructokinase 2 (**PFK2**).

   (2) After a meal, fructose 2,6-bisphosphate is high, PFK1 is activated, and glycolysis is stimulated. The liver is using glycolysis to produce fatty acids for triacylglycerol synthesis.

   (3) PFK2 is phosphorylated in the **fasting** state (when glucagon is elevated) by protein kinase A, which is activated by cAMP.

   (a) Phosphorylated PFK2 converts fructose 2,6-bisphosphate to fructose 6-phosphate.

   (b) Fructose 2,6-bisphosphate levels fall, and PFK1 is less active.

   (4) In the **fed state,** insulin causes phosphatases to be stimulated. A phosphatase dephosphorylates PFK2, causing it to become more active in forming fructose 2,6-bisphosphate from fructose 6-phosphate. Fructose 2,6-bisphosphate levels rise, and PFK1 is more active.

   (5) Thus, PFK2 acts as a kinase (in the fed state when it is dephosphorylated) and as a phosphatase (in the fasting state when it is phosphorylated). PFK2 catalyzes two different reactions.

   b. PFK1 is activated by **AMP,** an important regulatory mechanism in muscle (see Figure 5-18).

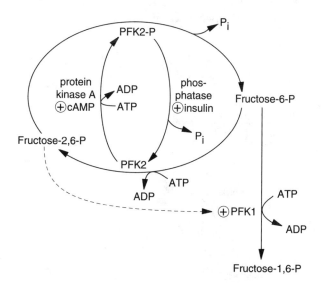

Figure 5-19. Regulation of fructose 2,6-bisphosphate levels. Fructose 2,6-P is an activator of phosphofructokinase 1 (*PFK1*), which converts fructose 6-phosphate to fructose 1,6-bisphosphate. Phosphofructokinase 2 (*PFK2*) acts as a kinase in the fed state and as a phosphatase during fasting.

(1) When muscle is exercising, AMP levels are high and ATP is low.

(2) Glycolysis is promoted by a more active PFK1, and ATP is generated.

c. PFK1 is inhibited by **ATP and citrate,** important regulatory mechanisms in muscle.

(1) When ATP is high, the cell does not need ATP, and glycolysis is inhibited.

(2) High levels of citrate indicate that adequate amounts of substrate are entering the TCA cycle. Therefore, glycolysis slows down.

4. Pyruvate kinase

a. Pyruvate kinase is inhibited by phosphorylation in the liver during **fasting** when glucagon levels are high (see Figure 5-18).

(1) Glucagon via cAMP activates protein kinase A, which phosphorylates and inactivates pyruvate kinase.

(2) The inhibition of pyruvate kinase promotes gluconeogenesis.

b. Pyruvate kinase is activated in the **fed state**.

(1) Insulin stimulates phosphatases that dephosphorylate and activate pyruvate kinase.

(2) Pyruvate kinase may also be induced in the fed state; that is, the quantity of the enzyme may increase.

E. The fate of pyruvate (Figure 5-20)

1. Conversion to lactate

–Pyruvate can be reduced in the cytosol by NADH, forming the corresponding α-hydroxy acid, **lactate,** and regenerating NAD^+.

a. **NADH,** which is produced by glycolysis, must be reconverted to NAD^+ so that carbons of glucose can continue to flow through glycolysis.

b. Lactate dehydrogenase (**LDH**) converts pyruvate to lactate. LDH consists of four subunits that can be either of the muscle (M) or heart (H) type.

(1) Five isozymes occur (MMMM, MMMH, MMHH, MHHH, and HHHH), which may be separated by electrophoresis.

(2) Different tissues have different mixtures of these isozymes.

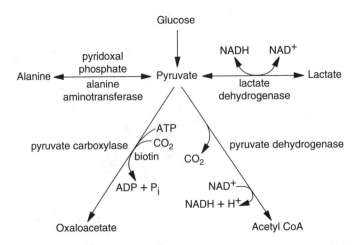

Figure 5-20. The fate of pyruvate.

    c. Lactate is produced by tissues such as red blood cells or exercising muscle.

    d. Lactate may be utilized by the liver for gluconeogenesis, or by tissues such as the heart and kidney where it is converted to pyruvate and oxidized for energy.

    –The LDH reaction is reversible.

2. Conversion to acetyl CoA

–Pyruvate can enter mitochondria and be converted by pyruvate dehydrogenase to acetyl CoA, which can enter the TCA cycle.

3. Conversion to oxaloacetate

–Pyruvate may be converted to oxaloacetate by **pyruvate carboxylase,** an enzyme found in tissues such as the liver and brain but not in muscle.

–This reaction serves to replenish intermediates of the TCA cycle.

4. Conversion to alanine

–Pyruvate may be transaminated to form the amino acid **alanine**.

F. Generation of ATP by glycolysis

1. Production of ATP and NADH in the glycolytic pathway

–Overall, when one mole of glucose is converted to two moles of pyruvate, two moles of ATP are utilized in the process, and four moles of ATP are produced, for a net yield of two moles of ATP. In addition, two moles of cytosolic NADH are generated

2. Energy generated by conversion of glucose to lactate (Figure 5-21)

–If the NADH generated by glycolysis is utilized to reduce pyruvate to lactate, the net yield is two moles of ATP per mole of glucose converted to lactate.

3. Energy generated by conversion of glucose to CO_2 and H_2O (Figure 5-22)

–When glucose is oxidized completely to CO_2 and H_2O, approximately 36 or 38 moles of ATP are generated.

a. Energy generated exclusive of NADH

–If the energy produced from the NADH generated by glycolysis is not included, the yield is about 32 moles of ATP.

(1) Two moles of ATP are generated from the conversion of one mole of glucose to two moles of pyruvate.

(2) The two moles of pyruvate enter mitochondria and are converted to two moles of acetyl CoA, producing two moles of NADH, which generate approximately six moles of ATP by oxidative phosphorylation.

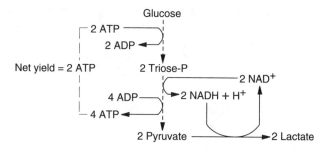

Figure 5-21. Conversion of glucose to lactate produces two molecules of ATP (net).

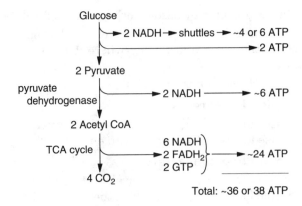

Glucose

2 NADH → shuttles → ~4 or 6 ATP

→ 2 ATP

2 Pyruvate

pyruvate
dehydrogenase

→ 2 NADH → ~6 ATP

2 Acetyl CoA

TCA cycle

6 NADH
2 FADH$_2$ → ~24 ATP
2 GTP

4 CO$_2$

Total: ~36 or 38 ATP

Figure 5-22. ATP produced by conversion of glucose to CO_2. The ATP produced by oxidative phosphorylation is approximate (indicated by ~).

(3) The 2 moles of acetyl CoA are oxidized in the TCA cycle, generating approximately 24 moles of ATP.

b. Energy generated from NADH

–NADH cannot directly cross the mitochondrial membrane; therefore, the electrons are passed to the mitochondrial electron transport chain by two shuttle systems.

(1) Glycerol phosphate shuttle (Figure 5-23, *left side*)

(a) Cytosolic DHAP is reduced to glycerol 3-phosphate by NADH.

(b) Glycerol 3-phosphate reacts with an FAD-linked dehydrogenase in the inner mitochondrial membrane. DHAP is regenerated and reenters the cytosol.

(c) Each mole of FADH$_2$ that is produced generates approximately two moles of ATP via oxidative phosphorylation.

(d) Because glycolysis produces two moles of NADH per mole of glucose, approximately **four moles of ATP are produced by this shuttle**.

(2) Malate aspartate shuttle (see Figure 5-23, *right side*)

(a) Cytosolic oxaloacetate is reduced to malate by NADH. The reaction is catalyzed by cytosolic malate dehydrogenase.

(b) Malate enters the mitochondrion and is reoxidized to oxaloacetate by the mitochondrial malate dehydrogenase, generating NADH in the matrix.

(c) Oxaloacetate cannot cross the mitochondrial membrane. In order to return carbon to the cytosol, oxaloacetate is transaminated to aspartate, which can be transported into the cytosol and reconverted to oxaloacetate by another transamination reaction.

(d) Each mole of NADH generates approximately three moles of ATP via oxidative phosphorylation.

(e) Because glycolysis produces two moles of NADH per mole of glucose, approximately six moles of ATP are produced by this shuttle.

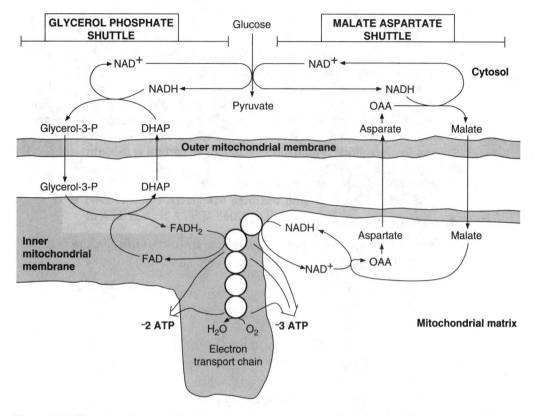

Figure 5-23. The glycerol phosphate and malate aspartate shuttles. *Left,* the glycerol phosphate shuttle produces $FADH_2$, each of which generates approximately 2 ATP by oxidative phosphorylation. *Right,* the malate aspartate shuttle produces NADH, each of which generates approximately 3 ATP.

c. Maximal ATP production

–Overall, when 1 mole of glucose is oxidized to CO_2 and H_2O, approximately 36 moles of ATP are produced if the glycerol phosphate shuttle is used, or 38 moles if the malate aspartate shuttle is used.

VI. Gluconeogenesis (Figure 5-24)

- Gluconeogenesis, which occurs mainly in the liver, is the synthesis of glucose from compounds that are not carbohydrates.
- The major precursors for gluconeogenesis are lactate, amino acids, and glycerol. Even-chain fatty acids do not produce any net glucose.
- Gluconeogenesis involves several enzymatic steps that do not occur in glycolysis; thus glucose is not generated by a simple reversal of glycolysis.
- Pyruvate carboxylase converts pyruvate to oxaloacetate in the mitochondrion. Oxaloacetate is converted to malate or aspartate, which travels to the cytosol and is reconverted to oxaloacetate.
- Phosphoenolpyruvate carboxykinase converts oxaloacetate to phosphoenolpyruvate. Phosphoenolpyruvate forms fructose 1,6-bisphosphate by reversal of the steps of glycolysis.

- Fructose 1,6-bisphosphatase converts fructose 1,6-bisphosphate to fructose 6-phosphate, which is converted to glucose 6-phosphate.
- Glucose 6-phosphatase converts glucose 6-phosphate to free glucose, which is released into the blood.
- Gluconeogenesis occurs under conditions in which pyruvate dehydrogenase, pyruvate kinase, phosphofructokinase 1, glucokinase, and hexokinase are relatively inactive. The low activity of these enzymes prevents futile cycles from occurring and ensures that, overall, pyruvate is converted to glucose.
- The synthesis of one mole of glucose from two moles of pyruvate requires energy equivalent to about six moles of ATP.

A. Reactions of gluconeogenesis

1. Conversion of pyruvate to phosphoenolpyruvate (Figure 5-25)

–In the liver, pyruvate is converted to phosphoenolpyruvate.

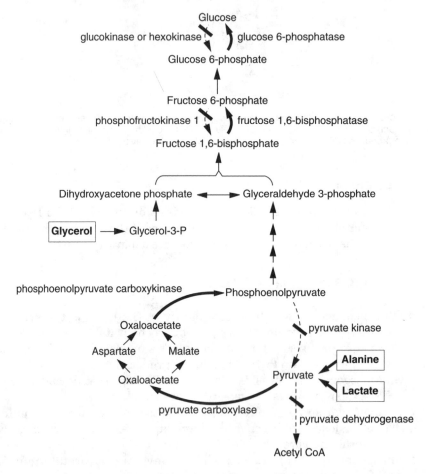

Figure 5-24. The key reactions of gluconeogenesis from the precursors alanine, lactate, and glycerol. *Heavy arrows* indicate steps that differ from those of glycolysis. *Broken arrows* are reactions that are inhibited (■) under conditions in which gluconeogenesis is occurring.

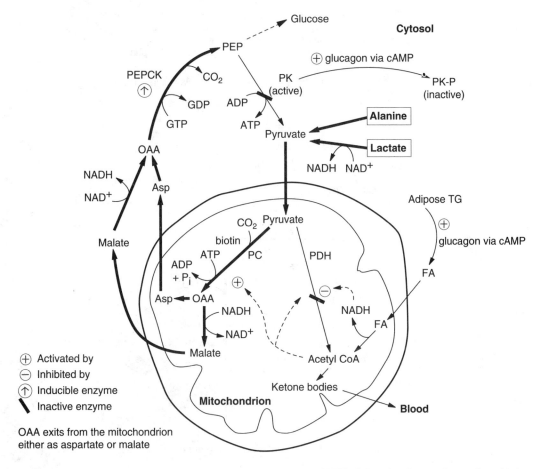

Figure 5-25. The conversion of pyruvate to phosphoenolpyruvate (*PEP*). Follow the diagram by starting with the precursors alanine and lactate. *OAA* = oxaloacetate; *FA* = fatty acid; *TG* = triacylglycerol; PDH = pyruvate dehydrogenase; PC = pyruvate carboxylase; PEPCK = phosphoenolpyruvate; PK = pyruvate kinase; PK-P = phosphorylated pyruvate kinase.

    a. Pyruvate is first converted to oxaloacetate (OAA) by pyruvate carboxylase, a mitochondrial enzyme that requires biotin and ATP.

    –**Oxaloacetate** cannot directly cross the mitochondrial membrane. Therefore, it is converted to malate or to aspartate, which can cross the mitochondrial membrane and be reconverted to OAA in the cytosol.

    b. Oxaloacetate is decarboxylated by phosphoenolpyruvate carboxykinase to form phosphoenolpyruvate. This reaction requires GTP.

    c. Phosphoenolpyruvate is converted to fructose 1,6-bisphosphate by reversal of the glycolytic reactions.

  2. Conversion of fructose 1,6-bisphosphate to fructose 6-phosphate (Figure 5-26)

    a. Fructose 1,6-bisphosphatase converts fructose 1,6-bisphosphate to fructose 6-phosphate in a reaction that releases inorganic phosphate.

    b. Fructose 6-phosphate is converted to glucose 6-phosphate by the same isomerase utilized in glycolysis.

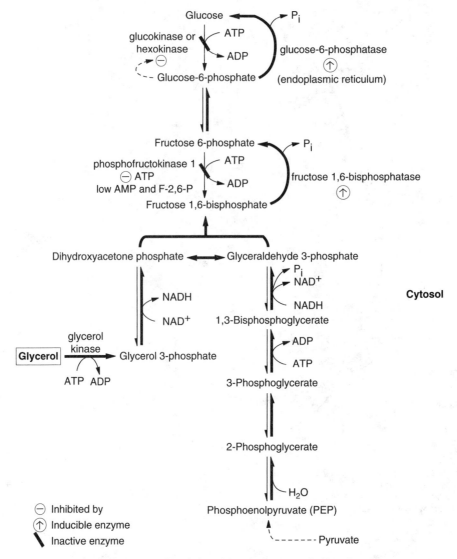

Figure 5-26. The conversion of phosphoenolpyruvate and glycerol to glucose.

3. Conversion of glucose 6-phosphate to glucose

    a. Glucose 6-phosphatase cleaves inorganic phosphate from glucose 6-phosphate, and free glucose is released into the blood.

    b. This enzyme catalyzes the release of free glucose from glucose 6-phosphate that is generated both from gluconeogenesis and glycogenolysis (see Figure 5-15).

B. Regulatory enzymes of gluconeogenesis

    –Under fasting conditions, **glucagon** is elevated and stimulates gluconeogenesis. Because of changes in the activity of certain enzymes, futile cycles are prevented from occurring, and the overall flow of carbon is from pyruvate to glucose (see Figures 5-25 and 5-26).

    –A futile cycle is the continuous recycling of substrates and products with the net consumption of energy and no useful result.

1. **Pyruvate dehydrogenase** (see Figure 5-25)

   a. **Glucagon stimulates the release of fatty acids** from adipose tissue.

   b. **Fatty acids** travel to the liver and **are oxidized,** producing acetyl CoA, NADH, and ATP, which cause inactivation of pyruvate dehydrogenase.

   c. Because **pyruvate dehydrogenase** is relatively **inactive,** pyruvate is converted to oxaloacetate, not to acetyl CoA.

2. **Pyruvate carboxylase**

   a. Pyruvate carboxylase is activated by acetyl CoA.

   b. Note that pyruvate carboxylase is active in both the fed and fasting states.

3. **Phosphoenolpyruvate carboxykinase (PEPCK)**

   a. PEPCK is an **inducible** enzyme.

   b. **Transcription** of the gene encoding PEPCK is stimulated by binding of cAMP–protein complexes and glucocorticoid–protein complexes to regulatory elements in the gene.

   c. Increased production of PEPCK mRNA leads to increased translation, resulting in higher PEPCK levels in the cell.

4. **Pyruvate kinase**

   a. **Glucagon,** via cAMP and protein kinase A, causes pyruvate kinase to be phosphorylated and **inactivated**.

   b. Because pyruvate kinase is relatively inactive, **phosphoenolpyruvate** formed from oxaloacetate is not reconverted to pyruvate but, in a series of steps, forms fructose 1,6-bisphosphate, which is converted to fructose 6-phosphate.

5. **Phosphofructokinase 1** (see Figure 5-26)

   –Phosphofructokinase 1 is relatively **inactive** because the concentrations of its activators, AMP and fructose 2,6-bisphosphate, are low and its inhibitor, ATP, is relatively high.

6. **Fructose 1,6-bisphosphatase**

   a. The level of **fructose 2,6-bisphosphate,** an inhibitor of fructose 1,6-bisphosphatase, is **low** during fasting. Therefore, fructose 1,6-bisphosphatase is **more active.**

   b. Fructose 1,6-bisphosphatase also is **induced** in the fasting state.

7. **Glucokinase and hexokinase**

   –Glucokinase and hexokinase are relatively inactive, so free glucose is not reconverted to glucose 6-phosphate.

   a. **Glucokinase** is relatively **inactive** because it has a high K_m for glucose, and under conditions that favor gluconeogenesis, the glucose concentration is low.

   b. **Hexokinase** is **inhibited by** glucose 6-phosphate.

C. **Precursors for gluconeogenesis**

   –Lactate, amino acids, and glycerol are the major precursors for gluconeogenesis in the human.

1. **Lactate** is oxidized by NAD^+ in a reaction catalyzed by lactate dehydrogenase to form pyruvate, which may be converted to glucose (see Figure 5-25).

   –Sources of lactate include red blood cells and exercising muscle.

2. **Amino acids** for gluconeogenesis come from **muscle**. They may be derived by degradation of muscle protein.

   a. Amino acids may be released directly into the blood from muscle, or carbons from amino acids may be converted to alanine and glutamine and released.

   –Alanine may also be formed by transamination of pyruvate that is derived by oxidation of glucose.

   –Glutamine may be converted to alanine by tissues such as gut and kidney.

   b. Amino acids travel to the liver and provide carbon for gluconeogenesis. Quantitatively, alanine is the major gluconeogenic amino acid.

   c. Amino acid nitrogen is converted to urea.

3. **Glycerol,** which is derived from **adipose** triacylglycerols, reacts with ATP to form glycerol 3-phosphate, which is oxidized to dihydroxyacetone phosphate and converted to glucose (see Figure 5-26).

D. **Role of fatty acids in gluconeogenesis**

   1. **Even-chain fatty acids**

      a. Fatty acids are oxidized to acetyl CoA, which enters the TCA cycle.

      b. For every two carbons of acetyl CoA that are converted to malate (which leaves the mitochondrion to form glucose), two carbons are released as CO_2. Therefore, there is **no net synthesis of glucose from acetyl CoA**.

      c. The pyruvate dehydrogenase reaction is irreversible, so acetyl CoA cannot be converted to pyruvate.

      d. Although even-chain fatty acids do not provide carbons for gluconeogenesis, β-oxidation of fatty acids does provide ATP.

   2. **Odd-chain fatty acids**

      –The three carbons at the ω-end of an odd-chain fatty acid are converted to propionate.

      a. **Propionate** enters the TCA cycle as succinyl CoA, which forms **malate,** an intermediate in glucose formation (see Figure 5-25).

      b. Certain amino acids also form glucose via propionate.

E. **Energy requirements for gluconeogenesis**

   1. **From pyruvate** (see Figures 5-25 and 5-26)

      a. Conversion of pyruvate to oxaloacetate by pyruvate carboxylase requires one ATP.

      b. Conversion of oxaloacetate to phosphoenolpyruvate by phosphoenolpyruvate carboxykinase requires one GTP (the equivalent of one ATP).

      c. Conversion of 3-phosphoglycerate to 1,3-bisphosphoglycerate by phosphoglycerate kinase requires one ATP.

d. Since two moles of pyruvate are required to form one mole of glucose, **six moles of high-energy phosphate are required for synthesis of 1 mole of glucose**.

2. From glycerol (see Figure 5-26)

—Glycerol enters the gluconeogenic pathway at the dihydroxyacetone phosphate (DHAP) level.

a. Conversion of glycerol to glycerol 3-phosphate, which is oxidized to DHAP, requires one ATP.

b. Since two moles of glycerol are required to form one mole of glucose, two moles of high-energy phosphate are required for synthesis of one mole of glucose.

VII. Fructose and Galactose Metabolism

- Although glucose is the most abundant monosaccharide derived from the diet, fructose and galactose are usually obtained in significant quantities, mainly from sucrose and lactose.
- After fructose and galactose enter cells, they are phosphorylated on carbon 1 and converted to intermediates in pathways of glucose metabolism.
- Fructose is metabolized mainly in the liver, where it is converted to fructose 1-phosphate and cleaved to produce dihydroxyacetone phosphate and glyceraldehyde, which may be phosphorylated to glyceraldehyde 3-phosphate. These two triose phosphates are intermediates of glycolysis.
- Fructose may be produced from sorbitol, which is generated from glucose.
- Galactose is phosphorylated to galactose 1-phosphate, which reacts with UDP-glucose. The products are glucose 1-phosphate and UDP-galactose, which is epimerized to UDP-glucose. The net result is that galactose is converted to the glucose moieties of UDP-glucose and glucose 1-phosphate, intermediates in pathways of glucose metabolism.
  —UDP-galactose may be used in the synthesis of glycoproteins, glycolipids, and proteoglycans.
  —UDP-galactose may react with glucose in the mammary gland to form the milk sugar lactose.
- Galactose may be reduced to galactitol.

A. Metabolism of fructose

—The major dietary source of fructose is the disaccharide sucrose (table sugar), but it is also present as the monosaccharide in fruit and in corn syrup, which is used as a sweetener.

1. Conversion of fructose to glycolytic intermediates (Figure 5-27)

a. Fructose is metabolized mainly in the **liver** where it is converted to pyruvate or, under fasting conditions, to glucose.

(1) Fructose is phosphorylated by ATP to form fructose 1-phosphate. The enzyme is fructokinase.

(2) Fructose 1-phosphate is cleaved by aldolase to form dihydroxyacetone phosphate (DHAP) and glyceraldehyde, which is phosphorylated by ATP to form glyceraldehyde 3-phosphate. DHAP and glyceraldehyde 3-phosphate are intermediates of glycolysis.

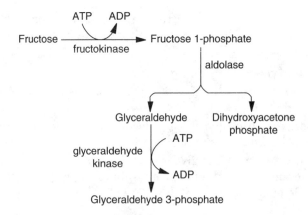

Figure 5-27. Conversion of fructose to intermediates of glycolysis.

    (3) Glyceraldehyde also may be reduced to glycerol, which is phosphorylated to glycerol 3-phosphate and used in lipid synthesis.
    b. In tissues other than liver, the major fate of fructose is phosphorylation by hexokinase to form fructose 6-phosphate, which enters glycolysis. Hexokinase has an affinity for fructose about one-twentieth of that for glucose.
  2. Production of fructose from glucose
    a. Glucose is reduced to sorbitol by aldose reductase, which reduces the aldehyde group to an alcohol (Figure 5-28).
    b. Sorbitol is then reoxidized at carbon 2 by sorbitol dehydrogenase to form fructose.
    c. Conversion of glucose to **fructose,** which serves as the major energy source for sperm cells, is particularly important in seminal vesicles.
B. Metabolism of galactose
  —The major dietary source of galactose is the disaccharide lactose found in milk or milk products.
  1. Conversion of galactose to intermediates of glucose pathways (Figure 5-29)
    a. Galactokinase uses ATP to phosphorylate galactose to galactose 1-phosphate.
    b. Galactose 1-phosphate reacts with UDP-glucose forming glucose 1-phosphate and UDP-galactose. The enzyme is galactose 1-phosphate uridyl transferase.
    c. UDP-galactose is epimerized to UDP-glucose in a reaction that is readily reversible.
    d. Repetition of reactions a–c results in a conversion of galactose to UDP-glucose and glucose 1-phosphate.

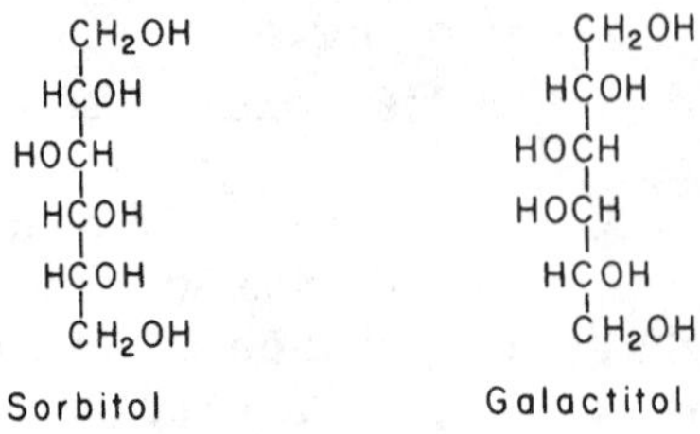

$$
\begin{array}{ll}
CH_2OH & CH_2OH \\
HCOH & HCOH \\
HOCH & HOCH \\
HCOH & HOCH \\
HCOH & HCOH \\
CH_2OH & CH_2OH \\
\text{Sorbitol} & \text{Galactitol}
\end{array}
$$

Figure 5-28. Reduced forms of sugars. Sorbitol is produced by reduction of glucose and reoxidized at carbon 2 to form fructose. Galactitol is produced by reduction of galactose.

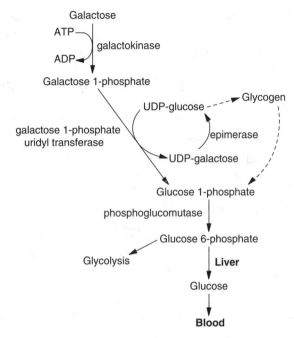

Figure 5-29. Conversion of galactose to intermediates of glucose metabolism. Galactose 1-phosphate uridyl transferase is deficient in classic galactosemia.

–These glucose derivatives may be converted to blood glucose or to glycogen by the liver. In various tissues, the glucose 1-phosphate may feed into glycolysis.

2. Other fates of UDP-galactose (Figure 5-30)

–UDP-galactose may be produced from galactose or from glucose via UDP-glucose and an epimerase.

a. UDP-galactose supplies galactose moieties for the synthesis of glyco-proteins, glycolipids, and proteoglycans.

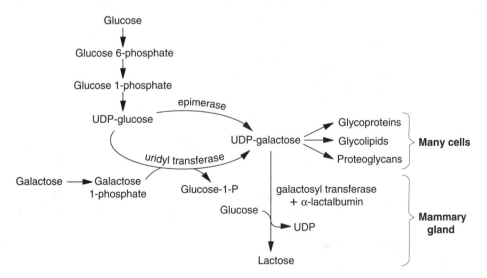

Figure 5-30. Metabolism of UDP-galactose. UDP-galactose can be produced from dietary glucose or galactose.

–The enzyme that adds galactose units to growing polysaccharide chains is galactosyl transferase.

b. UDP-galactose may react with glucose in the mammary gland to produce the milk sugar **lactose**.

–The modifier protein, **α-lactalbumin,** reacts with galactosyl transferase, causing its K_m for glucose to be lowered so that it adds galactose (from UDP-galactose) to glucose, forming lactose.

3. Conversion of galactose to galactitol

–Aldose reductase reduces the aldehyde of galactose to an alcohol, forming galactitol (see Figure 5-28).

VIII. Pentose Phosphate Pathway

● In the irreversible oxidative reactions of the pathway, one carbon of glucose 6-phosphate is released as CO_2, NADPH is generated, and ribulose 5-phosphate is produced (Figure 5-31).
  –NADPH is used for reductive biosynthesis (particularly of fatty acids) and for protection against oxidative damage.
  –Ribulose 5-phosphate provides ribose 5-phosphate for nucleotide biosynthesis or generates pentose phosphates, which enter the nonoxidative portion of the pathway.
● In the reversible nonoxidative reactions, pentose phosphates, produced from ribulose 5-phosphate, may be converted to the glycolytic intermediates fructose 6-phosphate and glyceraldehyde 3-phosphate.
● Because the nonoxidative reactions are reversible, they can be used to generate ribose 5-phosphate for nucleotide synthesis from intermediates of glycolysis.

A. Reactions of the pentose phosphate pathway

1. The oxidative reactions (Figure 5-32)

a. Glucose 6-phosphate is converted to 6-phosphogluconolactone, and $NADP^+$ is reduced to $NADPH + H^+$.

–Enzyme: **glucose 6-phosphate dehydrogenase**

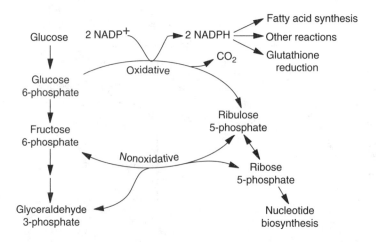

Figure 5-31. Overview of the pentose phosphate pathway.

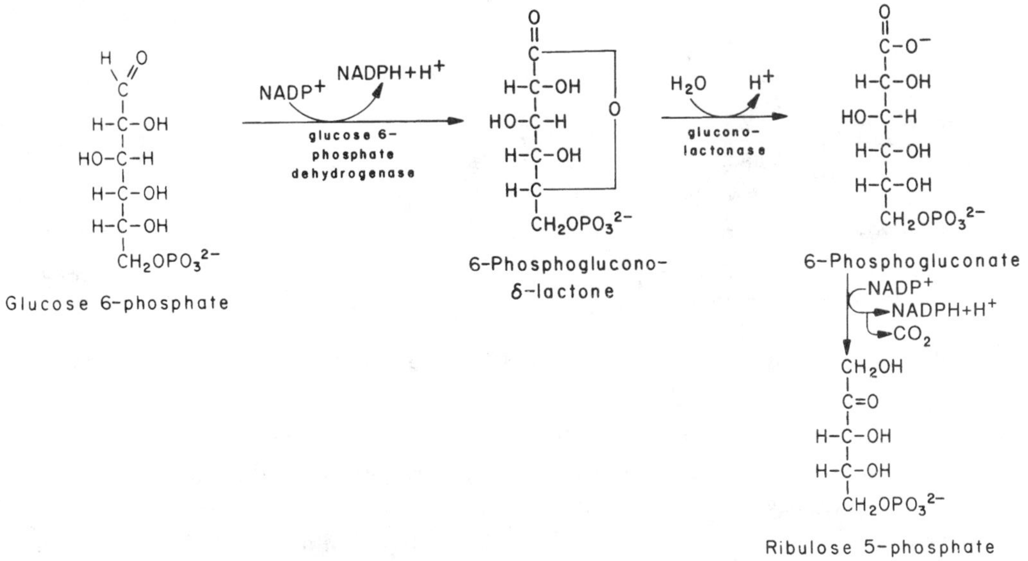

Figure 5-32. The oxidative reactions of the pentose phosphate pathway. These reactions are irreversible.

    b. 6-Phosphogluconolactone is hydrolyzed to 6-phosphogluconate.

    –Enzyme: **gluconolactonase**

    c. 6-Phosphogluconate is oxidatively decarboxylated. CO_2 is released and a second $NADPH + H^+$ is generated from $NADP^+$. The remaining carbons form ribulose 5-phosphate.

    –Enzyme: **6-phosphogluconate dehydrogenase**

2. The nonoxidative reactions (see Figure 5-31)

    a. Ribulose 5-phosphate may be isomerized to ribose 5-phosphate or epimerized to xylulose 5-phosphate.

    b. Ribose 5-phosphate and xylulose 5-phosphate may undergo reactions, catalyzed by transketolase and transaldolase, that transfer carbon units, ultimately forming fructose 6-phosphate and glyceraldehyde 3-phosphate.

        (1) Transketolase, which requires thiamine pyrophosphate, transfers two-carbon units (Figure 5-33).

        (2) Transaldolase transfers three-carbon units.

3. Overall reactions of the pentose phosphate pathway (Figure 5-34)

$$3 \text{ glucose-6-P} + 6 \text{ NADP}^+ \rightarrow 3 \text{ ribulose-5-P} + 3 \text{ CO}_2 + 6 \text{ NADPH}$$

$$3 \text{ ribulose-5-P} \rightarrow 2 \text{ xylulose-5-P} + \text{ribose-5-P}$$

$$2 \text{ xylulose-5-P} + \text{ribose-5-P} \rightarrow 2 \text{ fructose-6-P} + \text{glyceraldehyde-3-P}$$

B. Functions of NADPH (see Figure 5-31)

    1. The pentose phosphate pathway produces NADPH for **fatty acid synthesis**. Under these conditions, the fructose 6-phosphate and glyceraldehyde 3-phosphate generated in the pathway reenter glycolysis.

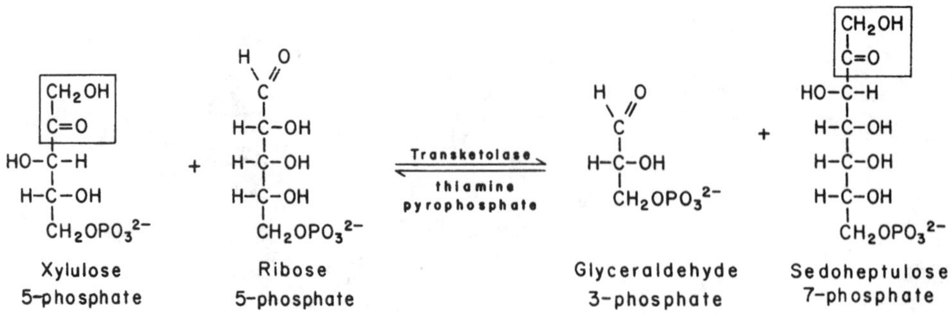

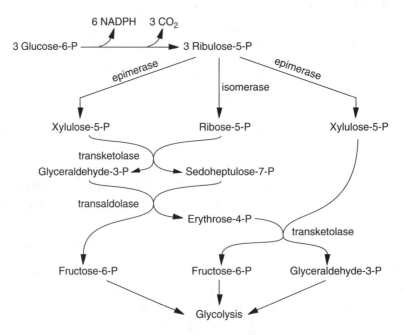

Figure 5-33. A two-carbon unit is transferred by transketolase. Thiamine pyrophosphate is a cofactor for this enzyme.

2. NADPH is also used to **reduce glutathione** (γ-glutamylcysteinyl-glycine).

   –Glutathione is used to reduce hydrogen peroxide and help prevent oxidative damage to membranes.

   –Glutathione is also used to transport amino acids across the membranes of certain cells by the γ-glutamyl cycle.

C. Generation of ribose 5-phosphate (see Figure 5-31)

1. When **NADPH levels are low,** the oxidative reactions of the pathway may be used to generate ribose 5-phosphate for nucleotide biosynthesis.

2. When **NADPH levels are high,** the reversible nonoxidative portion of the pathway can be used to generate ribose 5-phosphate for nucleotide biosynthesis from fructose 6-phosphate and glyceraldehyde 3-phosphate.

Figure 5-34. The reactions of the pentose phosphate pathway.

IX. Maintenance of Blood Glucose Levels

- Blood glucose levels are maintained within a very narrow range although the nature of the diet varies widely and the normal person eats periodically during the day and fasts between meals and at night. Even under circumstances when a person does not eat for extended periods of time, blood glucose levels decrease only slowly.
- The major hormones that regulate blood glucose are insulin and glucagon.
- After a meal, blood glucose is supplied by dietary carbohydrate.
- During fasting, blood glucose is maintained by the liver by the processes of glycogenolysis and gluconeogenesis.
  - Within the first few hours of fasting, glycogenolysis is primarily responsible for maintaining blood glucose levels.
  - As a fast progresses and glycogen stores decrease, gluconeogenesis becomes an important additional source of blood glucose.
  - After about 30 hours, when liver glycogen stores are depleted, gluconeogenesis becomes the only source of blood glucose.
- All cells use glucose for energy; however, the production of glucose during fasting is particularly important for tissues such as the brain and red blood cells.
- During exercise, blood glucose is also maintained by liver glycogenolysis and gluconeogenesis.

A. Blood glucose levels in the fed state

1. Changes in insulin and glucagon levels (Figure 5-35)

a. As a meal is digested, blood glucose levels rise, and, consequently, blood insulin levels increase.

  - Increases of blood glucose and of certain amino acids (particularly arginine and leucine) cause the release of insulin from β-cells of the pancreas.

b. The change in blood glucagon levels depends on the content of the meal. A high-carbohydrate meal causes glucagon levels to decrease. A high-protein meal causes **glucagon** to increase (see Figure 5-35).

c. On a normal mixed diet, glucagon will remain relatively constant after a meal while insulin increases.

2. Fate of dietary glucose in the liver

  - Glucose is oxidized for energy. Excess glucose is converted to glycogen and to the triacylglycerols of very-low-density lipoprotein (VLDL).

a. The enzyme glucokinase has a high K_m for glucose (about 10 mM), so its velocity increases after a meal when glucose levels are elevated. On a high-carbohydrate diet, glucokinase is induced.

b. Glycogen synthesis is prompted by insulin, which stimulates the phosphatase that dephosphorylates and activates glycogen synthetase.

c. Synthesis of triacylglycerols is also stimulated. The triacylglycerols are converted to VLDL and released into the blood.

3. Fate of dietary glucose in peripheral tissues

  - The brain, red blood cells, and all other cells oxidize glucose for energy.

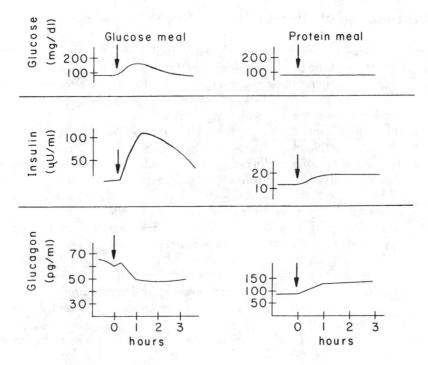

Figure 5-35. Changes in blood glucose, insulin, and glucagon levels in response to a glucose or a protein meal.

    a. Insulin stimulates the transport of glucose into adipose and muscle cells.

    b. In **muscle,** insulin stimulates the synthesis of glycogen.

    c. Adipose cells convert glucose to the glycerol moiety for synthesis of triacylglycerols.

  4. Return of blood glucose to fasting levels

    a. The **uptake of dietary glucose** by tissues, particularly liver, adipose, and muscle, causes blood glucose to decrease.

    b. By 2 hours after a meal, blood glucose has returned to the fasting level of 5 mM or 80–100 mg/dl (80–100 mg%).

B. Blood glucose levels in the fasting state (Figure 5-36)

  1. Changes in insulin and glucagon levels

    a. During fasting, insulin levels decrease, and glucagon levels increase.

    b. These hormonal changes promote **glycogenolysis** and **gluconeogenesis** in the liver so that blood glucose levels are maintained.

  2. Stimulation of glycogenolysis

    –Within a few hours after a meal, as **glucagon** levels increase, glycogenolysis is stimulated and begins to supply glucose to the blood (see Figure 5-15).

  3. Stimulation of gluconeogenesis

    a. By 4 hours after a meal, the liver is supplying glucose to the blood via gluconeogenesis and glycogenolysis (Figure 5-37).

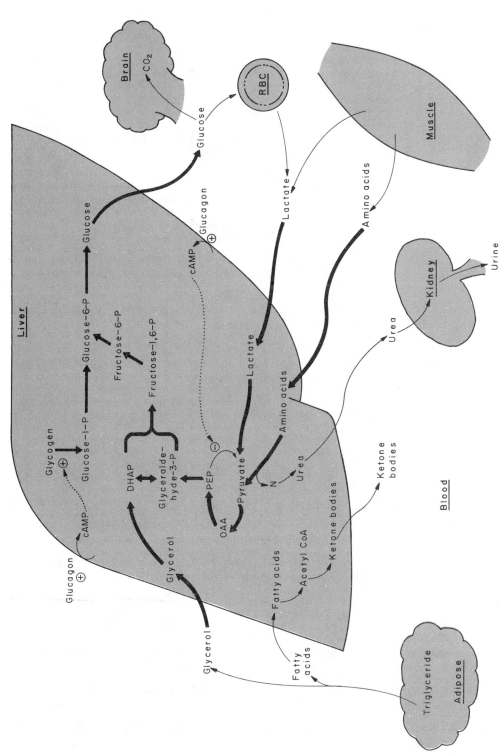

Figure 5-36. Tissue interrelationships in glucose production during fasting. Trace the precursors (shown in boxes) to blood glucose.

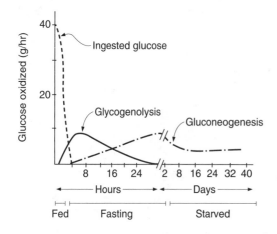

Figure 5-37. Sources of blood glucose in fed, fasting, and starved states. Note that the scale changes from hours to days. (Reproduced from Hanson RW and Mehlman MA (eds.): *Gluconeogenesis: Its Regulation in Mammalian Species.* New York, John Wiley & Sons, 1976, p 518.)

b. Regulatory mechanisms prevent futile cycles from occurring and promote the conversion of gluconeogenic precursors to glucose (see Figures 5-25 and 5-26).

4. Stimulation of lipolysis (see Figure 5-25)

　a. During fasting, the **breakdown of adipose triacylglycerols** is stimulated, and fatty acids and glycerol are released into the blood.

　b. Fatty acids are oxidized by certain tissues and converted to ketone bodies by the liver. The ATP and NADH produced by β-oxidation of fatty acids promotes gluconeogenesis.

　c. Glycerol is a source of carbon for gluconeogenesis in the liver.

5. Relative roles of glycogenolysis and gluconeogenesis in maintaining blood glucose (see Figure 5-37)

　a. As blood glucose falls to the fasting level after a meal, **glycogenolysis** is stimulated and is the main source of blood glucose for the next 8–12 hours.

　b. Gluconeogenesis is stimulated within a few (4) hours after a meal and supplies an increasingly larger share of blood glucose as the fasting state persists.

　c. By 16 hours of fasting, gluconeogenesis and glycogenolysis are approximately equal as sources of blood glucose.

　d. As liver glycogen stores become depleted, gluconeogenesis predominates.

　e. By about 30 hours of fasting, liver glycogen is depleted, and gluconeogenesis is the only source of blood glucose.

C. Blood glucose levels during prolonged fasting (starvation)

　–Even after 5–6 weeks of starvation, blood glucose levels are still in the range of 65 mg/dl.

　–Changes in fuel utilization by various tissues prevent blood glucose levels from decreasing abruptly during prolonged fasting.

　1. The levels of ketone bodies rise in the blood, and the **brain uses ketone bodies** for energy, decreasing its utilization of blood glucose.

　2. The rate of **gluconeogenesis** and, therefore, of urea production by the liver decreases.

3. Muscle protein is spared. There is less breakdown of muscle protein to provide amino acids for gluconeogenesis.

D. Blood glucose levels during exercise

–During exercise, blood glucose is maintained by essentially the same mechanisms that are used during fasting.

1. **Use of endogenous fuels**

   a. As the exercising muscle contracts, **ATP** is utilized.

   b. **ATP** is regenerated initially from creatine phosphate.

   c. Muscle **glycogen** is oxidized to produce ATP. The hormone epinephrine stimulates glycogen breakdown in muscle, causing the production of cAMP (see Figure 5-15).

2. **Use of fuels from the blood**

   a. As blood flow to the exercising muscle increases, **blood glucose** and **fatty acids** are taken up and **oxidized by muscle**.

   b. As blood glucose levels begin to decrease, the **liver,** by the processes of glycogenolysis and gluconeogenesis, acts to maintain blood glucose levels.

X. Clinical Correlations

A. Common problems associated with carbohydrate metabolism

1. **Intestinal lactase deficiency**

   Intestinal lactase deficiency is a common condition in which **lactose cannot be digested** and is oxidized by bacteria in the gut, producing gas, bloating, and watery diarrhea.

2. **Hypoglycemia**

   Low blood sugar is caused by the inability of the liver to maintain blood glucose levels. It can result from excessive insulin release, excessive cellular uptake of glucose, or an impairment of glycogenolysis or gluconeogenesis. Hypoglycemia is caused by liver disease, insulin-secreting tumors, and administration of inappropriately high doses of insulin or sulfonylureas. **Excessive alcohol ingestion** also can cause hypoglycemia. Metabolism of alcohol increases levels of NADH in the liver, which inhibits gluconeogenesis.

3. **Lactic acidosis**

   Lactate levels in the blood increase, producing an **acidosis**. This condition can be caused by **hypoxia** or by **alcohol ingestion**. Lack of oxygen results in increased NADH levels, and more pyruvate than normal is converted to lactate. High NADH levels from alcohol metabolism cause pyruvate to be converted to lactate.

4. **Glucose 6-phosphate dehydrogenase deficiency**

   A deficiency of glucose 6-phosphate dehydrogenase causes insufficient amounts of NADPH to be produced under certain conditions (e.g., when antimalarial drugs are being used). As a result, glutathione is not adequately reduced and, in turn, is not available to reduce compounds that are produced by the metabolism of these drugs. Red blood cells lyse and a **hemolytic anemia** may occur.

5. Diabetes mellitus

High blood glucose levels occur because of either a deficiency of insulin (insulin-dependent diabetes mellitus, IDDM) or the inability of tissues such as adipose and muscle to take up glucose in the presence of normal amounts of insulin (insulin resistance or noninsulin-dependent diabetes mellitus [NIDDM]). If insulin-deficiency diabetes mellitus is untreated, the body responds as if it is starving. Fuel stores are degraded in the face of high blood glucose, and ketoacidosis may occur. Many metabolic pathways are affected.

B. Rare problems associated with carbohydrate metabolism

1. Essential pentosuria

L-Xylulose reductase (**xylitol dehydrogenase**) is deficient in essential pentosuria. L-Xylulose (a pentose) appears in the urine and gives a positive reducing-sugar test. The condition is **benign**.

2. Glycogen storage diseases

Glycogen accumulates primarily in the liver or muscle, or both. Enzyme deficiencies occur mainly in glycogen degradation or conversion to glucose. In the **liver,** glycogen storage diseases may produce conditions ranging from mild hypoglycemia to liver failure. In **muscle,** they may cause problems ranging from difficulty in performing strenuous exercise to cardiorespiratory failure.

3. Pyruvate kinase deficiency

Deficiency of pyruvate kinase causes decreased production of ATP from glycolysis. Red blood cells have insufficient ATP for their sodium pumps; their membranes lyse, and a **hemolytic anemia** results.

4. Essential fructosuria

Fructokinase is deficient in essential fructosuria; therefore, fructose cannot be metabolized as rapidly as it normally would. Blood fructose levels rise, and fructose may appear in the urine. The condition is **benign**.

5. Fructose intolerance

The aldolase that cleaves fructose phosphates is deficient. Fructose 1-phosphate accumulates and inhibits glucose production, causing severe hypoglycemia if fructose is ingested.

6. Galactosemia

The appearance of high concentrations of galactose in the blood after lactose ingestion may be due to a galactokinase deficiency or to a uridyl transferase deficiency. In both conditions, excess galactose may be reduced to galactitol, which can produce **cataracts**. **Uridyl transferase deficiency** is more severe, causing elevation of galactose 1-phosphate, which inhibits phosphoglucomutase, interfering with glycogen synthesis and degradation.

7. Mucopolysaccharidoses and gangliosidoses (or sphingolipidoses)

A deficiency of lysosomal enzymes results in the inability to degrade the carbohydrate portions of proteoglycans or sphingolipids. Partially digested products accumulate in lysosomes. Tissues become engorged with these "residual bodies," and their function is impaired. These diseases are often **fatal**.

Review Test

Directions: Each of the numbered items or incomplete statements in this section is followed by answers or by completions of the statement. Select the **one** lettered answer or completion that is **best** in each case.

1. The sugars shown are

$$
\begin{array}{ccc}
\text{H} - \text{C} = \text{O} & & \text{H} - \text{C} = \text{O} \\
| & & | \\
\text{H} - \text{C} - \text{OH} & & \text{H} - \text{C} - \text{OH} \\
| & & | \\
\text{HO} - \text{C} - \text{H} & & \text{HO} - \text{C} - \text{H} \\
| & & | \\
\text{H} - \text{C} - \text{OH} & & \text{HO} - \text{C} - \text{H} \\
| & & | \\
\text{H} - \text{C} - \text{OH} & & \text{H} - \text{C} - \text{OH} \\
| & & | \\
\text{CH}_2\text{OH} & & \text{CH}_2\text{OH}
\end{array}
$$

(A) enantiomers
(B) ketoses
(C) hexoses of the L configuration
(D) epimers
(E) produced from a disaccharide by sucrase

2. The sugar shown

(A) contains a β-1,4 glycosidic bond
(B) is cleaved by lactase
(C) undergoes mutarotation
(D) contains a pentose sugar
(E) is sucrose (table sugar)

3. After digestion of a piece of cake that contains flour, milk, and sucrose as its primary ingredients, the major carbohydrate products entering the blood are

(A) glucose
(B) fructose and galactose
(C) galactose and glucose
(D) fructose and glucose
(E) glucose, fructose, and galactose

4. A patient has a genetic defect that causes intestinal epithelial cells to produce disaccharidases of much lower activity than normal. Compared to a normal person, after eating a bowl of milk and oatmeal, this patient will have higher levels of

(A) maltose, sucrose, and lactose in the stool
(B) starch in the stool
(C) galactose and fructose in the blood
(D) glycogen in the muscles
(E) insulin in the blood

5. The degradation of glycogen normally produces

(A) more glucose than glucose 1-phosphate
(B) more glucose 1-phosphate than glucose
(C) equal amounts of glucose and glucose 1-phosphate
(D) neither glucose nor glucose 1-phosphate
(E) only glucose 1-phosphate

6. Which of the following statements about liver phosphorylase kinase is true?

(A) It is present in an inactive form when epinephrine is elevated
(B) It phosphorylates phosphorylase to an inactive form
(C) It catalyzes a reaction that requires ATP
(D) It is phosphorylated in response to elevated insulin
(E) It is not affected by cAMP

7. A patient had large deposits of liver glycogen, which, after an overnight fast, had shorter than normal branches. This abnormality could be caused by

(A) a deficiency of phosphorylase
(B) a defect in the glucagon receptor
(C) an inability to produce glycogenin
(D) a deficiency of amylo-1,6-glucosidase (α-glucosidase)
(E) a genetic deficiency of amylo-4,6-transferase (4:6 transferase)

8. In which compartment of the cell does glycolysis occur?

(A) Mitochondrion
(B) Nucleus
(C) Soluble cytoplasm (cytosol)
(D) Rough endoplasmic reticulum
(E) Smooth endoplasmic reticulum

9. How many net molecules of ATP are generated in the conversion of glucose to pyruvate?

(A) 0
(B) 1
(C) 2
(D) 3
(E) 4

10. What type of bond is formed between phosphate and carbon 1 of 1,3-bisphosphoglycerate?

(A) Anhydride
(B) Ester
(C) Phosphodiester
(D) Amide
(E) Ether

11. During glycolysis, the conversion of compound I to compound II

$CH_2OPO_3{}^{2-}$ COO^-
| |
$C = O$ $C = O$
| |
$CH_2 - OH$ CH_3

Compound I   **Compound II**

(A) requires a dehydrogenase
(B) releases inorganic phosphate
(C) produces one molecule of ATP per molecule of product
(D) is catalyzed by a mutase
(E) requires two molecules of NADH

12. Which of the following statements about glycolysis is TRUE?

(A) Glucokinase catalyzes the conversion of glucose to glucose 6-phosphate in the liver
(B) Phosphofructokinase 1 catalyzes the conversion of fructose 1,6-bisphosphate to dihydroxyacetone phosphate
(C) When one molecule of glucose is converted to pyruvate via glycolysis, one molecule of NAD^+ is reduced
(D) When one molecule of glucose is converted to pyruvate via glycolysis, one carbon is lost as CO_2
(E) Hexokinase catalyzes the conversion of fructose 6-phosphate to fructose 1,6-bisphosphate

13. An adolescent patient with a deficiency of muscle phosphorylase was examined while exercising her forearm by squeezing a rubber ball. Compared to a normal person performing the same exercise, this patient

(A) could exercise for a longer period of time without fatigue
(B) had increased glucose levels in blood drawn from her forearm
(C) had decreased lactate levels in blood drawn from her forearm
(D) had lower levels of glycogen in biopsies of her forearm muscle

14. NADH is required for the one-step reaction by which pyruvate is converted to

(A) oxaloacetate
(B) acetyl CoA
(C) phosphoenolpyruvate
(D) lactate

15. Which of the following is NOT a regulatory mechanism of glycolysis?

(A) Activation of phosphofructokinase by AMP
(B) Inhibition of hexokinase by its product
(C) Inactivation of pyruvate kinase when glucagon levels are elevated
(D) Inhibition of aldolase by fructose 1,6-bisphosphate

16. Which of the following glycolytic enzymes is used in gluconeogenesis?

(A) Glucokinase
(B) Phosphofructokinase 1
(C) Pyruvate kinase
(D) Aldolase

17. Caffeine inhibits 3',5'-cAMP phosphodiesterase, which converts cAMP to AMP. Which of the following effects would be observed if cells were treated with caffeine?

(A) Decreased activity of liver protein kinase A
(B) Decreased activity of muscle protein kinase A
(C) Increased activity of liver pyruvate kinase
(D) Decreased activity of liver glycogen synthetase

18. In the conversion of pyruvate to glucose during gluconeogenesis

(A) biotin is required
(B) CO_2, added in one reaction, appears in the final product
(C) energy is utilized only in the form of GTP
(D) all of the reactions occur in the cytosol

19. In gluconeogenesis, both alanine and lactate are converted in a single step to

(A) oxaloacetate
(B) acetyl CoA
(C) phosphoenolpyruvate
(D) pyruvate
(E) aspartate

20. A common intermediate in the conversion of glycerol and lactate to glucose is

(A) pyruvate
(B) oxaloacetate
(C) malate
(D) glucose 6-phosphate
(E) phosphoenolpyruvate

21. In which of the following compounds do carbons derived from pyruvate leave mitochondria for the synthesis of glucose during fasting?

(A) Malate
(B) Acetyl CoA
(C) Oxaloacetate
(D) Lactate
(E) Glutamine

22. An infant with an enlarged liver has a glucose 6-phosphatase deficiency. This infant

(A) cannot maintain blood glucose levels either by glycogenolysis or by gluconeogenesis
(B) can use liver glycogen to maintain blood glucose levels
(C) can use muscle glycogen to maintain blood glucose levels
(D) can convert both alanine and glycerol to glucose to maintain blood glucose levels

23. Fructose and galactose enter the liver and are phosphorylated at carbon

(A) 1
(B) 2
(C) 3
(D) 5
(E) 6

24. Dietary fructose is phosphorylated in the liver and cleaved to form

(A) two molecules of dihydroxyacetone phosphate
(B) one molecule each of dihydroxyacetone phosphate and glyceraldehyde
(C) one molecule each of dihydroxyacetone phosphate and glyceraldehyde 3-phosphate
(D) one molecule each of dihydroxyacetone and glyceraldehyde 3-phosphate
(E) two molecules of glyceraldehyde 3-phosphate

25. Which of the following enzymes is not directly required in the sequence of reactions by which galactose is converted to UDP-glucose?

(A) Galactokinase
(B) An epimerase
(C) Phosphoglucomutase
(D) A uridyl transferase

26. Which of the following statements concerning lactose synthesis is TRUE?

(A) The reactions occur in most tissues
(B) α-Lactalbumin acts as a modifier of galactosyl transferase
(C) UDP-glucose reacts with galactose
(D) UDP-galactose requires dietary galactose for its synthesis

27. A pregnant woman who has a lactase deficiency and cannot tolerate milk in her diet is concerned that she will not be able to produce milk of sufficient caloric value to nourish her baby. She should be advised that

(A) she must eat pure galactose in order to produce the galactose moiety of lactose
(B) she will not be able to breast-feed her baby because she cannot produce lactose
(C) the production of lactose by the mammary gland does not require the ingestion of milk or milk products
(D) she can produce lactose by degrading α-lactalbumin

28. A mother with a deficiency of galactosyl 1-phosphate uridyl transferase

(A) can convert galactose to UDP-glucose for lactose synthesis during lactation
(B) can form galactose 1-phosphate from galactose
(C) can convert galactose to blood glucose
(D) can convert galactose to liver glycogen
(E) will have lower than normal blood galactose levels after ingestion of milk

29. The pentose phosphate pathway generates each of the following products EXCEPT

(A) NADPH, which may be used for fatty acid synthesis
(B) ribose 5-phosphate, which may be used for the biosynthesis of ATP
(C) compounds that are intermediates of glycolysis
(D) xylulose 5-phosphate by one of the nonoxidative reactions
(E) glucose from ribose 5-phosphate and CO_2

30. In the pentose phosphate pathway, thiamine pyrophosphate is required for the action of

(A) an epimerase
(B) transaldolase
(C) an isomerase
(D) transketolase
(E) a dehydrogenase

31. The major process responsible for maintaining blood glucose 4 hours after the last meal is

(A) glycolysis
(B) glycogenolysis
(C) the pentose phosphate pathway
(D) gluconeogenesis

32. The major process responsible for maintaining blood glucose 40 hours after the last meal is

(A) glycolysis
(B) glycogenolysis
(C) the pentose phosphate pathway
(D) gluconeogenesis

33. The transport of glucose across the cell membrane is stimulated by insulin in

(A) brain
(B) liver
(C) red blood cells
(D) skeletal muscle

34. Positive allosteric activators of phosphofructokinase 1 in the liver include

(A) ADP
(B) acetyl CoA
(C) fructose 2,6-bisphosphate
(D) ATP
(E) citrate

35. In an individual at rest who has fasted for 12 hours

(A) gluconeogenesis is the major process by which blood glucose is maintained
(B) adenyl cyclase is inactivated in liver
(C) liver glycogen stores are depleted
(D) phosphorylase, pyruvate kinase, and glycogen synthetase are phosphorylated in liver

36. When a normal individual in the basal metabolic state ingests a high-carbohydrate meal, there is

(A) enhanced glycogen synthetase activity in liver
(B) an increased ratio of phosphorylase a to phosphorylase b in liver
(C) an increased rate of lactate formation by erythrocytes
(D) inhibition of glycogen synthetase phosphatase activity in liver

37. An insulin infusion causes blood glucose to decrease because insulin promotes the dephosphorylation of each of the following enzymes EXCEPT

(A) liver glycogen phosphorylase
(B) liver pyruvate kinase
(C) liver glycogen synthetase
(D) liver protein kinase A
(E) muscle phosphorylase kinase

38. A 16-year-old insulin-dependent diabetic patient was admitted to the hospital with a blood glucose of 400 mg/dl. (The reference range for blood glucose is 80–100 mg/dl). One hour after an insulin infusion was begun, her blood glucose had decreased to 320 mg/dl. One hour later, it was 230 mg/dl. The patient's glucose level decreased because insulin

(A) stimulates the transport of glucose across the cell membranes of liver and brain
(B) stimulates the conversion of glucose to glycogen and triacylglycerol in liver
(C) inhibits the synthesis of ketone bodies from blood glucose
(D) stimulates glycogenolysis in liver
(E) inhibits the conversion of muscle glycogen to blood glucose

39. Which of the following statements concerning glycosaminoglycans is TRUE?

(A) They contain repeating disaccharides
(B) They are usually positively charged
(C) They contain short oligosaccharide chains
(D) They rarely contain sulfate groups
(E) They contain branches of *N*-acetylneuraminic acid

40. Which of the following statements concerning glycoproteins is TRUE?

(A) They are usually positively charged
(B) They never contain branched oligosaccharide chains
(C) They contain oligosaccharides that are synthesized on dolichol phosphate and transferred to serine residues
(D) They are degraded by lysosomal enzymes
(E) They are all secreted into the blood

Questions 41–43

A patient presented with a bacterial infection that produced an endotoxin that inhibits phosphoenolpyruvate carboxykinase.

41. In this patient, inhibition of phosphoenolpyruvate carboxykinase would cause

(A) inhibition of glucose production from alanine
(B) inhibition of glucose production from glycerol
(C) inhibition of glucose production from even-chain fatty acids
(D) inhibition of glucose oxidation to phosphoenolpyruvate

42. Administration of a high dose of glucagon to this patient 2–3 hours after a high-carbohydrate meal

(A) would result in a substantial increase in blood glucose levels
(B) would decrease blood glucose levels
(C) would have little effect on blood glucose levels

43. Administration of a high dose of glucagon to this patient 30 hours after a high-carbohydrate meal

(A) would result in a substantial increase in blood glucose levels
(B) would decrease blood glucose levels
(C) would have little effect on blood glucose levels

Directions: Each group of items in this section consists of lettered options followed by a set of numbered items. For each item, select the **one** lettered option that is most closely associated with it. Each lettered option may be selected once, more than once, or not at all.

Questions 44–47

Match each enzyme below with the protein that most directly alters its activity.

(A) cAMP-dependent protein kinase
(B) Phosphorylase kinase
(C) Glucagon receptor
(D) Phosphodiesterase

44. Phosphorylase b
45. Glycogen synthetase
46. Adenyl cyclase
47. Phosphorylase kinase

Questions 48–50

Match the products with the enzymes that catalyze their formation.

(A) Glucose 6-phosphate dehydrogenase
(B) 6-Phosphogluconate dehydrogenase
(C) Transaldolase
(D) Transketolase

48. NADPH and a lactone
49. CO_2
50. Glyceraldehyde 3-phosphate in a reaction requiring thiamine pyrophosphate

Questions 51–57

Match each description below with the most appropriate enzyme (A–E) indicated in the following diagram.

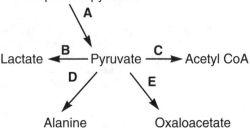

51. Inhibited by NADH and acetyl CoA
52. Requires thiamine pyrophosphate
53. Activated by acetyl CoA
54. Requires biotin
55. Requires pyridoxal phosphate
56. Phosphorylated and inactivated by protein kinase A
57. Forms product (as indicated by the *arrow*) when NADH levels are elevated

Answers and Explanations

1–D. Glucose (left) and galactose (right) are not mirror images (enantiomers). They contain hydroxyl groups on different sides of carbon 4 (that is, they are epimers). They are aldoses, not ketoses. They are hexoses (containing six carbons) in the D configuration.

2–E. This sugar is sucrose. It contains glucose and fructose joined by their anomeric carbons, so it is not a reducing sugar and does not mutarotate.

3–E. The cake contains starch, lactose (milk sugar), and sucrose (table sugar). Digestion of starch produces glucose. Lactase cleaves lactose to galactose and glucose, and sucrase cleaves sucrose to fructose and glucose.

4–A. Starch will be digested to small oligosaccharides and maltose, but a lower than normal amount of glucose will be produced because of the deficiency of the brush border disaccharidases, which include sucrase and lactase. Sucrose and lactose will not be cleaved; hence, there will be more maltose, sucrose, and lactose in the stool and less monosaccharides in the blood and tissues. Insulin levels will be low.

5–B. Phosphorylase produces glucose 1-phosphate from glucose residues linked α-1,4. Free glucose is produced from α-1,6-linked residues at branch points by an α-1,6-glucosidase. Degradation of glycogen produces glucose 1-phosphate and glucose in about a 10:1 ratio.

6–C. Glucagon in the liver and epinephrine in both the liver and muscle cause cAMP to rise, activating protein kinase A. Protein kinase A phosphorylates and activates phosphorylase kinase, which in turn phosphorylates and activates phosphorylase. These phosphorylation reactions require ATP.

7–D. If after fasting, the branches were shorter than normal, phosphorylase must be functional and capable of being activated by glucagon. The branching enzyme (the 4:6 transferase) must be normal because branches are present. The protein glycogenin must be present in order for large amounts of glycogen to be synthesized and deposited. The defect has to be in the debranching enzyme (α-glucosidase). If the debrancher is defective, phosphorylase would break the glycogen down to the branch points, but further degradation would not occur. So, short branches would be present in the glycogen.

8–C. All of the reactions of glycolysis occur in the cytosol.

9–C. One ATP is used to convert glucose to glucose 6-phosphate and a second to convert fructose 6-phosphate to the bisphosphate. Two triosephosphates are produced by cleavage of fructose 1,6-bisphosphate. As the two triosephosphates are converted to pyruvate, four ATPs are generated, two by phosphoglycerate kinase and two by pyruvate kinase. Net, two ATPs are produced.

10–A. The carboxylic acid (carbon 1) reacts with phosphoric acid, splitting out H_2O and forming an anhydride. Cleavage of this bond in the next step of glycolysis generates enough energy to produce one ATP from ADP and P_i.

11–A. Dihydroxyacetone phosphate (compound I) is isomerized to glyceraldehyde 3-phosphate and converted in a series of steps to pyruvate (compound II). One of the reactions requires glyceraldehyde 3-phosphate dehydrogenase, which utilizes inorganic phosphate and produces one NADH. In the conversion of one 1,3-bisphosphoglycerate to one pyruvate, two ATP are produced. A mutase is not required.

12–A. Glucokinase, a liver enzyme, converts glucose to glucose 6-phosphate. Phosphofructokinase 1 converts fructose 6-phosphate to fructose 1,6-bisphosphate. In glycolysis, one molecule of glucose is converted to two molecules of pyruvate. Two molecules of NADH are produced, and no carbons are lost as CO_2.

13–C. This patient has a glycogen storage disease (McArdle's disease). Because she cannot degrade glycogen to produce energy for muscle contraction, she becomes fatigued more readily than a normal person. Because she is not degrading glycogen normally, the glycogen levels in her muscle will be higher than normal and her blood lactate levels will be lower. She will use more blood glucose, so her blood glucose levels will not be increased, but may be decreased.

14–D. All of these compounds can be produced from pyruvates, but only lactate can be produced in a single reaction that requires NADH.

15–D. Aldolase is not inhibited by its substrate.

16–D. During gluconeogenesis, glucokinase, phosphofructokinase, and pyruvate kinase are not active and, thus, futile cycles are prevented. Aldolase is utilized both in glycolysis and gluconeogenesis.

17–D. If the phosphodiesterase that degrades cAMP were inhibited, cAMP levels would rise. Protein kinase A would become more active in the liver and muscle, pyruvate kinase would become less active, and glycogen synthetase activity would be decreased.

18–A. In mitochondria, CO_2 is added to pyruvate to form oxaloacetate (OAA). The enzyme is pyruvate carboxylase, which requires biotin and ATP. OAA leaves the mitochondrion as malate or aspartate and is regenerated in the cytosol. OAA is converted to phosphoenolpyruvate by a reaction that utilizes GTP and releases the same CO_2 that was added in the mitochondrion. The remainder of the reactions occur in the cytosol.

19–D. Alanine may be transaminated and lactate may be oxidized to form pyruvate. The other compounds are not produced in a single step from alanine and lactate.

20–D. The only intermediate included on the list that glycerol has in common with lactate is glucose 6-phosphate. Glycerol enters gluconeogenesis as dihydroxyacetone phosphate. Therefore, it bypasses the other compounds.

21–A. Pyruvate is converted in mitochondria to malate, which can cross the mitochondrial membrane. Oxaloacetate and acetyl CoA cannot. Lactate is produced from pyruvate in the cytosol. The reverse reaction is involved in gluconeogenesis. Glutamine is not derived from pyruvate during gluconeogenesis.

22–A. Glucose 6-phosphatase deficiency is a glycogen storage disease (von Gierke's disease) in which neither liver glycogen nor gluconeogenic precursors (such as alanine and glycerol) can be used to maintain normal blood glucose levels. The last step (conversion of glucose 6-phosphate to glucose) is deficient for both glycogenolysis and gluconeogenesis. Muscle glycogen cannot be used to maintain blood glucose because muscle does not contain glucose 6-phosphatase.

23–A. Fructokinase and galactokinase phosphorylate their substrates at carbon 1.

24–B. Fructose 1-phosphate is cleaved to dihydroxyacetone phosphate and glyceraldehyde.

25–C. Galactose is phosphorylated by galactokinase to galactose 1-phosphate, which reacts with UDP-glucose in a reaction catalyzed by uridyl transferase to form UDP-galactose and glucose 1-phosphate. An epimerase converts UDP-galactose to UDP-glucose. Phosphoglucomutase interconverts glucose 1-phosphate and glucose 6-phosphate.

26–B. UDP-galactose reacts with glucose to form lactose only in the mammary gland. α-Lactalbumin acts as a modifier of the enzyme galactosyl transferase, lowering its K_m for glucose. Glucose may be converted to UDP-glucose and epimerized to form the UDP-galactose used in lactose synthesis; therefore; dietary galactose is not required.

27–C. She will be able to breast-feed her baby because she can produce lactose. However, she does not have to eat pure galactose. Glucose can be converted to UDP-galactose (Glucose → glucose 6-phosphate → glucose 1-phosphate → UDP-glucose → UDP-galactose). UDP-galactose reacts with glucose to form lactose. α-Lactalbumin is a protein that serves as the modifier of galactosyl transferase, which catalyzes this reaction.

28–B. A person with a uridyl transferase deficiency (classical galactosemia) can phosphorylate galactose but will not be able to react the galactose 1-phosphate with UDP-glucose to form UDP-galactose and glucose 1-phosphate. Therefore, she will not be able to convert galactose to UDP-glucose, liver glycogen, or blood glucose. Cellular galactose 1-phosphate and blood galactose levels will be elevated if she consumes galactose in the lactose of milk.

29–E. In the first three reactions of the pentose phosphate pathway, glucose is converted to ribulose 5-phosphate and CO_2, with the production of NADPH. These reactions are *not* reversible. Ribose 5-phosphate and xylulose 5-phosphate may be formed from ribulose 5-phosphate. A series of reactions catalyzed by transketolase and transaldolase produce the glycolytic intermediates fructose 6-phosphate and glyceraldehyde 3-phosphate.

30–D. Transketolase requires thiamine pyrophosphate.

31–B. The breakdown of liver glycogen is the major process for maintenance of blood glucose in the first few hours after a meal. Gluconeogenesis becomes a significant source of blood glucose after about 4 hours. Glycolysis and the pentose phosphate pathway do not supply glucose for maintenance of blood levels.

32–D. By 40 hours after a meal, liver glycogen is depleted, and gluconeogenesis is the only pathway that supplies glucose to maintain blood levels.

33–D. Insulin stimulates glucose transport into muscle and adipose cells. It is not required by the brain, liver, and red blood cells.

34–C. Phosphofructokinase 1 is activated by AMP and fructose 2,6-bisphosphate. It is inhibited by ATP and citrate and not affected by acetyl CoA or ADP.

35–D. After 12 hours of fasting, liver glycogen stores remain substantial. Glycogenolysis is stimulated by glucagon, which activates adenyl cyclase. cAMP activates protein kinase A, which phosphorylates phosphorylase, pyruvate kinase, and glycogen synthetase. As a result, phosphorylase is activated, whereas glycogen synthetase and pyruvate kinase are inactivated. Gluconeogenesis does not become the major process for maintaining blood glucose until 18–20 hours of fasting. After about 30 hours, liver glycogen is depleted.

36–A. After a high-carbohydrate meal, glycogen synthetase is activated by a phosphatase. The ratio of phosphorylase a to phosphorylase b is decreased by a phosphatase, so glycogen degradation decreases. Red blood cells continue to use glucose and form lactate at their normal rate.

37–D. Insulin causes activation of phosphatases, which dephosphorylate all of the enzymes that were phosphorylated in the fasting state. Protein kinase A is not activated by phosphorylation during fasting but by the binding of cAMP to regulatory subunits.

38–B. Blood glucose decreases because insulin stimulates the transport of glucose into muscle and adipose cells and the conversion of glucose to glycogen and triacylglycerols in liver. Ketone bodies are not made from blood glucose. During fasting, when liver is producing ketone bodies, it is also synthesizing glucose. Carbon for ketone body synthesis comes from fatty acids. Insulin stimulates glycogenesis, not glycogenolysis. Muscle glycogen is not converted to blood glucose.

39–A. Glycosaminoglycans are long, linear carbohydrate chains that contain repeating disaccharide units, which usually contain a hexosamine and a uronic acid. They often contain sulfate groups. The uronic acid and sulfate residues cause them to be negatively charged. They are unbranched and do not contain N-acetylneuraminic acid.

40–D. Glycoproteins contain branched oligosaccharide chains. These chains may be synthesized by addition of sugars to serine or threonine residues of the protein, or they may be synthesized on dolichol phosphate and transferred to asparagine residues on the protein. They are not positively charged. They are synthesized in the RER and Golgi and may be secreted from cells, anchored in the cell membrane, or segregated into lysosomes. They are internalized by endocytosis and degraded by lysosomal enzymes.

41–A. Phosphoenolpyruvate carboxykinase converts oxaloacetate to phosphoenolpyruvate. It is a gluconeogenic enzyme required for the conversion of alanine and lactate (but not glycerol) to glucose. Acetyl CoA from oxidation of fatty acids is not converted to glucose.

42–A. By 2–3 hours after a high-carbohydrate meal, the patient's glycogen stores would be filled. Glucagon would stimulate glycogenolysis, and blood glucose levels would rise.

43–C. Thirty hours after a meal, liver glycogen is normally depleted, and blood glucose is maintained solely by gluconeogenesis after this time. However, in this case, a key gluconeogenic enzyme is inhibited by an endotoxin. Therefore, gluconeogenesis will not occur at a normal rate and glycogen stores will be depleted more rapidly than normal. Blood glucose levels will not change significantly if glucagon is administered after 30 hours of fasting.

44–B. Phosphorylase kinase phosphorylates phosphorylase b, converting it to the more active phosphorylase a.

45–A. Glycogen synthetase is phosphorylated and inactivated by a cAMP-dependent protein kinase.

46–C. Glucagon combines with its membrane receptor, and the complex activates adenyl cyclase in the cell membrane, causing the conversion of ATP to cAMP.

47–A. A cAMP-dependent protein kinase phosphorylates phosphorylase kinase, causing it to become active.

48–A. Both A and B produce NADPH, but only A produces 6-phosphogluconolactone.

49–B. 6-Phosphogluconate is decarboxylated to form ribulose 5-phosphate and CO_2.

50–D. Both transaldolase and transketolase produce glyceraldehyde 3-phosphate, but only transketolase requires thiamine pyrophosphate.

51–C. Pyruvate dehydrogenase is inhibited by NADH and acetyl CoA.

52–C. Pyruvate dehydrogenase requires thiamine pyrophosphate.

53–E. Pyruvate carboxylase is activated by acetyl CoA.

54–E. Pyruvate carboxylase requires biotin, CO_2, and ATP.

55–D. Alanine aminotransferase (transaminase) requires pyridoxal phosphate.

56–A. Pyruvate kinase is inactivated by protein kinase A. Pyruvate dehydrogenase is inactivated by a kinase that is a subunit of the enzyme complex.

57–B. Lactate dehydrogenase produces lactate from pyruvate when NADH levels are high.

6

Lipid Metabolism

Overview

- Lipids are a diverse group of compounds that are related by their insolubility in water.
- Membranes contain lipids, particularly phosphoglycerides, sphingolipids, and cholesterol.
- Triacylglycerols provide the body with its major source of energy and are obtained from the diet or synthesized in the liver. They are transported in the blood as lipoproteins and are stored in adipose tissue (Figure 6-1).
- The major classes of blood lipoproteins include chylomicrons, VLDL, IDL, LDL, and HDL.
- Chylomicrons are produced in intestinal cells from dietary lipid, and VLDL (very-low-density lipoprotein) is produced in the liver, mainly from dietary carbohydrate.
- IDL (intermediate-density lipoprotein) and LDL (low-density lipoprotein) are produced in blood capillaries by digestion of the triacylglycerols of VLDL.
- HDL (high-density lipoprotein) transfers an activator of lipoprotein lipase to chylomicrons and VLDL. HDL also carries cholesterol from peripheral tissues to the liver.
- The triacylglycerols of chylomicrons and VLDL are hydrolyzed by lipoprotein lipase to fatty acids and glycerol. The fatty acids may be oxidized by various tissues. In adipose cells, the fatty acids are converted to triacylglycerols and stored.
- During fasting, fatty acids (derived from adipose triacylglycerol stores) can be oxidized by various tissues to produce energy (Figure 6-2).
- The class of lipids includes cholesterol, the bile salts and steroid hormones (which are derived from cholesterol), the fat-soluble vitamins, and compounds such as the prostaglandins.

I. Lipid Structure

- Lipids have diverse structures but are similar in that they are insoluble in water.

A. Fatty acids may exist free or esterified to glycerol (Figure 6-3).

    1. In humans, fatty acids usually have an even number of carbon atoms, are 16 to 20 carbon atoms in length, and may be saturated or unsaturated (contain double bonds). They are described by the number of carbons

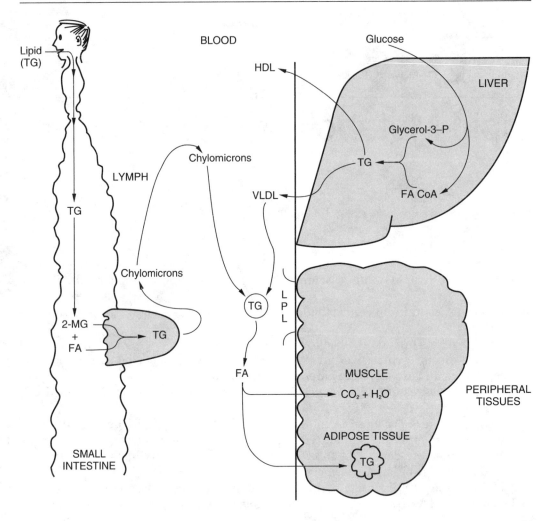

Figure 6-1. Overview of lipid metabolism in the fed state. *TG* = triacylglycerol; *2-MG* = 2-monoacylglycerol; *FA* = fatty acid; *LPL* = lipoprotein lipase; *VLDL* = very-low-density lipoprotein; *HDL* = high-density lipoprotein; *circled TG* = triacylglycerols of VLDL and chylomicrons.

and the positions of the double bonds (e.g., arachidonic acid, which has 20 carbons and 4 double bonds, is $20{:}4,\Delta^{5,8,11,14}$).

2. **Polyunsaturated fatty acids** are often classified according to the position of the first double bond from the ω-end (the carbon furthest from the carboxyl group; e.g., ω-3 or ω-6).

B. **Monoacylglycerols** (monoglycerides), **diacylglycerols** (diglycerides), and **triacylglycerols** (triglycerides) contain one, two, and three fatty acids esterified to glycerol, respectively.

C. **Phosphoglycerides** contain fatty acids esterified to positions 1 and 2 of the glycerol moiety and a phosphoryl group at position 3 (e.g., phosphocholine).

D. **Sphingolipids** contain ceramide with a variety of groups attached.

1. **Sphingomyelin** contains phosphocholine.

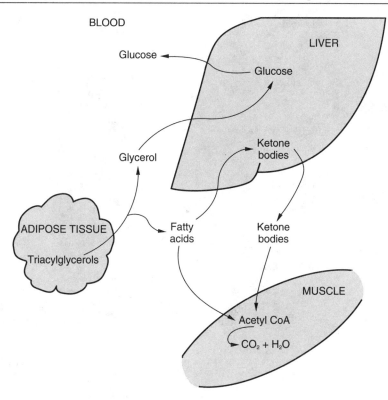

Figure 6-2. Overview of lipid metabolism in the fasting state.

   2. Cerebrosides contain a sugar residue.

   3. Gangliosides contain a number of sugar residues.

E. Cholesterol contains four rings and an aliphatic side chain (see Figure 6-17).
   –Bile salts and steroid hormones are derived from cholesterol (see Figure 6-18).

F. Prostaglandins and **leukotrienes** are derived from polyunsaturated fatty acids such as arachidonic acid (see Figure 6-16).

G. The **fat-soluble vitamins** include vitamins A, D, E, and K (see Figure 4-10).

II. Membranes

● The cell (plasma) membrane is a fluid mosaic of lipids and proteins.
● The proteins serve as transporters, enzymes, receptors, and mediators that allow extracellular compounds, such as hormones, to exert intracellular effects.

A. Membrane structure

   1. Membranes are composed mainly of lipids and proteins (Figure 6-4).

   2. The major membrane lipids are **phosphoglycerides,** but sphingolipids and cholesterol are also present.

Fatty acids

O
‖
R– C – OH
General structure

Glycerol

CH₂OH
|
H– C– OH
|
CH₂OH

$$CH_3-(CH_2)_{14}-\overset{\overset{\displaystyle O}{\|}}{C}-O^-$$
Palmitate (16:0)

$$CH_3-(CH_2)_7-CH=CH-(CH_2)_7-\overset{\overset{\displaystyle O}{\|}}{C}-O^-$$
Oleate (18:1, Δ⁹)

$$CH_3-(CH_2)_{16}-\overset{\overset{\displaystyle O}{\|}}{C}-O^-$$
Stearate (18:0)

Monoacylglycerol (monoglyceride)

¹CH₂– O – C– R₁
|
H–²C–OH
|
³CH₂OH

Diacylglycerol (diglyceride)

¹CH₂– O – CR₁
|
R₂C– O –²C–H
|
³CH₂OH

Triacylglycerol (triglyceride)

¹CH₂– O – CR₁
|
R₂C– O –²C–H
|
³CH₂– O – CR₃

Figure 6-3. The structures of fatty acids, glycerol, and the acylglycerols. *R* indicates a linear aliphatic chain. Fatty acids are identified by the number of carbons and the number of double bonds and their position (e.g., 18:1,Δ⁹).

–**Phospholipids** form a bilayer, with their hydrophilic head groups interacting with water on both the extracellular and intracellular surfaces, and their hydrophobic fatty acyl chains in the central portion of the membrane.

3. **Peripheral proteins** are embedded at the periphery; **integral proteins** span from one side to the other.

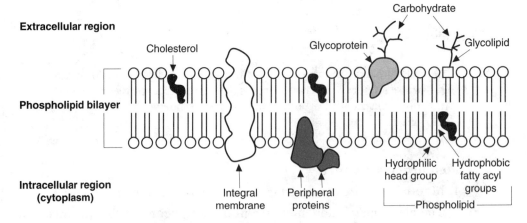

Figure 6-4. The structure of the cell membrane.

4. **Carbohydrates** may be attached to proteins or lipids on the exterior side of the cell membrane. They extend into the extracellular space.

5. **Lipids and proteins** can diffuse laterally within the plane of the membrane. Therefore, the membrane is a fluid mosaic.

B. Membrane function

1. Membranes serve as **barriers** that separate the contents of a cell from the external environment or the contents of organelles from the remainder of the cell.

2. The proteins in the cell membrane have many functions.

   a. Some are involved in the **transport** of substances across the membrane.

   b. Some are **enzymes** that catalyze biochemical reactions.

   c. Those on the exterior surface may function as **receptors** that bind external ligands such as hormones or growth factors.

   d. Others are **mediators** that aid the ligand–receptor complex in triggering a sequence of events; as a consequence, **second messengers** that alter metabolism are produced inside the cell. Therefore, an external agent, such as a hormone, can elicit intracellular effects without entering the cell.

III. Digestion of Dietary Triacylglycerol

- The major dietary fat is triacylglycerol, which is obtained from the fat stores of the plants and animals in the food supply.
- The dietary triacylglycerols, which are water-insoluble, are digested in the small intestine to fatty acids and 2-monoacylglycerols. These digestive products are resynthesized to triacylglycerols in intestinal epithelial cells and are secreted in chylomicrons via the lymph into the blood.

A. **Dietary triacylglycerols** are digested in the **small intestine** by a process that requires bile salts and secretions from the pancreas (Figure 6-5).

1. **Bile salts** are synthesized in the liver from cholesterol and are secreted into the bile. They pass into the intestine, where they emulsify the dietary lipids.

2. The **pancreas** secretes bicarbonate, which neutralizes the stomach acid, raising the pH into the optimal range for the digestive enzymes.

3. **Pancreatic lipase,** with the aid of colipase, digests the triacylglycerols to 2-monoacylglycerols and free fatty acids, which are packaged into micelles. The micelles, which are tiny microdroplets emulsified by bile salts, also contain other dietary lipids such as cholesterol and the fat-soluble vitamins.

4. The **micelles** travel to the microvilli of the intestinal epithelial cells, which absorb the fatty acids, 2-monoacylglycerols, and other dietary lipids.

5. The **bile salts are resorbed** and secreted into the gut during subsequent digestive cycles.

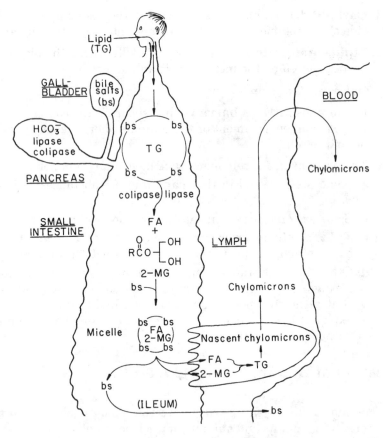

Figure 6-5. Digestion of triacylglycerols. *TG* = triacylglycerols; *bs* = bile salts; *FA* = fatty acid; *2-MG* = 2-monoacylglycerol.

B. Synthesis of chylomicrons

1. In intestinal epithelial cells, the fatty acids from micelles are activated by fatty acyl CoA synthetase (thiokinase) to form fatty acyl CoA.

2. A **fatty acyl CoA** reacts with a 2-monoacylglycerol to form a **diacylglycerol**. Then another fatty acyl CoA reacts with the diacylglycerol to form a **triacylglycerol**.

3. The triacylglycerols pass into the lymph, packaged in **nascent chylomicrons,** which eventually enter the blood.

IV. Fatty Acid and Triacylglycerol Synthesis

- Lipogenesis, the synthesis of fatty acids and their esterification to glycerol to form triacylglycerols, occurs mainly in the liver in humans, with dietary carbohydrate as the major source of carbon.
- The de novo synthesis of fatty acids from acetyl CoA occurs in the cytosol on the fatty acid synthase complex.
- Acetyl CoA, derived from glucose or from other sources, is converted by acetyl CoA carboxylase to malonyl CoA.

- The growing fatty acid chain on the fatty acid synthase complex is elongated, two carbons at a time, by the addition of the three-carbon compound, malonyl CoA, which is subsequently decarboxylated. With each two-carbon addition, the growing chain, which initially contains a β-keto group, is reduced in a series of steps that require NADPH. NADPH is produced by the pentose phosphate pathway and by the reaction catalyzed by the malic enzyme.
- Palmitate, the product released by the fatty acid synthase complex, is converted to a series of other fatty acyl CoAs by elongation and desaturation reactions.
- Fatty acyl CoA combines with glycerol 3-phosphate in the liver to form triacylglycerols by a pathway in which phosphatidic acid serves as an intermediate.
- The triacylglycerols, packaged in VLDL, are secreted into the blood.

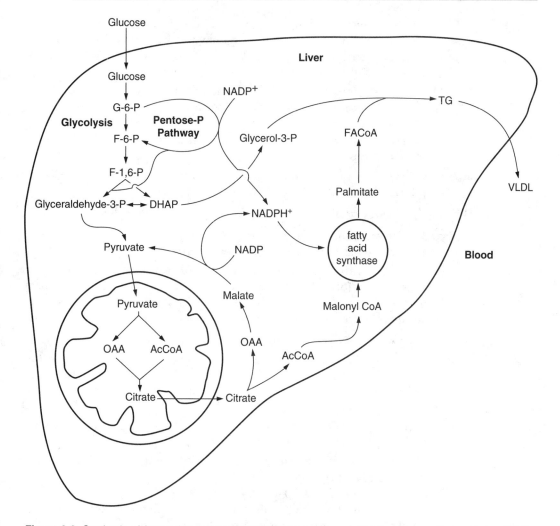

Figure 6-6. Synthesis of fatty acids and triacylglycerols from glucose. *G-6-P* = glucose 6-phosphate; *F-6-P* = fructose 6-phosphate; *F-1,6-P* = fructose 1,6-bisphosphate; *DHAP* = dihydroxyacetone phosphate; *AcCoA* = acetyl CoA; *VLDL* = very-low-density lipoprotein.

A. Conversion of glucose to acetyl CoA for fatty acid synthesis (Figure 6-6)

1. **Glucose** enters liver cells and is converted via glycolysis to pyruvate, which enters mitochondria.

2. **Pyruvate** is converted to acetyl CoA by pyruvate dehydrogenase and to **oxaloacetate** by pyruvate carboxylase.

3. Because acetyl CoA cannot directly cross the mitochondrial membrane and enter the cytosol to be used for the process of fatty acid synthesis, acetyl CoA and oxaloacetate condense to form **citrate,** which can cross the mitochondrial membrane.

4. In the cytosol, **citrate is cleaved** to oxaloacetate and acetyl CoA by citrate lyase, an enzyme that requires ATP and is induced by insulin.

   a. Oxaloacetate from the citrate lyase reaction is reduced in the cytosol by NADH, producing NAD^+ and malate. The enzyme is cytosolic malate dehydrogenase.

   b. In a subsequent reaction, malate is converted to pyruvate; NADPH is produced, and CO_2 is released. The enzyme is the malic enzyme (also known as decarboxylating malate dehydrogenase or $NADP^+$-dependent malate dehydrogenase).

      (1) **Pyruvate** may reenter the mitochondrion and be reutilized.

      (2) **NADPH** supplies reducing equivalents for reactions that occur on the fatty acid synthase complex.

      (3) **NADPH** is produced by the malic enzyme and by the pentose phosphate pathway.

5. **Acetyl CoA** (from the citrate lyase reaction or from other sources) supplies carbons for fatty acid synthesis in the cytosol.

B. Synthesis of fatty acids by the fatty acid synthase complex (Figure 6-7)

1. **Fatty acid synthase** is a multienzyme complex located in the cytosol that has two identical subunits with seven catalytic activities.

   –This enzyme contains a **phosphopantetheine residue,** derived from the vitamin pantothenic acid, and a **cysteine residue;** both can form thioesters with acyl groups. The growing fatty acyl chain moves from one to the other of these sulfhydryl residues as it is elongated.

2. **Addition of two-carbon units**

   a. Initially, **acetyl CoA** reacts with the phosphopantetheinyl residue and then the acetyl group is transferred to the cysteinyl residue. This acetyl group provides the **ω-carbon** of the fatty acid produced by the fatty acid synthase complex.

   b. A malonyl group from **malonyl CoA** forms a **thioester** with the phosphopantetheinyl sulfhydryl group.

      (1) Malonyl CoA is formed from acetyl CoA by a carboxylation reaction that requires biotin and ATP.

      (2) The enzyme is **acetyl CoA carboxylase,** a regulatory enzyme that is inhibited by phosphorylation, activated by dephosphorylation and by citrate, and induced by insulin.

   c. The **acetyl group** on the fatty acid synthase complex condenses with malonyl CoA, the CO_2 that was added to the malonyl group by acetyl

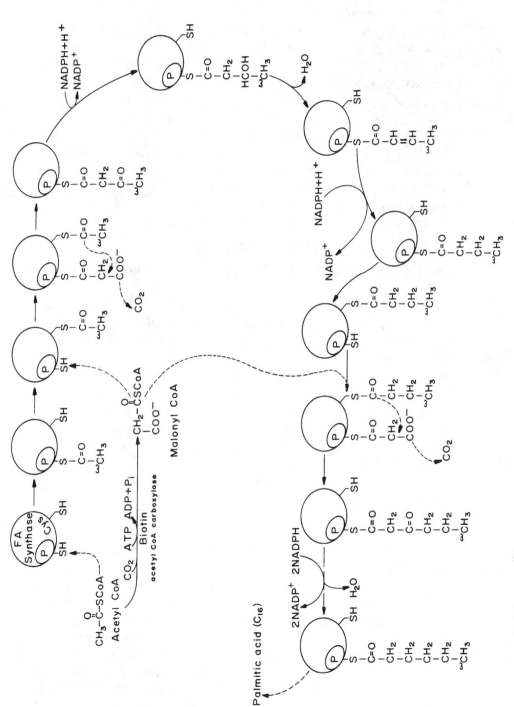

Figure 6-7. Fatty acid synthesis. Malonyl CoA provides the 2-carbon units that are added to the growing fatty acyl chain. The addition and reduction steps are repeated until palmitic acid is produced. P = a phosphopantetheinyl group attached to the fatty acid synthase complex; *Cys-SH* = a cysteinyl residue.

CoA carboxylase is released, and a **β-ketoacyl group,** containing four carbons, is produced.

3. Reduction of the β-ketoacyl group

   a. The β-keto group is reduced by NADPH to a β-hydroxy group.

   b. Then **dehydration** occurs, producing an **enoyl group** with the double bond between carbons 2 and 3.

   c. Finally, the double bond is reduced by NADPH, and a **four-carbon acyl group** is generated. The **NADPH** for these reactions is produced by the pentose phosphate pathway and by the malic enzyme.

4. Elongation of the growing fatty acyl chain

   a. The acyl group is transferred to the cysteinyl sulfhydryl group, and the malonyl CoA reacts with the phosphopantetheinyl group. Condensation of the acyl and malonyl groups occurs with the release of CO_2, followed by the three reactions that reduce the β-keto group.

   b. The chain is now longer by two carbons. This sequence of reactions repeats until the growing chain is 16 carbons in length.

   c. Palmitate, a 16-carbon saturated fatty acid, is the final product released by hydrolysis from the fatty acid synthase complex.

C. Elongation and desaturation of fatty acids

   1. Palmitate can be elongated and desaturated to form a **series of fatty acids**.

   2. These newly synthesized fatty acids are prevented from undergoing β-oxidation because malonyl CoA inhibits the carnitine acyltransferase that is involved in transporting fatty acids into mitochondria where β-oxidation occurs.

   3. Elongation of long-chain fatty acids occurs in the endoplasmic reticulum.

   a. Malonyl CoA provides the two-carbon units that add to palmitoyl CoA or to longer-chain fatty acyl CoAs.

   b. Malonyl CoA condenses with the carbonyl group of the fatty acyl residue and CO_2 is released.

   c. The β-keto group is reduced by NADPH to a β-hydroxy group, dehydration occurs, and a double bond is formed, which is reduced by NADPH.

   d. The net result is that fatty acyl CoA is elongated, using reactions similar but not identical to those that occur on the fatty acid synthase complex.

   4. Desaturation of fatty acids is a complex process that requires O_2, NADPH, and cytochrome b_5.

   –In humans, desaturases may add double bonds at the 9-10 position of a fatty acyl CoA and at 3-carbon intervals between carbon 9 and the carboxyl group.

   a. Plants can introduce double bonds between carbon 9 and the ω-carbon but animals cannot. Therefore, certain unsaturated fatty acids from plants are required in the human diet.

    b. Linoleate ($18{:}2,\Delta^{9,12}$) and **linolenate** ($18{:}3,\Delta^{9,12,15}$) are essential fatty acids required in the diet. Linoleate is used for the synthesis of **arachidonic acid,** from which compounds such as prostaglandins are produced.

D. Synthesis of triacylglycerols (Figure 6-8)

  1. In **intestinal epithelial** cells, triacylglycerol synthesis occurs by a different pathway than in other tissues.

    a. Fatty acids and **2-monoacylglycerols** produced by digestion enter the cells.

    b. The fatty acids are activated to fatty acyl CoA and react with a 2-monoacylglycerol, forming a **triacylglycerol.**

    c. The triacylglycerol is released as part of a **chylomicron.**

  2. In the **liver and in adipose tissue,** glycerol 3-phosphate provides the glycerol moiety that reacts with two fatty acyl CoAs to form phosphatidic acid. The phosphate group is cleaved to form a diacylglycerol, which reacts with another fatty acyl CoA to form a triacylglycerol.

    a. The **liver** may use glycerol to produce glycerol 3-phosphate by a reaction that requires ATP and is catalyzed by glycerol kinase.

    b. Adipose tissue, which lacks glycerol kinase, cannot generate glycerol 3-phosphate from glycerol.

    c. Both liver and adipose tissue can convert glucose, through glycolysis, to dihydroxyacetone phosphate (DHAP), which may be reduced by NADH to glycerol 3-phosphate.

    d. Triacylglycerol is stored in adipose tissue; in the liver, it is incorporated into **VLDL,** which enters the blood. Ultimately, the fatty acyl residues in the triacylglycerols of VLDL may be stored as triacylglycerols in adipose tissue.

E. Regulation of triacylglycerol synthesis from carbohydrate

  1. Synthesis of triacylglycerols from carbohydrate occurs in the liver in the **fed state.**

  2. Key regulatory enzymes in the pathway are activated, and a high-carbohydrate diet causes their induction.

    a. The glycolytic enzymes **glucokinase, phosphofructokinase 1,** and **pyruvate kinase** are active (see Chapter 5 for mechanisms).

    b. Pyruvate dehydrogenase is dephosphorylated and active.

    c. Pyruvate carboxylase is activated by acetyl CoA.

    d. Citrate lyase is inducible.

    e. Acetyl CoA carboxylase is induced, activated by citrate, and converted to its active, dephosphorylated state by a phosphatase that is stimulated by insulin.

    f. The **fatty acid synthase complex** is inducible.

  3. NADPH, which provides the reducing equivalents for fatty acid synthesis, is produced by the inducible malic enzyme and by the inducible enzymes of the pentose phosphate pathway, glucose 6-phosphate dehydrogenase and 6-phosphogluconate dehydrogenase.

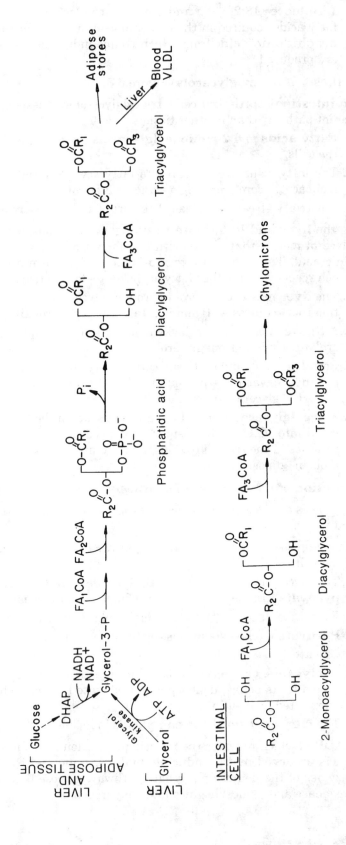

Figure 6-8. Synthesis of triacylglycerols in liver, adipose tissue, and intestinal cells. *R* = fatty acyl moiety. *VLDL* = very-low-density-lipoprotein.

4. **Malonyl CoA,** the product of acetyl CoA carboxylase, inhibits carnitine acyltransferase I, thereby preventing newly synthesized fatty acids from entering mitochondria and undergoing β-oxidation.

V. Regulation of Triacylglycerol Stores in Adipose Tissue

- The triacylglycerol stores of adipose tissue serve as a major source of fuel for the human body. The average 70-kg man has about 15 kg of fat.
- After a meal, triacylglycerols are stored in adipose cells. They are synthesized from fatty acids (derived by the action of lipoprotein lipase on chylomicrons and VLDL) and from a glycerol moiety (derived from glucose).
- The storage of triacylglycerols in adipose tissue is mediated by insulin, which stimulates adipose cells to secrete lipoprotein lipase and to take up glucose, the source of glycerol for triacylglycerol synthesis.
- During fasting, fatty acids and glycerol are released from adipose triacylglycerol stores and serve as a source of fuel for other tissues.
- Glucagon rises during fasting and causes the activation of hormone-sensitive lipase by a cAMP-dependent mechanism. Hormone-sensitive lipase initiates the conversion of adipose triacylglycerols to fatty acids and glycerol, which are released into the blood.
- Fatty acids are transported in the blood complexed with albumin, taken up by various tissues, and oxidized for energy. In the liver, fatty acids are converted to ketone bodies, and glycerol is converted to glucose. These fuels serve as energy sources for other tissues.

A. **Storage of triacylglycerols in adipose tissue** (Figure 6-9)

1. **After a meal,** the triacylglycerol stores of adipose tissue increase.

2. The triacylglycerols of chylomicrons and VLDL are hydrolyzed to fatty acids and glycerol by lipoprotein lipase, which is attached to membranes of cells in the walls of capillaries in adipose tissue.

   a. **Lipoprotein lipase** is synthesized in adipose cells and is secreted when insulin is elevated after a meal.

   b. **Apoprotein C$_{II}$,** which is transferred from HDL to chylomicrons and VLDL, is an activator of lipoprotein lipase.

3. **Fatty acids** released by lipoprotein lipase are taken up by adipose cells and converted to triacylglycerols, but glycerol is not used because adipose tissue lacks glycerol kinase (see IV D 2 for pathway).

   a. The **transport of glucose** into adipose cells is stimulated by insulin, which is elevated after a meal.

   b. Glucose is converted to **DHAP,** which is reduced by NADH to form **glycerol 3-phosphate,** which is used to produce the glycerol moiety.

4. The triacylglycerols are stored in large fat globules in adipose cells.

B. **Lipolysis of adipose triacylglycerols**

1. In the **fasting state,** lipolysis of adipose triacylglycerols occurs.

2. **Glucagon,** which is elevated during fasting, **stimulates lipolysis.** (Epinephrine and other hormones promote lipolysis by the same mechanism under different conditions.)

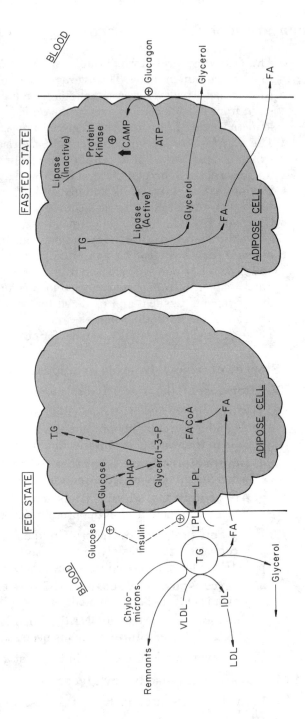

Figure 6-9. Regulation of triacylglycerol stores in adipose tissue. *Left* = in the fed state. *Right* = in the fasted state. *TG* = triacylglycerol; *FA* = fatty acid; *LPL* = lipoprotein lipase; *DHAP* = dihydroxyacetone phosphate; ⊕ = stimulated by; circled TG = triacylglycerol of chylomicrons and VLDL.

    a. Glucagon binds to receptors on the cell membrane and activates **adenyl cyclase**.

    b. Adenyl cyclase produces **cAMP,** which activates protein kinase A.

    c. Protein kinase A phosphorylates and thus activates the hormone-sensitive lipase of adipose tissue.

3. Hormone-sensitive lipase initiates lipolysis, and fatty acids and glycerol are released.

4. The **fatty acids** are carried on albumin in the blood

    a. In tissues such as **muscle** and **kidney,** the fatty acids are oxidized for energy.

    b. In the **liver,** fatty acids are converted to **ketone bodies** that are oxidized by tissues such as muscle and kidney. During starvation (after fasting has lasted about 3 or more days), the brain uses ketone bodies for energy.

5. Glycerol is used by the liver as a source of carbon for **gluconeogenesis,** which produces glucose for tissues such as the brain and red blood cells.

VI. Blood Lipoproteins

- The blood lipoproteins serve to transport water-insoluble triacylglycerols and cholesterol from one tissue to another.
- The major carriers of triacylglycerols are chylomicrons and VLDL.
- The triacylglycerols of the chylomicrons and VLDL are digested in capillaries by lipoprotein lipase. The fatty acids that are produced are either oxidized for energy or converted to triacylglycerols and stored. The glycerol is used for triacylglycerol synthesis or converted to DHAP and oxidized for energy, either directly or after conversion to glucose in the liver.
- The remnants of the chylomicrons are taken up by liver cells by the process of endocytosis and are degraded by lysosomal enzymes, and the products are reused by the cell.
- VLDL is converted to IDL, which is degraded by the liver or converted in blood capillaries to LDL by further digestion of triacylglycerols.
- LDL is taken up by various tissues and provides cholesterol, which the tissues utilize.
- HDL, which is synthesized by the liver, transfers apoproteins, including apo C_{II} and apo E to chylomicrons and VLDL.
- HDL picks up cholesterol from cell membranes or from other lipoproteins. Cholesterol is converted to cholesterol esters by the lecithin:cholesterol acyltransferase (LCAT) reaction. The cholesterol esters may be transferred to other lipoproteins or carried by HDL to the liver, where they are hydrolyzed to free cholesterol, which is used for synthesis of VLDL or converted to bile salts.

A. Composition of the blood lipoproteins (Table 6-1)

    –The major components of lipoproteins are triacylglycerols, cholesterol, cholesterol esters, phospholipids, and proteins. Purified proteins (apoproteins) are designated A, B, C, and E.

Table 6-1. Composition of the Blood Lipoproteins

Component	Chylomicrons	VLDL	IDL	LDL	HDL
Triacylglycerol	85%	55%	26%	10%	8%
Protein	2%	9%	11%	20%	45%
Type	B, C, E	B, C, E	B, E	B	A, C, E
Cholesterol	1%	7%	8%	10%	5%
Cholesterol ester	2%	10%	30%	35%	15%
Phospholipid	8%	20%	23%	20%	25%

1. **Chylomicrons** are the least dense of the blood lipoproteins because they have the most triacylglycerol and the least protein.

2. **VLDL** is more dense than chylomicrons but still has a high content of triacylglycerol.

3. **IDL,** which is derived from VLDL, is more dense than chylomicrons but still has a high content of triacylglycerol.

4. **LDL** has less triacylglycerol and more protein and, therefore, is more dense than the IDL from which it is derived. LDL has the highest content of cholesterol and its esters.

5. **HDL** is the most dense lipoprotein. It has the lowest triacylglycerol and the highest protein content.

B. **Metabolism of chylomicrons** (Figure 6-10)

1. Chylomicrons are synthesized in **intestinal epithelial cells.** Their triacylglycerols are derived from dietary lipid, and their major apoprotein is apo B-48.

2. Chylomicrons travel through the lymph into the blood. **Apo C$_{II}$,** the activator of lipoprotein lipase, and **apo E** are transferred to nascent chylomicrons **from HDL,** and mature chylomicrons are formed.

3. In peripheral tissues, particularly adipose and muscle, the triacylglycerols are digested by **lipoprotein lipase.**

4. The chylomicron remnants interact with receptors on liver cells and are taken up by **endocytosis.** The contents are degraded by **lysosomal enzymes,** and the products (amino acids, fatty acids, glycerol, and cholesterol) are released into the cytosol and reutilized.

C. **Metabolism of VLDL**

1. **VLDL** is synthesized in the **liver,** particularly after a high-carbohydrate meal. It is formed from triacylglycerols that are packaged with cholesterol, apoproteins (particularly apo B-100), and phospholipids, and it is released into the blood.

2. In **peripheral tissues,** particularly adipose and muscle, VLDL triacylglycerols are digested by lipoprotein lipase, and VLDL is converted to IDL.

3. **IDL** returns to the liver, is taken up by endocytosis, and is degraded by **lysosomal enzymes**.

4. **IDL** may also be further degraded by lipoprotein lipase, forming LDL.

5. **LDL** reacts with receptors on various cells, is taken up by endocytosis, and is digested by **lysosomal enzymes.**

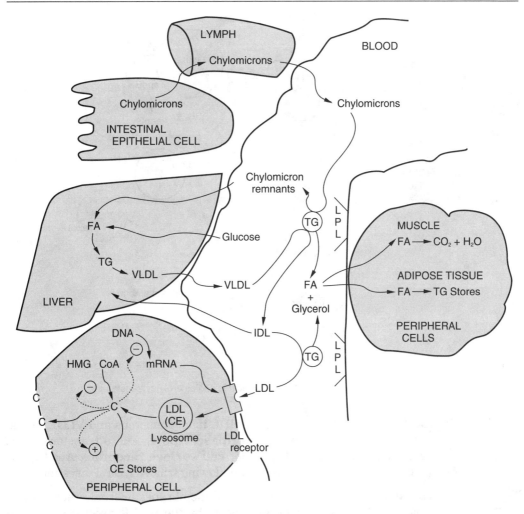

Figure 6-10. Metabolism of chylomicrons and VLDL. *LPL* = lipoprotein lipase; *TG* = triacylglycerol; *CE* = cholesterol esters; *circled TG* = triacylglycerol of chylomicrons and VLDL; ⊖= inhibits; ⊕= stimulates.

    a. Cholesterol, released from cholesterol esters by a lysosomal esterase, can be used for the synthesis of cell membranes or for the synthesis of bile salts in the liver or steroid hormones in endocrine tissue.

    b. Cholesterol inhibits **HMG CoA reductase,** a key enzyme in cholesterol biosynthesis, which regulates the rate of cholesterol synthesis by the cell.

    c. Cholesterol inhibits synthesis of LDL receptors (down-regulation), which subsequently reduces the amount of cholesterol taken up by the cells.

    d. Cholesterol activates acyl:cholesterol acyltransferase (**ACAT**), which converts cholesterol to cholesterol esters for storage in the cells.

D. Metabolism of HDL (Figure 6-11)

    1. HDL is synthesized by the **liver** and released into the blood as disk-shaped particles. The major **protein** of HDL is **apo A.**

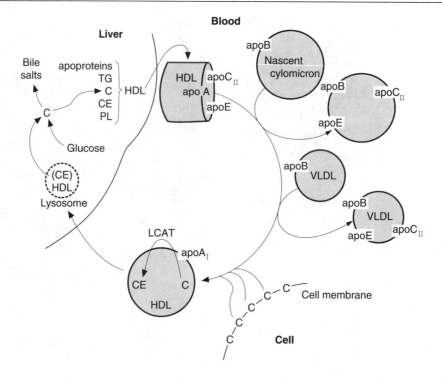

Figure 6-11. HDL metabolism. *LCAT* = lecithin:cholesterol acyltransferase.

2. **Apo C$_{II}$**, which is transferred by HDL to chylomicrons and VLDL, serves as an **activator of lipoprotein lipase. Apo E** is also transferred and serves as a **recognition factor** for **cell surface receptors**. Apo C$_{II}$ and apo E are transferred back to HDL following triacylglycerol digestion.

3. **HDL cholesterol,** obtained from cell membranes or from other lipoproteins, is converted to cholesterol esters by the **LCAT reaction,** which is activated by apo A$_I$. A fatty acid from position 2 of lecithin (phosphatidylcholine) forms an ester with the 3-hydroxyl group of cholesterol, producing lysolecithin and a cholesterol ester. As cholesterol esters accumulate in the core of the lipoprotein, HDL particles become spheroids.

4. **HDL particles** are taken up by the liver by endocytosis and **hydrolyzed by lysosomal enzymes**. Cholesterol, released from cholesterol esters, may be packaged by the liver in VLDL and released into the blood or converted to bile salts and secreted into the bile.

VII. Fatty Acid Oxidation

- Fatty acids, which are the major source of energy in the human, are oxidized by β-oxidation.
- Prior to oxidation, long-chain fatty acids are activated, forming fatty acyl CoA, which is transported into mitochondria by a carnitine carrier system.
- The process of β-oxidation occurs in mitochondria. In four steps that produce FADH$_2$ and NADH, two carbons are cleaved from a fatty acyl CoA

and are released as acetyl CoA. This series of steps is repeated until the fatty acid is completely converted to acetyl CoA.

- ATP is obtained when $FADH_2$ and NADH interact with the electron transport chain or when acetyl CoA is oxidized further.
- In tissues such as skeletal and heart muscle, acetyl CoA enters the TCA cycle and is oxidized to CO_2 and H_2O. In the liver, acetyl CoA is converted to ketone bodies.
- Fatty acid oxidation is regulated by the mechanisms that control oxidative phosphorylation—by the demand for ATP.

A. Activation of fatty acids

1. In the cytosol of the cell, long-chain fatty acids are activated by **ATP and coenzyme A,** and fatty acyl CoA is formed (Figure 6-12). Short-chain fatty acids are activated in mitochondria.

2. The **ATP** is converted to **AMP and pyrophosphate** (PP_i), which is cleaved by pyrophosphatase to two inorganic phosphates ($2 P_i$). Because two high-energy phosphate bonds are cleaved, the equivalent of two molecules of ATP are used for fatty acid activation.

B. Transport of fatty acyl CoA from the cytosol into mitochondria

1. **Fatty acyl CoA** from the cytosol reacts with **carnitine** in the outer mitochondrial membrane, forming fatty acyl carnitine, which passes through both membranes and re-forms fatty acyl CoA on the matrix side.

2. **Carnitine acyltransferase I,** which catalyzes the transfer of acyl groups from coenzyme A to carnitine, is **inhibited by malonyl CoA,** an intermediate in fatty acid synthesis. Therefore, when fatty acids are being synthesized in the cytosol, malonyl CoA inhibits their transport into mitochondria and, thus, prevents a futile cycle—synthesis followed by immediate degradation.

3. Inside the mitochondrion, fatty acyl CoA may undergo **β-oxidation**.

C. Oxidation of fatty acids

−**β-Oxidation** (all reactions involve the β-carbon of fatty acyl CoA) is a spiral consisting of four sequential steps, the first three of which are similar to those in the TCA cycle between succinate and oxaloacetate. These steps are repeated until all the carbons of an even-chain fatty acyl CoA are converted to acetyl CoA (see Figure 6-12).

1. In the first step, **FAD accepts hydrogens** from a fatty acyl CoA. A double bond is produced between the α- and β-carbons, and an enoyl CoA is formed. The $FADH_2$ that is produced interacts with the electron transport chain, generating ATP.

   −Enzyme: **acyl CoA dehydrogenase**

2. **H_2O** adds across the double bond, and a β-hydroxyacyl CoA is formed.

   −Enzyme: **enoyl CoA hydratase**

3. **β-Hydroxyacyl CoA** is oxidized by NAD^+ to a β-ketoacyl CoA. The NADH that is produced interacts with the electron transport chain, generating ATP.

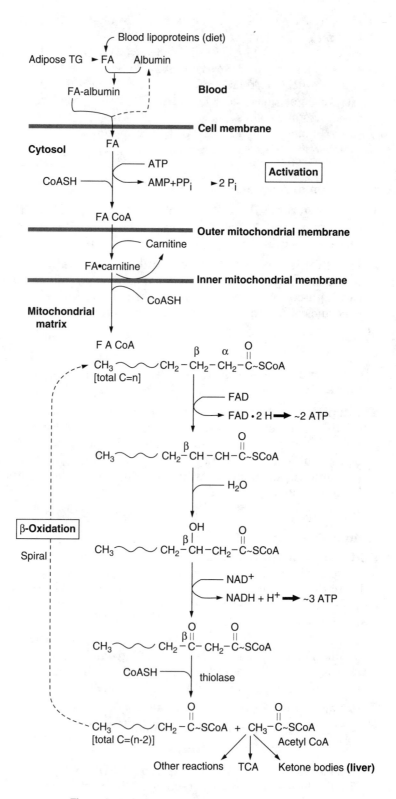

Figure 6-12. Activation and oxidation of fatty acids.

–Enzyme: **L-3-hydroxyacyl CoA dehydrogenase** (which is specific for the L-isomer of the β-hydroxyacyl CoA)

4. The **bond between the α- and β-carbons** of the β-ketoacyl CoA is cleaved by a **thiolase** that requires coenzyme A. Acetyl CoA is produced from the two carbons at the carboxyl end of the original fatty acyl CoA, and the remaining carbons form a fatty acyl CoA that is two carbons shorter than the original fatty acyl CoA.

   –Enzyme: **β-ketothiolase**

5. The shortened **fatty acyl CoA** repeats these four steps. Repetitions continue until all the carbons of the original fatty acyl CoA are converted to acetyl CoA.

   a. The 16-carbon palmitoyl CoA undergoes seven repetitions.

   b. In the last repetition, a four-carbon fatty acyl CoA (butyryl CoA) is cleaved to two acetyl CoAs.

6. **Energy is generated** from the products of β-oxidation.

   a. When one palmitoyl CoA is oxidized, seven $FADH_2$, seven NADH, and eight acetyl CoA are formed.

      (1) The 7 $FADH_2$ each generate approximately 2 ATP, a total of about 14 ATP.

      (2) The 7 NADH each generate about 3 ATP, a total of about 21 ATP.

      (3) The 8 acetyl CoA can enter the TCA cycle, each producing about 12 ATP, a total of about 96 ATP.

      (4) From the oxidation of palmitoyl CoA to CO_2 and H_2O, a total of about 131 ATP are produced.

   b. The **net ATP** produced from palmitate that enters the cell from the blood is about 129 because palmitate must undergo activation (a process that requires the equivalent of 2 ATP) before it can be oxidized (131 ATP – 2 ATP = 129 ATP).

   c. **Oxidation of other fatty acids** will yield different amounts of ATP. For instance, stearoyl CoA, an 18-carbon fatty acid, will produce 8 $FADH_2$, 8 NADH, and 9 acetyl CoA and, as a result, about 146 ATP, if the acetyl CoA is oxidized to CO_2 and H_2O and the activation of the fatty acid is considered.

7. **Oxidation of odd-chain and unsaturated fatty acids**

   a. **Odd-chain fatty acids** produce acetyl CoA and propionyl CoA.

      (1) These fatty acids repeat the four steps of the spiral, producing **acetyl CoA** until the last cleavage when the three remaining carbons are released as propionyl CoA.

      (2) **Propionyl CoA,** but not acetyl CoA, can be converted to glucose. (See Chapter 5 VI and Figure 7-10.)

   b. **Unsaturated fatty acids,** which comprise about half the fatty acid residues in human lipids, require enzymes in addition to the four that catalyze the repetitive steps of the β-oxidation spiral.

      (1) **β-Oxidation** occurs until a double bond of the unsaturated fatty acid is near the carboxyl end of the fatty acyl chain.

      (2) If the double bond is not between the α and β carbons in a *trans* configuration, it is moved so that it is in the appropriate position. The normal steps of β-oxidation can then proceed.

8. ω-Oxidation of fatty acids

a. The ω (omega)-**carbon** (the methyl carbon) of fatty acids may be oxidized to a carboxyl group in the endoplasmic reticulum.

b. β-Oxidation can then occur in mitochondria at this end of the fatty acid as well as from the original carboxyl end. **Dicarboxylic acids** are produced.

9. Process of oxidation of very-long-chain fatty acids in peroxisomes

a. The process differs from β-oxidation in that **molecular O_2** is used and **hydrogen peroxide** (H_2O_2) is formed.

b. The shorter-chain fatty acids that are produced travel to mitochondria, where they undergo β-oxidation.

VIII. Ketone Body Synthesis and Utilization

- The ketone bodies, acetoacetate and 3-hydroxybutyrate, serve as a source of fuel. They are synthesized mainly in liver mitochondria whenever fatty acid levels are high in the blood.
- Fatty acids are activated in liver cells and converted to acetyl CoA, generating ATP. As NADH and ATP levels rise, acetyl CoA accumulates.
- Acetyl CoA reacts with acetoacetyl CoA to form HMG CoA, which is cleaved to form acetoacetate.
- Acetoacetate may be reduced to a second ketone body, 3-hydroxybutyrate (β-hydroxybutyrate), by NADH.
- Acetone is produced by spontaneous (nonenzymatic) decarboxylation of acetoacetate.
- The liver cannot use ketone bodies because it lacks the thiotransferase enzyme that activates acetoacetate.
- Ketone bodies are used as fuels by tissues such as muscle and kidney. During starvation (after about 3–5 days of fasting), the brain also oxidizes ketone bodies.
- Acetoacetate may enter a cell, or it can be formed in the cell by oxidation of 3-hydroxybutyrate, a reaction that produces NADH.
- Acetoacetate is activated to acetoacetyl CoA by reacting with succinyl CoA. Acetoacetyl CoA is cleaved by β-ketothiolase to 2 acetyl CoAs, which enter the TCA cycle and are oxidized to CO_2 and H_2O, generating ATP.

A. Synthesis of ketone bodies (Figure 6-13) occurs in **liver mitochondria** when fatty acids are in high concentration in the blood (during fasting, starvation, or as a result of a high-fat diet).

−**β-Oxidation** produces NADH and ATP and results in the accumulation of acetyl CoA. The liver is producing glucose, using oxaloacetate, so there is decreased condensation of acetyl CoA with OAA to form citrate.

1. Two molecules of acetyl CoA condense to produce acetoacetyl CoA. This reaction is catalyzed by thiolase or an isoenzyme of thiolase.

2. Acetyl CoA and acetoacetyl CoA form **HMG CoA** in a reaction catalyzed

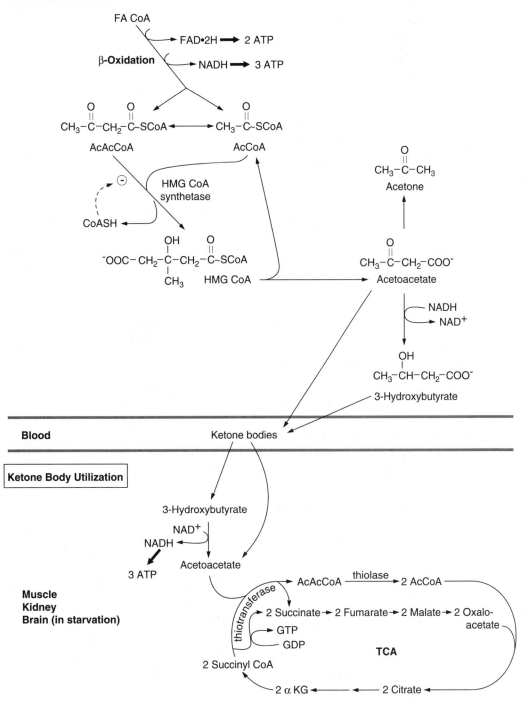

Figure 6-13. Ketone body synthesis and utilization. *FA* = fatty acid; *AcCoA* = acetyl CoA; *AcAcCoA* = acetoacetyl CoA; *αKG* = α-ketoglutarate; *OAA* = oxaloacetate; *HMG CoA* = hydroxymethylglutaryl CoA. The thiotransferase is succinyl CoA-acetoacetate-CoA transferase.

a. **HMG CoA synthetase** is the regulatory enzyme for ketone body synthesis. The enzyme is inducible. It is synthesized when fats are high in the blood.

b. When fats are high in the blood, there is a higher level of substrates for the enzyme.

c. HMG CoA synthetase is inhibited by high levels of one of its products, coenzyme A (CoASH).

(1) When fatty acids flow to the liver, CoASH is used for activation and for the thiolase reaction. CoASH levels are reduced, and HMG CoA synthetase is active.

(2) When the flow of fatty acids to the liver decreases, CoASH accumulates and inhibits the enzyme.

3. **HMG CoA** is cleaved by HMG CoA lyase to form acetyl CoA and acetoacetate.

4. **Acetoacetate** may be reduced by an NAD-requiring dehydrogenase (3-hydroxybutyrate dehydrogenase) to **3-hydroxybutyrate.** This is a reversible reaction.

5. Acetoacetate may also be spontaneously **decarboxylated,** in a nonenzymatic reaction, forming **acetone** (the source of the odor on the breath of ketotic diabetics).

6. The **liver** lacks succinyl CoA-acetoacetate-CoA transferase (a thiotransferase) so it **cannot use ketone bodies.** Therefore, acetoacetate and 3-hydroxybutyrate are released into the blood by the liver.

B. **Utilization of ketone bodies** (see Figure 6-13)

1. When ketone bodies are released from the liver into the blood, they are taken up by peripheral tissues such as **muscle and kidney,** where they are oxidized for energy.

–During **starvation,** ketone bodies in the blood increase to a level that permits entry into **brain** cells, where they are oxidized.

2. **Acetoacetate** may enter cells directly, or it may be produced from the oxidation of 3-hydroxybutyrate by 3-hydroxybutyrate dehydrogenase. NADH is produced by this reaction and can generate ATP.

3. Acetoacetate is activated by reacting with succinyl CoA to form **acetoacetyl CoA** and succinate. The enzyme is succinyl CoA-acetoacetate-CoA transferase (a thiotransferase).

4. Acetoacetyl CoA is cleaved by **thiolase** to form two acetyl CoAs, which enter the TCA cycle and are oxidized to CO_2 and H_2O.

5. **Energy is produced** from the oxidation of ketone bodies.

a. One acetoacetate produces 2 acetyl CoAs, each of which generates about 12 ATP, or a total of about 24 ATP.

b. However, activation of acetoacetate results in the loss of 1 ATP (because GTP, the equivalent of ATP, is not produced in the conversion of succinyl CoA to succinate as it would be in the TCA cycle reaction). Therefore, the oxidation of acetoacetate produces a net yield of only 23 ATP.

c. When **3-hydroxybutyrate** is oxidized, three additional ATP are formed because the oxidation of 3-hydroxybutyrate to acetoacetate produces NADH.

IX. Phospholipid and Sphingolipid Metabolism

- Phospholipids and sphingolipids are major components of cell membranes. They are amphipathic molecules; that is, one portion of the molecule is hydrophilic and associates with H_2O, and another portion contains the hydrocarbon chains derived from fatty acids, which are hydrophobic and associate with lipids (see Figure 6-4).
- Phosphoglycerides (the major phospholipids) contain glycerol, fatty acids, and phosphate. The phosphate is esterified to choline, serine, ethanolamine, or inositol.
- The phosphoglycerides may be synthesized from compounds containing cytosine nucleotides, or they may be interconverted (e.g., phosphatidylcholine [lecithin] is synthesized from a diacylglycerol and CDP-choline, or it may be produced by methylation of phosphatidylethanolamine).
- Degradation of phosphoglycerides involves phospholipases, which are each specific for one of the ester linkages of the phosphodiester bonds.
- The sphingolipids include sphingomyelin (also a phospholipid), the cerebrosides, and gangliosides. These compounds are major components of cell membranes in nervous tissue.
- The sphingolipids are synthesized from ceramide. Ceramide is produced from serine and palmitoyl CoA via the intermediate, sphingosine, which forms an amide with another fatty acyl group. Ceramide reacts with CDP-choline to form sphingomyelin, with a UDP-sugar to form a cerebroside, and with UDP-sugars and CMP-NANA (*N*-acetylneuraminic acid) to form gangliosides.
- During degradation, the phosphocholine and sugar units of the sphingolipids are removed by lysosomal enzymes.

A. Synthesis and degradation of phosphoglycerides (Figure 6-14)

–The phosphoglycerides are synthesized by a process similar in its initial steps to triacylglycerol synthesis—glycerol 3-phosphate combines with two fatty acyl CoAs to form **phosphatidic acid**.

1. Synthesis of phosphatidylinositol

a. Phosphatidic acid reacts with CTP to form CDP-diacylglycerol, which reacts with inositol to form phosphatidylinositol.

b. Phosphatidylinositol may be further phosphorylated to form phosphatidylinositol 4,5-bisphosphate, which is cleaved in response to various stimuli to form the compounds inositol 1,4,5-trisphosphate (IP_3) and diacylglycerol (DAG), which serve as second messengers (see Chapter 8).

2. Synthesis of phosphatidylethanolamine, phosphatidylcholine, and phosphatidylserine

a. Inorganic phosphate is released from phosphatidic acid, and diacyl-

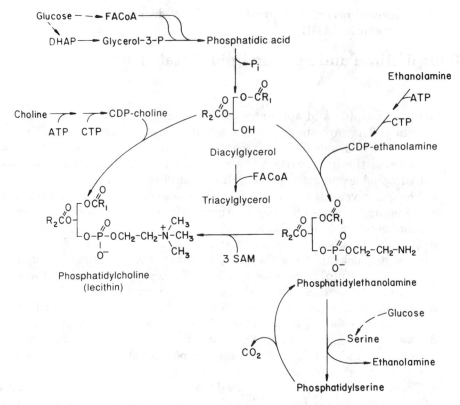

Figure 6-14. Synthesis of phospholipids. *FA* = fatty acid; *DHAP* = dihydroxyacetone phosphate; *SAM* = *S*-adenosylmethionine.

glycerol is produced.

b. Diacylglycerol reacts with compounds containing cytosine nucleotides to form phosphatidylethanolamine and phosphatidylcholine.

(1) Diacylglycerol reacts with CDP-ethanolamine to form phosphatidyl-ethanolamine.

(2) Phosphatidylethanolamine can also be formed by **decarboxylation of phosphatidylserine**.

(3) Diacylglycerol reacts with CDP-choline to form phosphatidyl-choline (lecithin).

(4) Phosphatidylcholine can also be formed by methylation of phos-phatidylethanolamine.

—In addition to being an important component of cell membranes and the blood lipoproteins, phosphatidylcholine provides the fatty acid for the synthesis of cholesterol esters in HDL by the LCAT reaction and, as the dipalmitoyl derivative, serves as lung surfactant. If choline is deficient in the diet, phosphatidyl-choline can be synthesized de novo from glucose (see Figure 6-14).

3. Degradation of phosphoglycerides

a. Phosphoglycerides are hydrolyzed by **phospholipases**.

b. Phospholipase A_1 releases the fatty acid at position 1 of the glycerol moiety, phospholipase A_2 releases the fatty acid at position 2, phospholipase C releases the phosphorylated base at position 3, and phospholipase D releases the free base.

B. Synthesis and degradation of sphingolipids (Figure 6-15)

–Sphingolipids are derived from **serine** rather than glycerol.

1. **Serine** condenses with palmitoyl CoA in a reaction in which the serine is decarboxylated by a pyridoxal phosphate-requiring enzyme.

2. The product is converted to **sphingosine** by a reduction and then an oxidation reaction.

3. A fatty acyl CoA adds onto the nitrogen of sphingosine, forming an amide. The resulting compound is **ceramide**.

4. The hydroxymethyl moiety of ceramide combines with various compounds to form **sphingolipids**.

   a. **CDP-choline** can react with ceramide to form **sphingomyelin**.

   b. **UDP-galactose,** or **UDP-glucose,** can react with ceramide to form **galactocerebrosides** or **glucocerebrosides**.

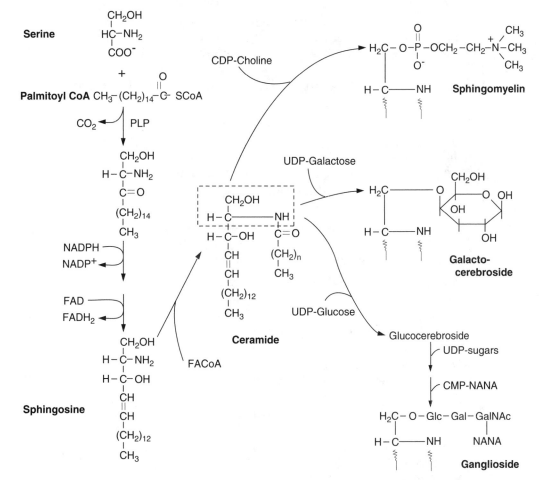

Figure 6-15. Synthesis of sphingolipids. *NANA* = *N*-acetylneuraminic acid; *Glc* = glucose; *Gal* = galactose; *GalNAc* = *N*-acetylgalactosamine; *PLP* = pyridoxal phosphate; *FA* = fatty acyl groups derived from fatty acids; ⌇ = hydrophobic chains of ceramide. The *dashed box* contains the portion of ceramide derived from serine.

c. A series of sugars can add to ceramide, UDP-sugars serving as precursors. **CMP-NANA** (*N*-acetylneuraminic acid, a sialic acid) can form branches from the carbohydrate chain. These ceramide–oligosaccharide compounds are **gangliosides**.

5. Sphingolipids are degraded by **lysosomal enzymes**.

- Prostaglandins, thromboxanes, and leukotrienes are synthesized from 20-carbon polyunsaturated fatty acids (e.g., arachidonic acid) that are released from membrane phospholipids by phospholipase A_2, which is inhibited by glucocorticoids and other steroidal anti-inflammatory agents.
- For prostaglandin synthesis, the polyunsaturated fatty acid is cyclized and oxidized by a cyclooxygenase, which is inhibited by aspirin and the nonsteroidal anti-inflammatory agents. Further oxidations and rearrangements can occur that produce a series of prostaglandins, including the prostacyclins. Thromboxanes can be produced from certain prostaglandins.
- Leukotrienes can be produced from arachidonic acid by a pathway that differs from that for prostaglandin synthesis.

X. Metabolism of the Prostaglandins and Related Compounds

A. Prostaglandins, prostacyclins, and thromboxanes (Figure 6-16)

1. **Polyunsaturated fatty acids** containing 20 carbons and three to five double bonds (e.g., arachidonic acid) are usually esterified to position 2 of the glycerol moiety of phospholipids in cell membranes. These fatty acids may require **dietary linoleic acid** ($18:2,\Delta^{9,12}$), an essential fatty acid, for their synthesis.

2. The polyunsaturated fatty acid is cleaved from the membrane **phospholipid** by phospholipase A_2, which is inhibited by the steroidal anti-inflammatory agents.

3. Oxygen is added and a 5-carbon ring is formed by a **cyclooxygenase** that produces the initial prostaglandin, which is converted to other classes of prostaglandins and the thromboxanes. Aspirin, indomethacin, and other nonsteroidal anti-inflammatory agents inhibit this cyclooxygenase.

   a. The **prostaglandins** have a multitude of effects that differ from one tissue to another and include inflammation, pain, fever, and reproduction. These compounds are known as **autocoids** because they exert their effects primarily in the tissue in which they are produced.

   b. Certain prostacyclins (**PGI₂**), produced by vascular endothelial cells, inhibit platelet aggregation, while certain thromboxanes (**TXA₂**) promote platelet aggregation.

4. Inactivation of the prostaglandins occurs when the molecule is oxidized from the carboxyl and ω-methyl ends to form **dicarboxylic acids** that

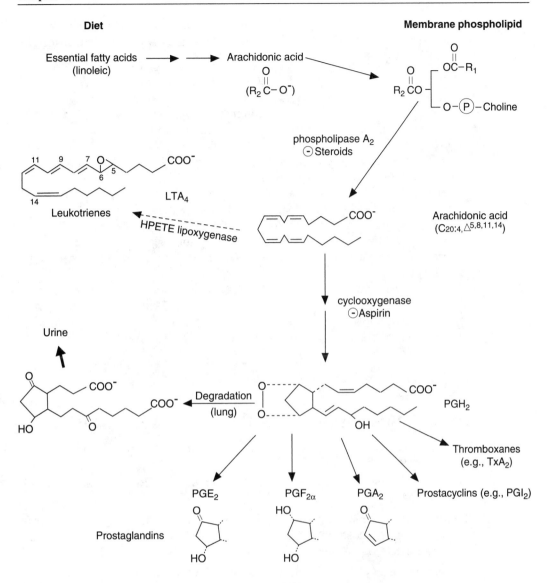

Figure 6-16. Prostaglandins, thromboxanes, and leukotrienes. For each of the classes of prostaglandins (H, E, F, A), the *ring* contains hydroxyl and keto groups at different positions. Only the ring portions of PGE_2, $PGF_{2\alpha}$, and PGA_2 are shown. The remainder of the molecule (not shown) is the same as PGH_2. The subscript refers to the number of double bonds in the nonring portion. The class with two double bonds is derived from arachidonate. Other classes (with one or three double bonds) are derived from other polyunsaturated fatty acids.

are more water-soluble than the prostaglandins and can be excreted in the urine.

B. Leukotrienes

–**Arachidonic acid,** derived from membrane phospholipids, is the precursor for synthesis of the leukotrienes.

1. In the first step, oxygen is added by lipoxygenases, and a family of linear molecules, hydroperoxyeicosatetraenoic acids (**HPETEs**), is formed.

2. A series of compounds, comprising the family of leukotrienes, is produced from these HPETEs. (The leukotrienes are involved in **allergic**

- Although cholesterol is synthesized in most tissues of the body where it serves as a component of cell membranes, it is produced mainly in the liver and intestine.
- Cholesterol is stored in tissues as cholesterol esters.
- Cholesterol and cholesterol esters are transported in blood lipoproteins.
- All the carbons of cholesterol are derived from acetyl CoA.
- Key intermediates in cholesterol biosynthesis are HMG CoA, mevalonic acid, isopentenyl pyrophosphate, and squalene.
- In the liver, bile salts are formed from cholesterol by hydroxylation of the sterol ring, oxidation of the side chain, and conjugation of the carboxylic acid group with glycine or taurine.
- The bile salts are stored in the gallbladder and released during a meal to aid in lipid digestion. Ninety-five percent of the bile salts are resorbed and recycled.
- In certain endocrine tissues, cholesterol is converted to steroid hormones.
- A cholesterol precursor may be converted to 1,25-dihydroxycholecalciferol, the active form of vitamin D_3.
- The steroid nucleus cannot be degraded. It is excreted intact, mainly as unresorbed bile salts.

reactions.)

XI. Cholesterol, Bile Salt, and Steroid Hormone Metabolism

A. Cholesterol (Figure 6-17)

1. Cholesterol is synthesized from **cytosolic acetyl CoA** by a sequence of reactions.

2. Glucose is a major source of carbon for acetyl CoA. Acetyl CoA is produced from glucose by the same sequence of reactions used to produce cytosolic acetyl CoA for fatty acid biosynthesis (see Figure 6-6).

3. Cytosolic acetyl CoA forms **acetoacetyl CoA,** which condenses with another acetyl CoA to form **HMG CoA.**

–Acetyl CoA undergoes similar reactions in the mitochondrion, where HMG CoA is used for ketone body synthesis.

4. Cytosolic **HMG CoA** is a key intermediate in cholesterol biosynthesis. In the endoplasmic reticulum, it is reduced to **mevalonic acid** by the regulatory enzyme HMG CoA reductase.

–**HMG CoA reductase** is inhibited by cholesterol and bile salts in the liver and is induced when blood insulin levels are elevated.

5. Mevalonic acid is phosphorylated and decarboxylated to form the 5-carbon (C-5) isoprene compound, **isopentenyl pyrophosphate**.

6. Two C-5 units condense, forming a C-10 compound, geranyl pyrophosphate, which reacts with another C-5 unit to form a C-15 compound, farnesyl pyrophosphate. This C-15 compound is converted to **squalene**,

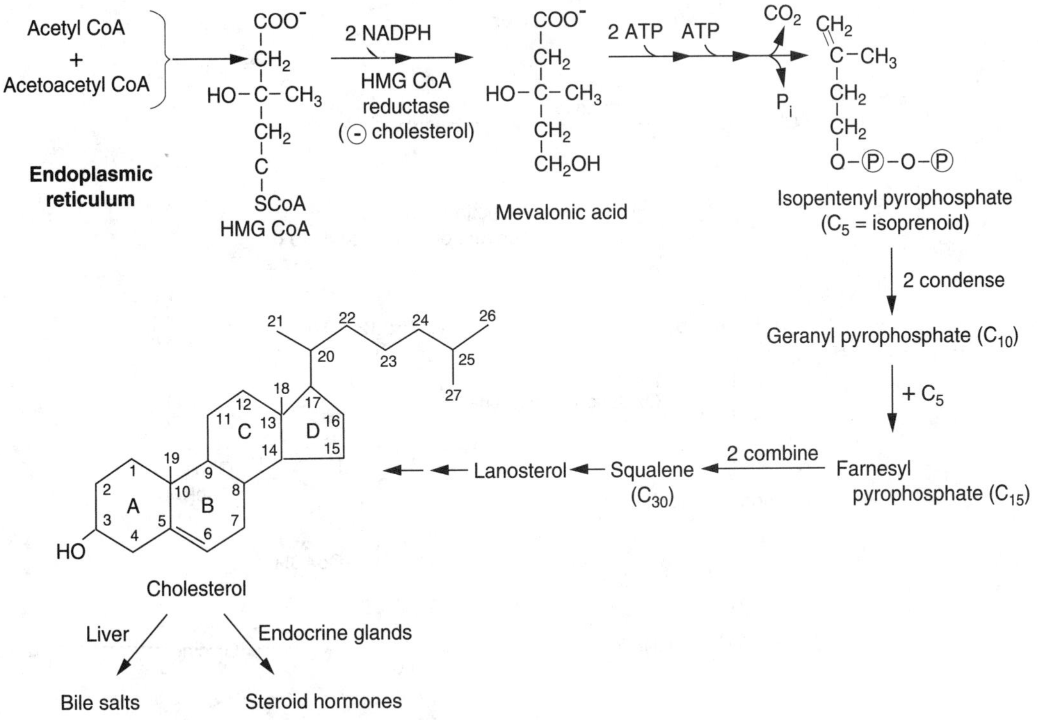

Figure 6-17. Cholesterol biosynthesis. *HMG CoA* = hydroxymethylglutaryl CoA; ⊖ = inhibited by: ℗ = phosphate group.

which is oxidized and cyclized, forming lanosterol.

7. In a series of steps, **lanosterol** loses three methyl groups as CO_2, two double bonds are reduced, one double bond is generated between carbons 5 and 6, and **cholesterol is formed**.

8. The **ring structure** of cholesterol **cannot be degraded** in the body. The bile salts excreted in the feces are the major form in which the steroid nucleus is excreted.

B. Bile salts are synthesized in the liver from cholesterol (Figure 6-18).

1. An **α-hydroxyl group** is added to carbon 7 of cholesterol. A 7 α-hydroxylase, which is inhibited by bile salts, catalyzes this rate-limiting step.

2. The double bond is reduced and further hydroxylations occur, resulting in two compounds. One has α-hydroxyl groups at positions 3 and 7, and the other at positions 3, 7, and 12.

3. The bile acids are produced when the **side chain is oxidized** and converted to a branched, 5-carbon chain, containing a carboxylic acid at the end.

   a. The bile acid with hydroxyl groups at positions 3 and 7 is **chenocholic acid**. The bile acid with hydroxyl groups at positions 3, 7, and 12 is **cholic acid**.

   b. These bile acids each have a **pK of about 6**. Therefore, at pH 6, half

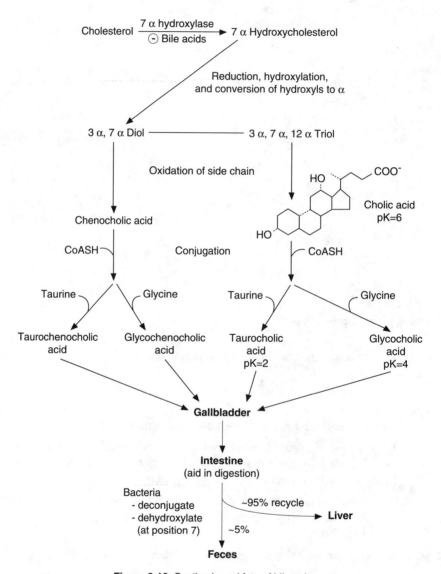

Liver

Cholesterol $\xrightarrow[\ominus \text{ Bile acids}]{7\,\alpha \text{ hydroxylase}}$ 7 α Hydroxycholesterol

Reduction, hydroxylation,
and conversion of hydroxyls to α

3 α, 7 α Diol ——————————— 3 α, 7 α, 12 α Triol

Oxidation of side chain

Chenocholic acid

Cholic acid
pK=6

CoASH   Conjugation   CoASH

Taurine   Glycine   Taurine   Glycine

Taurochenocholic   Glycochenocholic   Taurocholic   Glycocholic
acid   acid   acid   acid
pK=2   pK=4

Gallbladder

Intestine
(aid in digestion)

Bacteria
- deconjugate   ~95% recycle
- dehydroxylate   **Liver**
(at position 7)   ~5%

Feces

Figure 6-18. Synthesis and fate of bile salts.

of the molecules are ionized and carry a negative charge. Below pH 6, the molecules become protonated, and their charge decreases as the pH is lowered. Above pH 6, the molecules ionize and carry a negative charge.

4. Conjugation of the bile acids

a. The bile acids may be activated by ATP and coenzyme A, forming their CoA derivatives, which can form conjugates with either **glycine** or **taurine**.

b. Glycine, an amino acid, forms an amide with the carboxyl group of a bile acid, forming **glycocholic acid or glycochenocholic acid**. These bile salts each have a pK of about 4, lower than the unconjugated bile acids, so they are more completely ionized at pH 6 in the gut and serve as better detergents.

   c. **Taurine,** which is derived from the amino acid cysteine, forms an amide with the carboxyl group of a bile acid. Because of the sulfite group on the taurine moiety, the **taurocholic** and **taurochenocholic acids** have a pK of about 2. They ionize very readily in the gut and are the best detergents among the bile salts.

5. **Fate of the bile salts**

   a. Cholic acid, chenocholic acid, and their conjugates are known as the primary bile salts. They are made in the liver and secreted via the bile through the gallbladder into the **intestine,** where, because they are amphipathic (contain both hydrophobic and hydrophilic regions), they aid in **lipid digestion.**

   b. In the intestine, bile salts may be deconjugated and dehydroxylated (at position 7) by intestinal bacteria.

   c. Bile salts are resorbed in the ileum and return to the liver, where they may be reconjugated with glycine or taurine. However, they are not rehydroxylated. Those that lack the 7 α-hydroxyl group are called secondary bile salts.

   d. The liver recycles about 95% of the bile salts each day; 5% are lost in the feces.

C. **Steroid hormones** are synthesized from cholesterol, and 1,25-dihydroxycholecalciferol (active **vitamin D$_3$**) is synthesized from a precursor of cholesterol (see Chapter 8).

XII. Clinical Correlations

A. Hyperlipidemias

In the hyperlipidemias, the blood levels of cholesterol or triacylglycerols, or both, are elevated resulting from overproduction of lipoproteins or defects in various stages of their degradation. In **familial hypercholesterolemia,** cellular receptors for LDL are defective. Therefore, LDL cannot be taken up normally by cells and degraded by lysosomal enzymes. The consequent increase of blood LDL is associated with **xanthomas** (lipid deposits often found under the skin) and **coronary artery disease**. Treatment may involve diets low in saturated fat and cholesterol, HMG CoA reductase inhibitors, bile-acid binding resins, and nicotinic acid. In hypertriglyceridemia due to deficiencies in lipoprotein lipase (LPL) or apo C$_{II}$ (the LPL activator), triacylglycerol levels rise markedly. These deficiencies are associated with characteristic xanthomas and an intolerance to fatty foods. Low-fat diets may be effective.

B. Atherosclerosis

Atherosclerosis involves the formation of **lipid-rich plaques** in the intima of arteries. The plaques begin as fatty streaks containing foam cells, which initially are macrophages filled with lipids, particularly cholesterol esters. These early lesions develop into fibrous plaques that may occlude an artery and cause a **myocardial infarct or a cerebral infarct**. Formation of these plaques is often associated with abnormalities in plasma lipoprotein metabolism. In contrast to the other lipoproteins, HDL may have a protective effect.

C. Diabetic ketoacidosis

If an **insulin-dependent diabetic** who has failed to take insulin, is suffering from an illness, or is subjected to stress, blood glucose may rise markedly. Elevated glucagon levels cause adipose tissue to release increased amounts of fatty acids, which are converted to ketone bodies by the liver. Ketone body levels may become extremely high, causing a **metabolic acidosis** that, if not treated rapidly and effectively, may lead to **coma and death**.

D. Fatty liver related to alcoholism

Oxidation of ingested alcohol produces acetaldehyde, acetate, and NADH. A high NADH/NAD$^+$ ratio slows the TCA cycle and promotes the synthesis of glycerol 3-phosphate from dihydroxyacetone phosphate. Fatty acid synthesis is stimulated and, because of the effects of ethanol on mitochondria, fatty acid oxidation is decreased. The net result is that fatty acids react with glycerol 3-phosphate to form triacylglycerols, which accumulate in the liver. Impairment of protein synthesis due to **chronic liver dysfunction** results in an inability to produce and secrete VLDL and adds to the hepatic buildup of fats.

E. Malabsorption of fats

Blockage of the bile duct caused by problems such as cholesterol-containing gallstones or duodenal or pancreatic tumors can lead to an inadequate concentration of bile salts in the intestine. Digestion and absorption of dietary lipids is diminished. Certain diseases that affect the pancreas can lead to a decrease in bicarbonate and digestive enzymes in the intestinal lumen. (Bicarbonate is required to raise the intestinal pH so that bile salts and digestive enzymes can function.) If dietary fats are not adequately digested, **steatorrhea** may result. Malabsorption of fats can lead to caloric deficiencies and lack of fat-soluble vitamins and essential fatty acids.

F. Use of nonsteroidal anti-inflammatory drugs (NSAIDs)

NSAIDs, such as aspirin, indomethacin, and ibuprofen, inhibit the cyclooxygenase involved in prostaglandin synthesis. These drugs reduce pain, inflammation, and fever associated with the action of the prostaglandins. Aspirin irreversibly acetylates the enzyme in platelets, inhibiting thromboxane (TXA$_2$) formation, thus reducing platelet aggregation and preventing constriction of vascular and bronchial smooth muscle.

G. Sphingolipidoses

Sphingolipids are normally degraded by lysosomal enzymes. If these enzymes are deficient, partially degraded sphingolipids accumulate in cells, compromising cell function. Death may result. An α-galactosidase is deficient in **Fabry's disease,** a β-glucosidase in **Gaucher's disease,** a sphingomyelinase in **Neimann-Pick disease,** and a hexosaminidase in **Tay-Sachs disease**.

H. Respiratory distress syndrome in the newborn

Dipalmitoylphosphatidylcholine serves as the lung surfactant in adults, allowing the lungs to function normally. This phospholipid develops in the fetus after week 30 of gestation. Premature infants do not have an adequate amount of this phospholipid. As a result, **acute respiratory distress syndrome** is a leading cause of morbidity and death in premature infants.

I. Jamaican vomiting sickness

Jamaican vomiting sickness is caused by a toxin (hypoglycin) from the unripe fruit of the akee tree. This toxin inhibits an acyl CoA dehydrogenase of β-oxidation; consequently, more glucose must be oxidized to compensate for the decreased ability to use fatty acids as a fuel, and severe **hypoglycemia** can occur. ω-Oxidation of fatty acids is increased, and dicarboxylic acids are excreted in the urine. Unwary children are usually the victims of this frequently fatal disease.

Review Test

Directions: Each of the numbered items or incomplete statements in this section is followed by answers or by completions of the statement. Select the **one** lettered answer or completion that is **best** in each case.

1. The process by which dietary lipids are digested and absorbed requires all of the following EXCEPT

(A) the production of chylomicrons
(B) bile salts secreted by the gallbladder
(C) the hydrolysis of ester bonds in triacylglycerols
(D) glycerol 3-phosphate in the intestinal epithelial cell

2. Which one of the following statements about acetyl CoA carboxylase is correct?

(A) Requires thiamine for the carboxylation of acetyl CoA
(B) Utilizes citrate as a substrate
(C) Produces malonyl CoA, which is subsequently decarboxylated
(D) Is located mainly in the matrix of liver mitochondria

3. In the pathway for triacylglycerol synthesis in the liver

(A) fatty acids react with glycerol 3-phosphate
(B) coenzyme A is not required
(C) phosphatidic acid is an intermediate
(D) a 2-monoacylglycerol is an intermediate

4. Which of the following is involved in the synthesis of triacylglycerols in adipose tissue?

(A) Fatty acids are obtained from chylomicrons and VLDL
(B) Glycerol 3-phosphate is derived from blood glycerol
(C) Coenzyme A is not required
(D) A 2-monoacylglycerol is an intermediate
(E) Lipoprotein lipase catalyzes the formation of ester bonds

5. All of the following occur during fasting EXCEPT

(A) hormone-sensitive lipase is activated because the glucagon level in blood is increased and insulin is decreased
(B) glycerol, released from adipose tissue, is utilized for glucose synthesis in liver
(C) liver is converting fatty acids, derived from adipose tissue, to ketone bodies
(D) liver is actively synthesizing chylomicrons from dietary lipids

6. Which of the following is characteristic of HDL?

(A) It is digested by liver lysosomes
(B) It carries cholesterol that is converted to cholesterol esters by the ACAT reaction
(C) It carries apoprotein E, an activator of lipoprotein lipase
(D) It is produced by the action of hormone-sensitive lipase on VLDL

7. Which one of the following statements about fatty acids is TRUE?

(A) Fatty acids are very soluble in water and need no carrier in the blood
(B) When fatty acids are activated in the cytosol, ATP is converted to ADP
(C) Fatty acyl groups are covalently linked to carnitine by an enzyme inhibited by malonyl CoA
(D) Fatty acids may be oxidized to CO_2 and H_2O in the mitochondria of red blood cells

8. The synthesis of fatty acids from glucose in the liver

(A) occurs in mitochondria
(B) requires a covalently bound derivative of pantothenic acid
(C) utilizes NADPH derived solely from the pentose phosphate pathway
(D) is regulated mainly by isocitrate
(E) does not require biotin

9. Which one of the following is a characteristic of the product of the fatty acid synthase complex in the liver?

(A) May be elongated to stearic acid
(B) May be reduced to form oleic acid
(C) May be oxidized directly to palmitic acid
(D) May be converted to arachidonic acid
(E) Is converted into LDL and secreted into the blood

10. Which one of the following sequences places the lipoproteins in the order of most dense to least dense?

(A) HDL/VLDL/chylomicrons/LDL
(B) HDL/LDL/VLDL/chylomicrons
(C) LDL/chylomicrons/HDL/VLDL
(D) VLDL/chylomicrons/LDL/HDL
(E) LDL/chylomicrons/VLDL/HDL

11. Which one of the following occurs during β-oxidation?

(A) FAD is required to form a double bond in fatty acyl CoA
(B) Carbon 2 of the fatty acid is oxidized to form a β-hydroxy compound
(C) NAD$^+$ removes water from the β-hydroxy fatty acyl CoA intermediate
(D) Thiolase removes one carbon from the β-keto intermediate
(E) Two acetyl CoA molecules are produced in each turn of the β-oxidation spiral

12. If 1 mole of the compound shown is oxidized to CO_2 and H_2O in muscle mitochondria, what will be the approximate net number of moles of ATP produced?

$$CH_3 - (CH_2)_9 - CH = CH - CH_2 - COOH$$

(A) 95
(B) 97
(C) 110
(D) 114
(E) 119

13. Newly synthesized fatty acids are not immediately degraded because

(A) tissues that synthesize fatty acids do not contain the enzymes that degrade fatty acids
(B) high NADPH levels inhibit β-oxidation
(C) transport of fatty acids into mitochondria is inhibited under conditions in which fatty acids are being synthesized
(D) in the presence of insulin, the key fatty acid degrading enzyme is not induced
(E) newly synthesized fatty acids cannot be converted to their CoA derivatives

14. Which one of the following occurs in the conversion of fatty acids to ketone bodies?

(A) Carnitine transports the fatty acid across the plasma membrane
(B) Activation of the fatty acid is driven by the conversion of ATP to ADP
(C) Thiolase cleaves HMG CoA
(D) Acetoacetate and acetyl CoA are produced by cleavage of HMG CoA
(E) The complete sequence of reactions occurs in all tissues of the body

15. Approximately how many net moles of ATP are generated when one mole of β-hydroxybutyrate is oxidized to carbon dioxide and water in skeletal muscle?

(A) 23
(B) 24
(C) 25
(D) 26
(E) 27

16. The complete oxidation of

$$CH_3 - \overset{\overset{O}{\|}}{C} - CH_2 - COOH$$

to CO_2 and H_2O in muscle requires

(A) elevated insulin levels
(B) thiamine pyrophosphate
(C) HMG CoA synthetase
(D) biotin
(E) cytosolic ATP for activation of the molecule

17. In the human, prostaglandins can be derived from

(A) glucose
(B) acetyl CoA
(C) arachidonic acid
(D) oleic acid
(E) leukotrienes

18. The compound shown below is

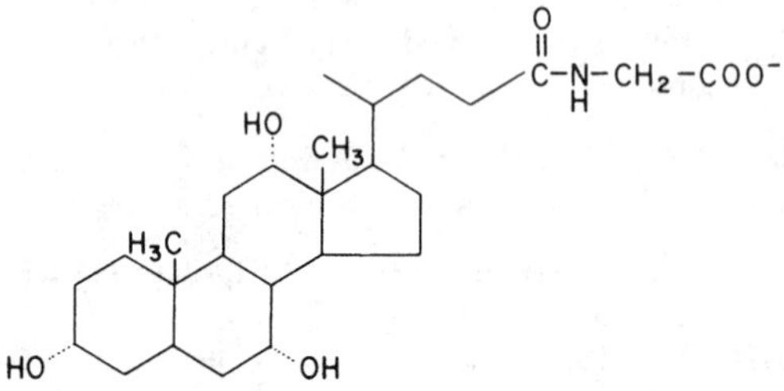

(A) a bile salt
(B) cholesterol
(C) a steroid hormone
(D) vitamin D_3
(E) a cholesterol ester

19. Which one of the following is a characteristic of phospholipids?

(A) Always contain choline and glycerol
(B) Are an important source of energy during fasting
(C) Are a major component of membranes
(D) Are not charged in the body
(E) Are not soluble in water

20. Which one of the following is a characteristic of sphingosine?

(A) Is not a precursor of gangliosides
(B) Is converted to ceramide by reacting with a UDP-sugar
(C) Has palmitoyl CoA and serine as precursors
(D) Contains a glycerol moiety

21. Cytosine nucleotides are involved in the biosynthesis of

(A) galactocerebroside
(B) ceramide
(C) phosphatidic acid
(D) phosphatidylcholine

22. Which one of the following is a characteristic of prostaglandins?

(A) Are derived from fatty acids with 22 carbons
(B) Contain ring structures with 8 carbons
(C) Do not contain keto or hydroxy groups
(D) Are synthesized from polyunsaturated fatty acids

23. A cyclooxygenase, which is inhibited by aspirin, is required for the production of

(A) thromboxanes from arachidonic acid
(B) leukotrienes from arachidonic acid
(C) phospholipids from arachidonic acid
(D) arachidonic acid from linoleic acid

24. Each of the following statements about the conversion of HMG CoA to mevalonic acid is correct EXCEPT

(A) it requires NADPH and H^+
(B) it is a key reaction in the synthesis of compounds that contain isoprenoid units
(C) it is regulated by cholesterol
(D) it is a step in the synthesis of ketone bodies

25. In the conversion of cholesterol to bile salts

(A) carbon 8 is hydroxylated
(B) the side chain is oxidized and may be conjugated with serine or taurine
(C) the double bond is reduced
(D) the hydroxyl group on carbon 3 remains in the β-position

26. After an overnight fast, the blood levels of which of the following will be higher in a person with a carnitine deficiency than in a normal person?

(A) Glucose
(B) Fatty acids
(C) Acetoacetate
(D) 3-Hydroxybutyrate

27. Pancreatic insufficiency may result in

(A) increased pH in the intestinal lumen
(B) increased absorption of fat-soluble vitamins
(C) decreased formation of bile salt micelles
(D) an increase in blood chylomicrons
(E) a decrease of fat in the stool

28. If intestinal pH decreases to 3 as a result of pancreatic insufficiency, which of the following will be most negatively charged?

(A) Glycocholic acid
(B) Taurocholic acid
(C) Palmitate
(D) Cholic acid
(E) Cholesterol

29. A person with type IIA hyperlipoproteinemia had a blood cholesterol level of 360 mg/dl (recommended level below 200 mg/dl) and blood triglyceride (triacylglycerol) levels of 140 mg/dl (recommended level below 160 mg/dl). This person most likely has

(A) a decreased ability for receptor-mediated endocytosis of LDL
(B) a decreased ability to degrade the triacylglycerols of chylomicrons
(C) an increased ability to produce VLDL
(D) an elevation of HDL in the blood
(E) a decreased ability to convert VLDL to IDL

30. A person with an LDL-receptor deficiency was treated with lovastatin. As a consequence of the action of this drug, the person should have

(A) fewer LDL receptors in cell membranes
(B) increased de novo cholesterol synthesis
(C) increased ACAT activity
(D) lower blood cholesterol levels
(E) higher blood triacylglycerol levels

31. A patient with a hyperlipoproteinemia would be most likely to benefit from a low-carbohydrate diet if the lipoproteins that are elevated in the blood are

(A) chylomicrons
(B) VLDL
(C) LDL
(D) HDL

32. Respiratory distress syndrome in premature newborns is caused by deficiency in the lungs of

(A) sphingomyelin
(B) gangliosides
(C) triacylglycerols
(D) phosphatidylcholine
(E) prostaglandins

33. Insulin-dependent diabetes mellitus (IDDM) is caused by a decreased ability of the β cells of the pancreas to produce insulin. A person with IDDM who has neglected to take insulin injections will have

(A) increased fatty acid synthesis from glucose in liver
(B) decreased conversion of fatty acids to ketone bodies
(C) increased stores of triacylglycerol in adipose tissue
(D) increased conversion of acetoacetate to acetone

34. The accumulation of the GM_2 ganglioside in Tay-Sachs disease is caused by

(A) an increased synthesis of the ganglioside precursor, ceramide
(B) an increased concentration of the UDP-sugars required for ganglioside synthesis
(C) a genetic deficiency of phospholipase A_2
(D) a deficiency of a lysosomal enzyme that degrades gangliosides

Directions: Each group of items in this section consists of lettered options followed by a set of numbered items. For each item, select the **one** lettered option that is most closely associated with it. Each lettered option may be selected once, more than once, or not at all.

Questions 35–38

Match the following descriptions with the appropriate protein.

(A) Lipoprotein lipase
(B) Pancreatic lipase
(C) Hormone-sensitive lipase
(D) Colipase

35. Does not cleave fatty acids from triacylglycerols
36. Produces 2-monoacylglycerols
37. Degrades the triacylglycerols of chylomicrons in blood capillaries
38. Is activated by protein kinase A

Questions 39–42

Match the following descriptions with the appropriate lipid.

(A) Palmitate
(B) Acetoacetate
(C) Cholesterol
(D) Bile salts

39. Oxidized by the brain during prolonged starvation
40. Produced from acetyl CoA by most cells in the body
41. Efficiently recycled by the liver
42. Produced by the liver during fasting

Questions 43–46

A molecule of palmitic acid, attached to carbon 1 of the glycerol moiety of a triacylglycerol, is ingested and digested. It passes into the blood, is stored in a fat cell, and ultimately is oxidized to CO_2 and H_2O in a muscle cell. Choose the molecular complex in the blood in which the palmitate residue is carried from the first site to the second.

(A) VLDL
(B) Chylomicron
(C) Fatty acid–albumin complex
(D) Bile salt micelle
(E) LDL

43. From the lumen of the gut to the surface of the gut epithelial cell
44. From the gut epithelial cell to the blood
45. From the blood in the intestine to a fat cell
46. From a fat cell to a muscle cell

Questions 47–50

Identify compounds A, B, C, and D in this figure.

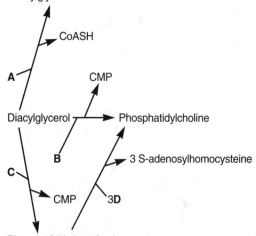

47. CDP-Choline
48. *S*-Adenosylmethionine
49. CDP-Ethanolamine
50. Fatty acyl CoA

Answers and Explanations

1–D. Bile salts, secreted by the gallbladder, emulsify dietary triacylglycerols, which contain ester bonds that are hydrolyzed by pancreatic lipase to produce fatty acids and 2-monoacylglycerols. These products are absorbed by intestinal epithelial cells, where they are reconverted to triacylglycerols (by a process that does not require glycerol 3-phosphate) and secreted into the lymph in chylomicrons.

2–C. Biotin is required for the acetyl CoA carboxylase reaction in which the substrate, acetyl CoA, is carboxylated by the addition of CO_2 to form malonyl CoA. This reaction occurs in the cytosol. Malonyl CoA provides the 2-carbon units that add to the growing fatty acid chain on the fatty acid synthase complex. As the growing chain is elongated, malonyl CoA is decarboxylated.

3–C. In the liver, 2 fatty acyl CoAs react with glycerol 3-phosphate to form phosphatidic acid, which releases inorganic phosphate to form a diacylglycerol. The diacylglycerol reacts with fatty acyl CoA to form a triacylglycerol.

4–A. Fatty acids, cleaved from the triacylglycerols of blood lipoproteins by the action of lipoprotein lipase, are taken up by adipose cells and react with coenzyme A to form fatty acyl CoA. Glucose is converted via dihydroxyacetone phosphate to glycerol 3-phosphate, which reacts with fatty acyl CoA to form phosphatidic acid. (Adipose tissue lacks glycerol kinase and cannot use glycerol.) After inorganic phosphate is released from phosphatidic acid, the resulting diacylglycerol reacts with another fatty acyl CoA to form a triacylglycerol, which is stored in adipose cells. (2-Monoacylglycerol is an intermediate only in intestinal cells.)

5–D. During fasting, the hormone-sensitive lipase of adipose tissue is activated by a mechanism involving glucagon, cAMP, and protein kinase. Triacylglycerols are degraded, and fatty acids and glycerol are released into the blood. In the liver, glycerol is converted to glucose and fatty acids to ketone bodies. These fuels are released into the blood and supply energy to various tissues. Under these conditions, the liver is not producing significant quantities of VLDL. Chylomicrons are produced in intestinal cells.

6–A. HDL is produced in the liver. It transfers apoprotein C_{II}, which activates lipoprotein lipase, to chylomicrons and VLDL. HDL picks up cholesterol from cell membranes. This cholesterol is converted to cholesterol esters by the LCAT reaction. Ultimately, HDL enters liver cells by endocytosis and is digested by lysosomal enzymes. Hormone-sensitive lipase degrades triacylglycerols stored in adipose cells.

7–C. Fatty acids cross the inner mitochondrial membrane on a carnitine carrier. This process is inhibited during fatty acid synthesis by malonyl CoA. Fatty acids are very insoluble in water and are transported in the blood by serum albumin. They cross the plasma membrane and are converted to fatty acyl CoA by CoASH and ATP. In the process, ATP is converted to AMP, so fatty acid activation utilizes the equivalent of 2 ATP. In mitochondria, fatty acids are oxidized to CO_2 and H_2O. They cannot be oxidized in red blood cells, which lack mitochondria.

8–B. The synthesis of fatty acids from glucose occurs in the cytosol, except for the mitochondrial reactions in which pyruvate is converted to citrate. Biotin is required for the conversion of pyruvate to oxaloacetate, which combines with acetyl CoA to form citrate. Biotin is also required by acetyl CoA carboxylase. Pantothenic acid is covalently bound to the fatty acid synthase complex as part of a phosphopantetheinyl residue. The growing fatty acid chain is attached to this residue during the sequence of reactions that produce palmitic acid. NADPH, produced by the malic enzyme as well as by the pentose phosphate pathway, provides reducing equivalents. Citrate, not isocitrate, is a key regulatory compound.

9–A. The 16-carbon, fully saturated fatty acid, palmitate (16:0), is the product of the fatty acid synthase complex. It may be elongated by two carbons to form stearic acid (18:0), or it may be oxidized to form palmitoleic acid ($16:1,\Delta^9$). Stearate can be oxidized to oleic acid ($18:1,\Delta^9$). Arachidonic acid ($20:4,\Delta^{5,8,11,14}$) can be synthesized from the essential fatty acid linoleate ($18:2,\Delta^{9,12}$). It cannot be produced from palmitate. Fatty acids synthesized in the liver are converted to triacylglycerols, packaged in VLDL, and secreted into the blood.

10–B. Because chylomicrons contain the most triacylglycerol, they are the least dense of the blood lipoproteins. VLDL is more dense than chylomicrons. Because LDL is produced by degradation of the triacylglycerols of VLDL, LDL is more dense than VLDL. HDL is the most dense of the blood lipoproteins.

11–A. During β-oxidation, a double bond is formed between the α and β carbons of a fatty acyl CoA, and FAD is reduced to $FADH_2$. Then water adds across the double bond, and a β-hydroxy compound is formed. The hydroxyl group on carbon 3 (the β-carbon) is oxidized to a keto group by NAD^+, which is converted to NADH + H^+. Finally, a cleavage catalyzed by thiolase releases one acetyl CoA (which contains two carbons).

12–C. This fatty acid contains 14 carbons. If it were fully saturated, it would undergo six spirals of β-oxidation, which would produce 6 $FADH_2$ and 6 NADH + H^+, which would generate about 2 x 6 and 3 x 6 ATP, respectively, or a total of 30 ATP. Seven acetyl CoA would be produced, which would enter the TCA cycle and generate about 7 x 12, or 84 ATP. Therefore, for a fully saturated, 14-carbon fatty acid, approximately 114 ATP would be produced. However, 2 ATP are required to activate the fatty acid, reducing the ATP yield to 112. Because this fatty acid is unsaturated (it contains one double bond), one fewer $FADH_2$ would be produced (two fewer ATP would be generated), and the net yield of ATP would be approximately 110.

13–C. During fatty acid synthesis (which occurs in the cytosol), malonyl CoA is produced. Malonyl CoA inhibits the carnitine acyltransferase that is involved in the transport of fatty acids into mitochondria (where β-oxidation occurs).

14–D. Ketone bodies are synthesized in the liver from fatty acids derived from the blood. During the cytosolic activation of the fatty acid, ATP is converted to AMP. Carnitine is required to carry the fatty acyl group across the mitochondrial membrane. In the mitochondrion, the fatty acid is oxidized. Acetyl CoA and acetoacetyl CoA are produced and react to form HMG CoA, which is cleaved by HMG CoA lyase to form the ketone body acetoacetate and acetyl CoA.

15–D. This reaction will produce 26 net moles of ATP, as follows: + 3 ATP produced from the NADH generated when β-hydroxybutyrate is oxidized to acetoacetate. – 1 ATP because, when succinyl CoA is converted to succinate via the thiotransferase reaction that converts acetoacetate to acetoacetyl CoA, no GTP is produced. + 24 ATP produced when 2 acetyl CoA are oxidized in the TCA cycle; therefore, + 26 ATP net.

16–B. This compound is acetoacetate, which is synthesized in the liver when blood insulin levels are low. HMG CoA synthetase is the key regulatory enzyme for synthesis, not oxidation. Acetoacetate is transported to tissues, such as muscle, where it is activated in the mitochondrion by succinyl CoA (not ATP), cleaved to 2 acetyl CoA, and oxidized via the TCA cycle, which requires the vitamin thiamine as thiamine pyrophosphate, a cofactor for α-ketoglutarate dehydrogenase. Biotin is not required.

17–C. Prostaglandins can be synthesized from arachidonic acid (which requires the essential fatty acid, linoleate, for its synthesis). They cannot be synthesized from glucose, so they cannot be made from acetyl CoA or oleic acid. Although leukotrienes are derived from arachidonic acid, they are not precursors of prostaglandins.

18–A. This compound is the bile salt glycocholic acid. During its synthesis, the ring structure of cholesterol is hydroxylated and reduced, and the side chain is oxidized and conjugated with glycine. Although cholesterol may be converted to steroid hormones, this is not one of them. The cholesterol ring structure opens when vitamin D_3 is formed. When cholesterol is converted to a cholesterol ester, the hydroxyl group at position 3 becomes esterified to a fatty acid.

19–C. Phospholipids are important components of membranes but are also found in blood lipoproteins and in lung surfactant. They are amphipathic molecules that are not involved in storing energy but in interfacing between body lipids and their aqueous environment. They are soluble in water because they contain a phosphate residue that is negatively charged and often contain either choline, ethanolamine, or serine residues that have a positive charge at physiologic pH. A serine residue will also contain a negative charge.

20–C. Palmitoyl CoA and serine react to form a precursor that is converted to sphingosine. Sphingosine does not contain a glycerol moiety. Formation of an amide with a fatty acyl CoA converts sphingosine to ceramide. Ceramide may be converted to sphingomyelin, cerebrosides, and gangliosides.

21–D. CDP-choline reacts with a diacylglycerol to form phosphatidylcholine. UDP-galactose reacts with ceramide to form a galactocerebroside. Cytosine nucleotides are not required for the synthesis of phosphatidic acid or ceramide.

22–D. Prostaglandins are synthesized from 20-carbon polyunsaturated fatty acids with three, four, or five double bonds. A cyclooxygenase converts the fatty acid to a compound that contains a 5-membered ring. In subsequent reactions, a series of prostaglandins is produced that contain various keto and hydroxy groups.

23–A. Arachidonic acid is produced from linoleic acid (an essential fatty acid) by a series of elongation and desaturation reactions. Arachidonic acid is stored in membrane phospholipids, released, and oxidized by a cyclooxygenase (which is inhibited by aspirin) in the first step in the synthesis of prostaglandins, prostacyclins, and thromboxanes. Leukotrienes require a lipoxygenase, rather than a cyclooxygenase, for their synthesis from arachidonic acid.

24–D. In the synthesis of cholesterol, but not of ketone bodies, HMG CoA is reduced by $NADPH + H^+$ to mevalonic acid. The enzyme, HMG CoA reductase, is highly regulated (it is inhibited by cholesterol and bile salts and induced by insulin). Mevalonic acid is converted to isopentenyl pyrophosphate, which provides isoprenoid units for the synthesis of cholesterol and its derivatives and for many other compounds.

25–C. During the conversion of cholesterol to bile salts, carbon 7 is hydroxylated. For the cholic acid of bile salts, carbon 12 is also hydroxylated. All hydroxyl groups, including the one on carbon 3, assume an α-configuration. The double bond is reduced and the side chain is oxidized and may be conjugated with glycine or taurine.

26–B. After an overnight fast, fatty acids, released from adipose tissue, serve as fuel for other tissues. Carnitine is required to transport the fatty acids into mitochondria for β-oxidation. In the liver, β-oxidation supplies acetyl CoA for ketone body (acetoacetate and 3-hydroxybutyrate) synthesis. In a carnitine deficiency, blood levels of fatty acids will be elevated and ketone bodies will be low. Consequently, the body will use more glucose, so glucose levels will be decreased.

27–C. The pancreas produces bicarbonate (which neutralizes stomach acid) and digestive enzymes (including the lipase that degrades dietary lipids). Decreased bicarbonate will lead to a decrease of intestinal pH. Decreased digestion of dietary triacylglycerols will lead to formation of fewer bile salt micelles. Intestinal cells will have less substrate for chylomicron formation, and less fat-soluble vitamins will be absorbed. More dietary fat will be excreted in the feces.

28–B. Taurocholic acid has the lowest pK (pK = 2) of these compounds. At pH 3, the ratio of negatively charged to uncharged taurocholate molecules would be 10:1.

29–A. Of the blood lipoproteins, LDL contains the highest concentrations of cholesterol and lowest concentrations of triacylglycerols. Elevation of LDL would result in high cholesterol levels and relatively normal triacylglycerol levels. Decreased ability to degrade triacylglycerol of microns or to convert VLDL to IDL, as well as an increased ability to produce VLDL, would all result in elevated triacylglycerol levels. Because HDL helps to transfer cholesterol from peripheral cells to the liver, high levels are associated with low cholesterol.

30–D. HMG CoA reductase inhibitors cause cells to decrease the rate of cholesterol synthesis, which causes decreased conversion of cholesterol to cholesterol esters (by the ACAT reaction) for storage and increased production of LDL receptors. An increased number of receptors will cause more LDL to be taken up by cells and degraded by lysosomes. Thus, blood cholesterol levels will decrease.

31–B. VLDL is produced mainly from dietary carbohydrate, LDL from VLDL, and chylomicrons from dietary triacylglycerol. Elevated HDL levels are desirable and not considered to be a lipid disorder.

32–D. Respiratory distress syndrome is caused by a deficiency of lung surfactant, which is composed mainly of dipalmitoylphosphatidylcholine.

33–D. Decreased insulin levels cause fatty acid synthesis to decrease and glucagon levels to increase. Adipose triacylglycerols are degraded. Fatty acids are converted to ketone bodies in liver; a ketoacidosis can occur. There is increased decarboxylation of acetoacetate to form acetone, which causes the odor associated with diabetic ketoacidosis.

34–D. Accumulation of gangliosides is not caused by increased synthesis, but rather by decreased degradation in lysosomes. Phospholipase A_2 cleaves fatty acids from position 2 of phospholipids in cell membranes.

35–D. Colipase aids pancreatic lipase. It does not have enzymatic activity.

36–B. Pancreatic lipase produces 2-monoacylglycerols.

37–A. Lipoprotein lipase, which is attached to cell membranes of blood capillary walls, degrades the triacylglycerols of chylomicrons and VLDL.

38–C. Hormone-sensitive lipase is phosphorylated and activated by protein kinase A in response to cAMP.

39–B. Ketone bodies such as acetoacetate are oxidized by the brain during prolonged starvation.

40–C. Most cells produce cholesterol.

41–D. Ninety-five percent of the bile salts secreted by the liver are returned from the gut to the liver.

42–B. The liver produces ketone bodies such as acetoacetate during fasting.

43–D. A palmitate residue attached to carbon 1 of a dietary triacylglycerol is released by pancreatic lipase and carried from the intestinal lumen to the gut epithelial cell in a bile salt micelle.

44–B. Palmitate is absorbed into the intestinal cell and utilized to synthesize a triacylglycerol, which is packaged in a nascent chylomicron and secreted via the lymph into the blood.

45–B. The chylomicron, containing the palmitate, matures in the blood by accepting proteins from HDL. It travels to a fat cell.

46–C. The chylomicron triacylglycerol is digested by lipoprotein lipase, and the palmitate enters a fat cell and is stored as triacylglycerol. It is released as free palmitate and carried, complexed with albumin, to a muscle cell, where it is oxidized.

47–B. Phosphatidylcholine (lecithin) may be produced when CDP-choline reacts with a diacylglycerol.

48–D. Phosphatidylcholine may also be produced when S-adenosylmethionine donates methyl groups to phosphatidylethanolamine.

49–C. Phosphatidylethanolamine can be produced from CDP-ethanolamine and a diacylglycerol.

50–A. A triacylglycerol can be produced from a diacylglycerol by reaction with fatty acyl CoA.

7

Nitrogen Metabolism

Overview

- Nitrogen is obtained mainly from protein in the diet, which is digested to amino acids by the combined action of proteases produced by the stomach, pancreas, and intestinal epithelial cells (Figure 7-1).
- Amino acids are absorbed by intestinal epithelial cells, pass into the blood, and are taken up by other cells of the body. The transport of amino acids into muscle cells is stimulated by insulin.
- Amino acids are used by cells for the synthesis of proteins, which is a dynamic process; proteins are constantly being synthesized and degraded.

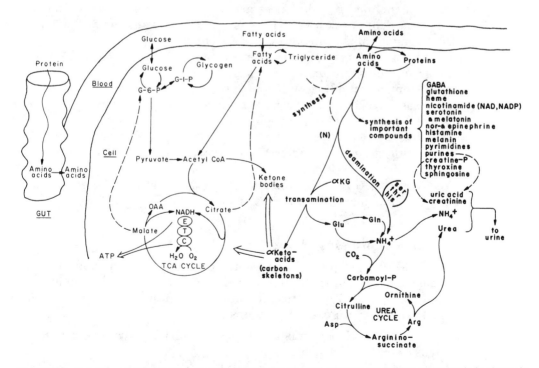

Figure 7-1. Overview of nitrogen metabolism. The metabolism of nitrogen-containing compounds is shown in bold face type on the *right,* and that of glucose and fatty acids is shown in light face on the *left*. αKG = α-ketoglutarate; *ETC* = electron transport chain; *OAA* = oxaloacetate; *G-6-P* = glucose 6-phosphate; *G-1-P* = glucose 1-phosphate.

231

- After nitrogen is removed from amino acids, the carbon skeletons may be oxidized for energy.
- The nitrogen of amino acids is converted to urea in the liver and ultimately excreted by the kidney.
- Although urea is the major nitrogenous excretory product, nitrogen may also be excreted as NH_4^+, uric acid, and creatinine.
- In the liver, in the fed state, amino acid carbons are converted to fatty acids and triacylglycerol. During fasting, amino acid carbons are converted to glucose or to ketone bodies.
- In muscle during fasting, protein is broken down to amino acids.
- The essential amino acids (histidine, isoleucine, leucine, lysine, methionine, phenylalanine, threonine, tryptophan, and valine) are required in the diet. Arginine and increased amounts of histidine are required during periods of growth.
- The nonessential amino acids can be synthesized in the body. The carbons of 10 of the nonessential amino acids may be derived from glucose.
  –Cysteine derives its sulfur from methionine.
  –Tyrosine is produced by hydroxylation of the essential amino acid phenylalanine.
- Amino acids are used for the synthesis of many nitrogen-containing compounds such as the purine and pyrimidine bases, heme, creatine, nicotinamide, thyroxine, epinephrine, melanin, and sphingosine.

I. Protein Digestion and Amino Acid Absorption

- Proteins are converted to amino acids by digestive enzymes.
- Many of the digestive proteases are produced and secreted as inactive zymogens. They are converted to their active forms by removal of a peptide fragment in the lumen of the digestive tract.
- The digestion of proteins begins in the stomach, where pepsin converts dietary proteins into smaller polypeptides.
- In the lumen of the small intestine, proteolytic enzymes produced by the pancreas (trypsin, chymotrypsin, elastase, and the carboxypeptidases) cleave the polypeptides into oligopeptides and amino acids.
- Digestive enzymes produced by the intestinal epithelial cells (aminopeptidases, dipeptidases, and tripeptidases) cleave the small peptides to amino acids.
- Amino acids, the final products of protein digestion, are absorbed through intestinal epithelial cells and enter the blood.

A. Digestion of proteins (Figure 7-2)

1. The 70–100 g of **protein** consumed each day and an equal or larger amount of protein that enters the digestive tract as digestive enzymes or in sloughed-off cells from the intestinal epithelium are converted to amino acids by **digestive enzymes**.

2. In the **stomach,** pepsin is the major proteolytic enzyme. It cleaves proteins to smaller polypeptides.

    a. **Pepsin** is produced and secreted by the chief cells of the stomach as the inactive zymogen **pepsinogen**.

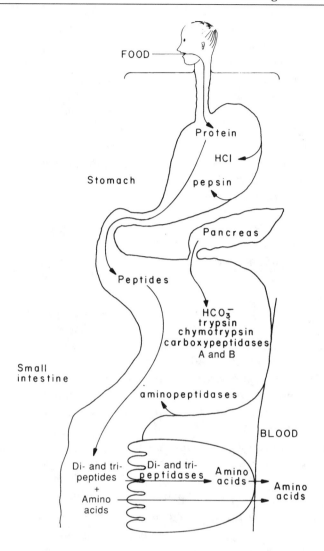

FOOD

Protein

HCl

Stomach

pepsin

Pancreas

Peptides

HCO_3^-
trypsin
chymotrypsin
carboxypeptidases
A and B

Small
intestine

aminopeptidases

BLOOD

Di- and tri-
peptides
+
Amino
acids

Di- and tri-
peptidases

Amino
acids

Amino
acids

Figure 7-2. Digestion of proteins.

b. HCl produced by the parietal cells of the stomach causes pepsinogen to be cleaved to pepsin. Pepsin also catalyzes the cleavage of pepsinogen to pepsin (autocatalysis).

c. Pepsin has a broad specificity but tends to cleave peptide bonds in which the carboxyl group is contributed by the aromatic amino acids or by **leucine**.

3. In the **intestine,** the partially digested material from the stomach encounters pancreatic secretions, which include bicarbonate and a group of proteolytic enzymes.

a. Bicarbonate neutralizes the stomach acid, raising the pH of the contents of the intestinal lumen into the optimal range for the digestive enzymes to act.

b. Endopeptidases from the pancreas cleave peptide bonds within protein chains.

(1) **Trypsin** cleaves peptide bonds in which the carboxyl group is contributed by **arginine** or **lysine**.

–Trypsin is secreted as the inactive zymogen **trypsinogen**.

–Trypsinogen is cleaved to form trypsin by the enzyme **enteropeptidase** (enterokinase), which is produced by intestinal cells. Trypsinogen may also undergo autocatalysis by trypsin.

(2) **Chymotrypsin** usually cleaves peptide bonds in which the carboxyl group is contributed by the aromatic **amino acids** or by **leucine. Chymotrypsinogen,** the inactive zymogen, is cleaved to form chymotrypsin by trypsin.

(3) **Elastase** cleaves at the carboxyl end of amino acid residues with small, uncharged side chains such as alanine, glycine, or serine. **Proelastase,** the inactive zymogen, **is cleaved** to elastase by trypsin.

c. **Exopeptidases** from the pancreas (**carboxypeptidases A** and **B**) cleave one amino acid at a time from the C-terminal end of the peptide.

(1) The carboxypeptidases are produced as **procarboxypeptidases,** which are cleaved to the active form by trypsin.

(2) Carboxypeptidase A cleaves aromatic amino acids from the C-terminal end of peptides.

(3) Carboxypeptidase B cleaves the basic amino acids, lysine and arginine, from the C-terminal end of peptides.

d. **Proteases** produced by intestinal epithelial cells complete the conversion of dietary proteins to amino acids.

(1) **Aminopeptidases** are exopeptidases produced by intestinal cells that cleave one amino acid at a time from the *N*-terminal end of peptides.

(2) **Dipeptidases and tripeptidases** associated with the intestinal cells produce amino acids from dipeptides and tripeptides.

B. Transport of amino acids from intestinal lumen into the blood

–Amino acids are absorbed by intestinal epithelial cells and released into the blood by two types of transport systems.

–There are at least seven different carrier proteins that transport different groups of amino acids.

1. Sodium-amino acid carrier system

a. The major transport system involves the uptake by the cell of a sodium ion and an amino acid by the same carrier protein on the luminal surface.

b. The **sodium ion** is pumped from the cell into the blood by the Na^+-K^+ ATPase, while the **amino acid** travels down its concentration gradient into the blood.

–Thus, the transport of amino acids from the intestinal lumen to the blood is driven by the hydrolysis of ATP.

2. γ-Glutamyl cycle

a. An amino acid in the lumen reacts with **glutathionine** (γ-glutamyl-cysteinyl-glycine) in the cell membrane, forming a γ-glutamyl amino acid and the dipeptide cysteinyl-glycine.

b. The amino acid is carried across the cell membrane attached to **γ-glutamate** and released into the cytoplasm. The γ-glutamyl moiety is used in the resynthesis of glutathione.

II. Addition and Removal of Amino Acid Nitrogen

- When amino acids are synthesized, nitrogen is added to the carbon precursors.
- When amino acids are oxidized to produce energy, the nitrogen is removed and converted to urea.
- Nitrogen may be transferred from one amino acid to another by transamination reactions, which always involve two different pairs of amino acids and their corresponding α-keto acids.
  –Glutamate and α-ketoglutarate usually serve as one of the pairs. Pyridoxal phosphate is the cofactor.
- Nitrogen may be removed from glutamate by glutamate dehydrogenase; from glutamine by glutaminase; from histidine by histidase; from serine and threonine by a dehydratase; and from asparagine by asparaginase. Ammonium ions also may be removed from amino acids by the purine nucleotide cycle.
- Glutamate is a pivotal compound in amino acid metabolism.

A. Transamination reactions (Figure 7-3)

–Transamination involves the transfer of an amino group from one amino acid (which is converted to its corresponding α-keto acid) to an **α-keto acid** (which is converted to its corresponding α-amino acid). Thus, the nitrogen from one amino acid appears in another amino acid.

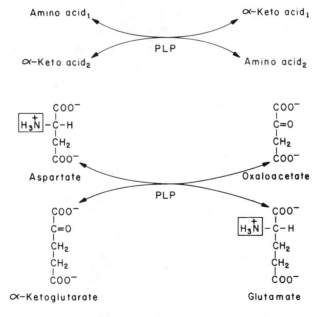

Figure 7-3. Transamination.

1. The enzymes that catalyze transamination reactions are known as **transaminases or aminotransferases**.

2. **Glutamate and α-ketoglutarate** are often involved in transamination reactions, serving as one of the amino acid/α-keto acid pairs.

3. Transamination reactions are readily reversible and can be used in the **synthesis** or the **degradation** of amino acids.

4. Most amino acids participate in transamination reactions. **Lysine** is an exception; it **is not transaminated**.

5. Pyridoxal phosphate (PLP) serves as the cofactor for transamination reactions. PLP is derived from vitamin B_6.

B. Removal of amino acid nitrogen as ammonia

–A number of amino acids undergo reactions in which their nitrogen is released as ammonia or ammonium ion (NH_4^+).

1. **Glutamate dehydrogenase** catalyzes the oxidative deamination of glutamate (Figure 7-4). Ammonium ion is released, and α-ketoglutarate is formed. The glutamate dehydrogenase reaction, which is readily reversible, requires NAD or NADP.

2. **Histidine** is deaminated by histidase to form NH_4^+ and urocanate.

3. **Serine** and **threonine** are deaminated by serine dehydratase, which requires pyridoxal phosphate. Serine is converted to pyruvate, and threonine to α-ketobutyrate; NH_4^+ is released.

4. The amide groups of **glutamine** and **asparagine** are released as ammonium ions by hydrolysis. Glutaminase converts glutamine to glutamate and NH_4^+. Asparaginase converts asparagine to aspartate and NH_4^+.

5. The **purine nucleotide cycle** serves to release NH_4^+ from amino acids, particularly in muscle.

    a. Glutamate collects nitrogen from other amino acids and transfers it to aspartate by a transamination reaction.

    b. Aspartate reacts with inosine monophosphate (IMP) to form AMP and generate fumarate.

    c. NH_4^+ is released from AMP, and IMP is re-formed.

C. The role of glutamate

1. **Glutamate plays a key role in removing nitrogen** from amino acids.

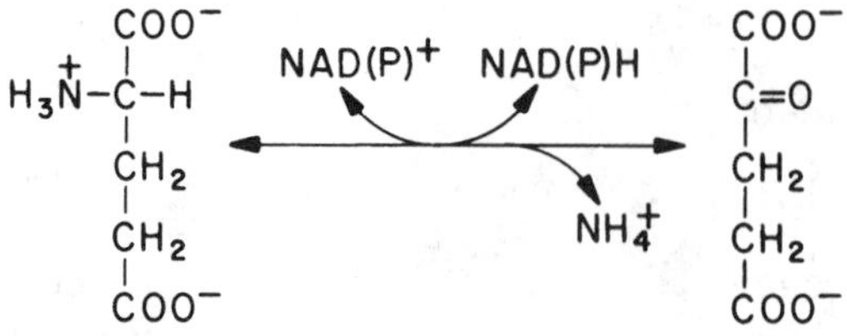

Figure 7-4. The reaction catalyzed by glutamate dehydrogenase. This reaction is readily reversible and can use either NAD^+ or $NADP^+$ as a cofactor.

   a. Glutamate collects nitrogen from other amino acids by means of transamination reactions.

   b. The nitrogen of glutamate may be released as NH_4^+ via the glutamate dehydrogenase reaction.

   c. NH_4^+ and aspartate (which may be produced from glutamate by transamination of oxaloacetate) provide nitrogen for urea synthesis via the urea cycle.

 2. Glutamate provides nitrogen for synthesis of many amino acids.

   a. NH_4^+ may provide the nitrogen for amino acid synthesis by reacting with α-ketoglutarate to form glutamate in the glutamate dehydrogenase reaction.

   b. Glutamate may transfer nitrogen by transamination reactions to α-keto acids to form their corresponding α-amino acids.

III. The Urea Cycle

- Ammonia, which is very toxic in humans, is converted to urea, which is nontoxic, very soluble, and readily excreted by the kidneys.
- Urea is formed in the urea cycle from NH_4^+, CO_2, and the nitrogen of aspartate. The cycle occurs mainly in the liver.
- NH_4^+, CO_2, and ATP react to form carbamoyl phosphate, which reacts with ornithine to form citrulline.
- Citrulline reacts with aspartate to form argininosuccinate, which releases fumarate to form arginine.
- The cleavage of arginine by arginase regenerates ornithine and releases urea.
- The enzymes of the urea cycle are induced if a high-protein diet is consumed for several days.
- When the nitrogen of amino acids is converted to urea in the liver, their carbon skeletons are converted either to glucose (in the fasting state) or to fatty acids (in the fed state).

A. Transport of nitrogen to the liver

  –**Ammonia** is very toxic, particularly affecting the central nervous system.

  1. The concentration of ammonia and ammonium ions in the blood is normally very low.

  2. Ammonia travels to the **liver** from other tissues mainly in the form of **alanine and glutamine**.

   a. NH_4^+ **and aspartate,** the forms in which nitrogen enters the urea cycle, are produced from amino acids in the liver by a series of transamination and deamination reactions.

   b. Glutamate dehydrogenase is a key enzyme in this process.

B. Reactions of the urea cycle (Figure 7-5)

  –NH_4^+ **and aspartate** provide the nitrogen that is used to produce urea, and CO_2 provides the carbon. Ornithine serves as a carrier that is regenerated by the cycle.

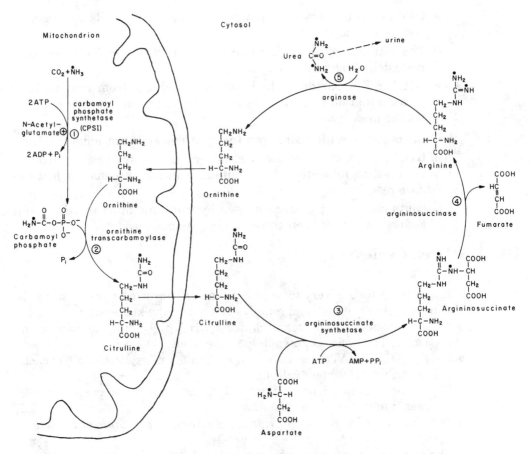

Figure 7-5. The urea cycle. The solid black dots (●) indicate the nitrogens from which urea is formed. *Numbers* correspond to the steps described in the text in section III B.

1. **Carbamoyl phosphate** is synthesized in the first reaction from NH_4^+, CO_2, and two ATP. Inorganic phosphate and two ADP are also produced.

   —Enzyme: **carbamoyl phosphate synthetase I,** which is located in mitochondria and is activated by *N*-acetylglutamate.

2. **Ornithine** reacts with carbamoyl phosphate to form citrulline. Inorganic phosphate is released.

   —Enzyme: **ornithine transcarbamoylase,** which is found in mitochondria. The product, citrulline, is transported to the cytosol.

3. **Citrulline** combines with aspartate to form argininosuccinate in a reaction that is driven by the hydrolysis of ATP to AMP and inorganic pyrophosphate.

   —Enzyme: **argininosuccinate synthetase**

4. **Argininosuccinate** is cleaved to form arginine and fumarate.

   —Enzyme: **argininosuccinase.** This reaction occurs in the cytosol.

   a. The carbons of fumarate, which are derived from the aspartate added in reaction 3, may be converted to malate.

   b. In the fasting state in the liver, malate may be converted to glucose.

   c. In the fed state in the liver, malate may be converted by the malic enzyme to **pyruvate,** which serves as a **carbon source** for the synthesis of fatty acids.

   d. Malate may also be converted to oxaloacetate, which is transaminated to regenerate the aspartate required for reaction 3.

5. Arginine is cleaved to form urea and regenerate ornithine.

   –Enzyme: **arginase,** which is located primarily in the liver and is inhibited by ornithine.

   a. Urea passes into the blood and is excreted by the kidneys.

   b. The urea excreted each day by a healthy adult (about 30 g) accounts for about 90% of the nitrogenous excretory products.

6. Ornithine is transported back into the mitochondrion where it can be utilized for another round of the cycle.

   a. When the cell requires additional **ornithine,** it may be synthesized from glucose via glutamate.

   b. Thus, **arginine** may be synthesized from glucose via ornithine and the first four reactions of the urea cycle.

C. Regulation of the urea cycle

   1. *N*-**Acetylglutamate** is an activator of carbamoyl phosphate synthetase I, the first enzyme of the urea cycle.

   2. The synthesis of *N*-**acetylglutamate** from acetyl CoA and glutamate is stimulated by arginine.

   3. Although the liver normally has a great capacity for urea synthesis, the enzymes of the urea cycle may be induced if a high-protein diet is consumed for 4 or more days.

IV. Synthesis and Degradation of Amino Acids

- Of the 20 amino acids commonly found in proteins, 11 are not essential in the adult diet because they can be synthesized in the body.
- Ten of the nonessential amino acids contain carbon skeletons that may be derived from glucose (Figure 7-6).
- Tyrosine is produced by hydroxylation of the essential amino acid phenylalanine.
- The major products obtained by degradation of the carbon skeletons of the amino acids are pyruvate, intermediates of the TCA cycle, acetyl CoA, and acetoacetate (Figure 7-7).

A. Synthesis of amino acids

   –Messenger RNA contains codons for 20 amino acids. Eleven of these amino acids can be synthesized in the body. The carbon skeletons of 10 of these amino acids can be derived from **glucose.**

   1. Amino acids derived from intermediates of glycolysis (Figure 7-8)

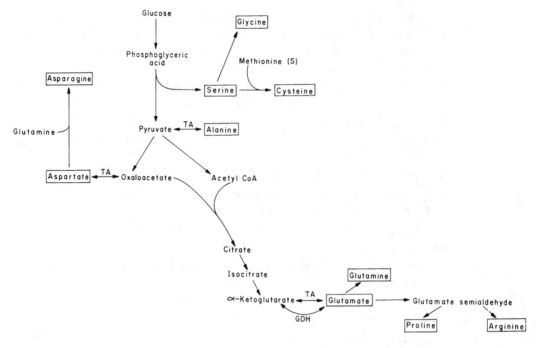

Figure 7-6. Overview of the synthesis of nonessential amino acids. Ten amino acids may be produced from glucose via intermediates of glycolysis or the TCA cycle. The eleventh amino acid, tyrosine, is synthesized by hydroxylation of the essential amino acid phenylalanine. *TA* = transamination; *GDH* = glutamate dehydrogenase.

   a. Intermediates of glycolysis serve as precursors for serine, glycine, cysteine, and alanine.

   b. Serine can be synthesized from the glycolytic intermediate 3-phosphoglyceric acid, which is oxidized, transaminated by glutamate, and dephosphorylated.

   c. Glycine and cysteine can be derived from serine.

      (1) Glycine can be produced from serine by a reaction in which a methylene group is transferred to tetrahydrofolate.

      (2) Cysteine derives its carbon and nitrogen from serine. The essential amino acid **methionine** supplies the sulfur.

   d. Alanine can be derived by transamination of pyruvate.

2. Amino acids derived from TCA cycle intermediates (see Figure 7-6)

   a. Aspartate can be derived from oxaloacetate by transamination.

   b. Asparagine is produced from aspartate by amidation.

   c. Glutamate is derived from α-ketoglutarate by the addition of NH_4^+ via the glutamate dehydrogenase reaction or by transamination (Figure 7-9).

   d. Glutamine, proline, and arginine can be derived from glutamate.

      (1) Glutamine is produced by amidation of glutamate.

      (2) Proline and arginine can be derived from **glutamate semialdehyde,** which is formed by reduction of glutamate.

        –**Proline** can be produced by cyclization of glutamate semialdehyde.

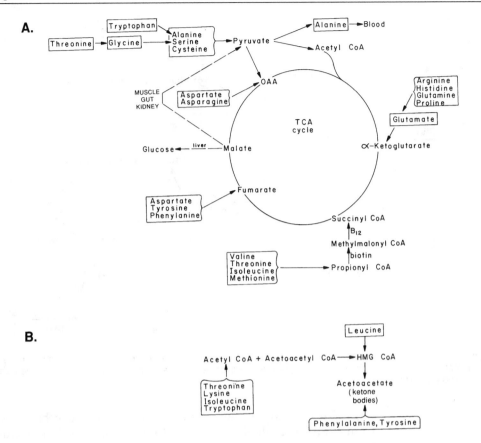

Figure 7-7. Degradation of amino acids. (*A*) Amino acids that produce pyruvate or intermediates of the TCA cycle. (*B*) Amino acids that produce acetyl CoA or ketone bodies. *OAA* = oxaloacetate; *HMG CoA* = hydroxymethylglutaryl CoA.

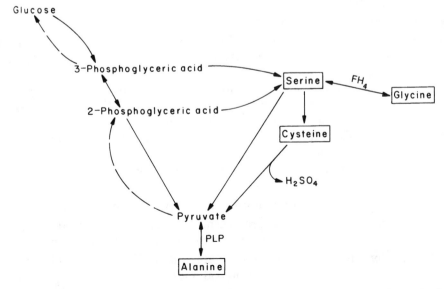

Figure 7-8. Amino acids derived from intermediates of glycolysis. These amino acids may be synthesized from glucose and can be reconverted to glucose in the liver.

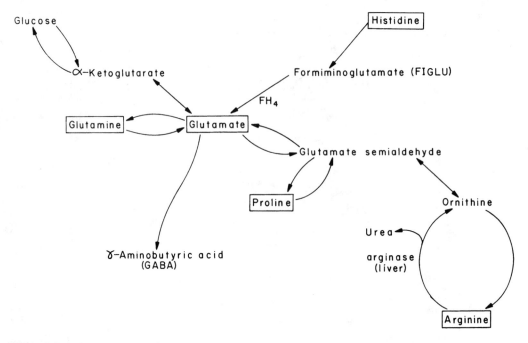

Figure 7-9. Amino acids related through glutamate. These amino acids contain carbons that can be converted to glutamate, which can be converted to glucose in the liver. All of these amino acids except histidine can be synthesized from glucose.

  –**Arginine,** via three reactions of the urea cycle, can be derived from ornithine, which is produced by transamination of glutamate semialdehyde.

  3. **Tyrosine,** the eleventh nonessential amino acid, is synthesized by hydroxylation of the essential amino acid phenylalanine in a reaction that requires tetrahydrobiopterin.

B. Degradation of amino acids

  –When the carbon skeletons of amino acids are degraded, the major products are pyruvate, intermediates of the TCA cycle, acetyl CoA, and acetoacetate (see Figure 7-7).

  –Amino acids that form pyruvate or intermediates of the TCA cycle in the liver are **glucogenic** (or gluconeogenic); that is, they provide carbon for the synthesis of glucose (see Figure 7-7 *A*).

  –Amino acids that form acetyl CoA or acetoacetate are **ketogenic;** that is, they form ketone bodies (see Figure 7-7 *B*).

  –Some amino acids (isoleucine, tryptophan, phenylalanine, and tyrosine) are both glucogenic and ketogenic.

  1. **Amino acids that are converted to pyruvate** (see Figure 7-8).

    –The amino acids that are synthesized from intermediates of glycolysis (serine, glycine, cysteine, and alanine) are degraded to form pyruvate.

    a. **Serine** is converted to pyruvate and NH_4^+ by serine dehydratase, an enzyme that requires pyridoxal phosphate.

    b. **Glycine,** in a reversal of the reaction utilized for its synthesis, reacts with methylene tetrahydrofolate to form serine.

(1) Glycine also reacts with tetrahydrofolate and NAD^+ to produce CO_2 and NH_4^+.

(2) Glycine can be converted to glyoxylate, which can be oxidized to CO_2 and H_2O or converted to oxalate.

c. Cysteine forms pyruvate. Its sulfur, which was derived from methionine, is converted to H_2SO_4.

d. Alanine can be transaminated to pyruvate.

2. Amino acids that are converted to intermediates of the TCA cycle (see Figure 7-7).

–Carbons from four groups of amino acids form the TCA cycle intermediates α-ketoglutarate, succinyl CoA, fumarate, and oxaloacetate.

a. Amino acids related to glutamate can form **α-ketoglutarate** (see Figure 7-9).

(1) Glutamine can be converted by glutaminase to glutamate with the release of its amide nitrogen as NH_4^+.

(2) Proline can be oxidized so that its ring opens and glutamate is formed.

(3) Arginine can be cleaved by arginase in the liver to form urea and ornithine. Ornithine can be transaminated to glutamate semialdehyde, which can be oxidized to glutamate.

(4) Histidine can be converted to formiminoglutamate (FIGLU). The formimino group is transferred to tetrahydrofolate (FH_4), and the remaining five carbons form glutamate.

(5) Glutamate can be deaminated by glutamate dehydrogenase or transaminated to form α-ketoglutarate.

b. Amino acids that form succinyl CoA (Figure 7-10)

–Four amino acids are converted to **propionyl CoA,** which is carboxylated in a biotin-requiring reaction to form methylmalonyl CoA, which is rearranged to form succinyl CoA in a reaction that requires vitamin B_{12}.

(1) Threonine is converted by a dehydratase to NH_4^+ and α-ketobutyrate, which is oxidatively decarboxylated to propionyl CoA.

–In a different reaction, threonine may be converted to glycine and acetyl CoA.

(2) Methionine provides methyl groups for the synthesis of various compounds; its sulfur is incorporated into **cysteine;** and the remaining carbons form **succinyl CoA.**

–Methionine and ATP form **S-adenosylmethionine (SAM),** which donates a methyl group and forms homocysteine.

–**Homocysteine** is reconverted to methionine by accepting a methyl group from the tetrahydrofolate pool via vitamin B_{12}.

–**Homocysteine** can also react with serine to form **cystathionine.** The cleavage of cystathionine produces cysteine, NH_4^+, and α-ketobutyrate, which is converted to propionyl CoA.

(3) Valine and isoleucine, two of the three branched-chain amino acids, form succinyl CoA (see Figure 7-10)

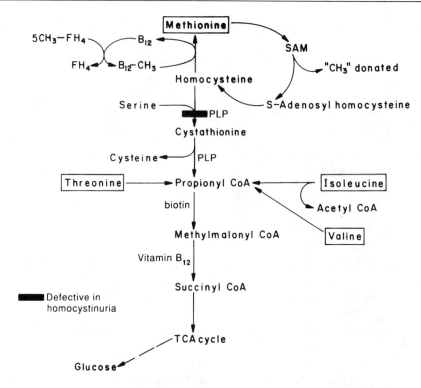

Figure 7-10. Amino acids that can be converted to succinyl CoA. The amino acids methionine, threonine, isoleucine, and valine, which form succinyl CoA via methylmalonyl CoA, are all essential. The carbons of serine are converted to cysteine and do not form succinyl CoA by this pathway. A defect in cystathionine synthase (■■) causes homocystinuria. *SAM* = *S*-adenosylmethionine; *PLP* = pyridoxal phosphate.

–Degradation of all three branched-chain amino acids begins with a transamination followed by an **oxidative decarboxylation** catalyzed by the branched-chain α-keto acid dehydrogenase complex. This enzyme, like α-ketoglutarate dehydrogenase, requires thiamine pyrophosphate, lipoic acid, coenzyme A, FAD, and NAD$^+$ (Figure 7-11).

–**Valine** is eventually converted to succinyl CoA via propionyl CoA and methylmalonyl CoA.

–**Isoleucine** also forms succinyl CoA after two of its carbons are released as acetyl CoA.

c. **Amino acids that form fumarate**

–Three amino acids (phenylalanine, tyrosine, and aspartate) are converted to fumarate (see Figure 7-7)

(1) **Phenylalanine is converted to tyrosine** by phenylalanine hydroxylase in a reaction requiring tetrahydrobiopterin and O_2.

(2) **Tyrosine,** obtained from the diet or by hydroxylation of phenylalanine, is converted to homogentisate, whose aromatic ring is opened and cleaved, **forming fumarate and acetoacetate** (Figure 7-12).

(3) **Aspartate** is converted to fumarate via reactions of the **urea cycle** and the **purine nucleotide cycle**.

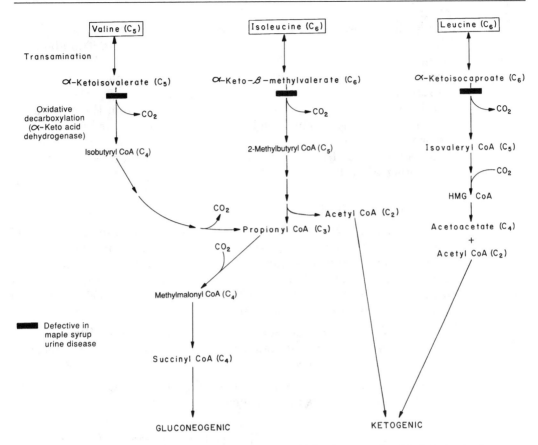

Figure 7-11. Degradation of the branched-chain amino acids. Valine forms propionyl CoA. Isoleucine forms propionyl CoA and acetyl CoA. Leucine forms acetoacetate and acetyl CoA.

–Aspartate reacts with IMP to form AMP and fumarate in the purine nucleotide cycle.

d. Amino acids that form oxaloacetate (see Figure 7-7)

   (1) Aspartate is transaminated to form oxaloacetate.

   (2) Asparagine loses its amide nitrogen as NH_4^+, forming aspartate in a reaction catalyzed by asparaginase.

3. Amino acids that are converted to acetyl CoA or acetoacetate (see Figure 7-12)

   –Four amino acids (lysine, threonine, isoleucine, and tryptophan) can form acetyl CoA, and phenylalanine and tyrosine form acetoacetate. Leucine is degraded to form both acetyl CoA and acetoacetate.

V. Interrelationships of Various Tissues in Amino Acid Metabolism

- During fasting, amino acids from muscle protein are released into the blood, predominantly as alanine and glutamine. (Figure 7-13).
- The branched-chain amino acids are oxidized by muscle to produce energy. Some of the carbons are converted to glutamine and alanine.

Alanine is also produced by the glucose–alanine cycle.

● The gut takes up glutamine from the blood and converts it to alanine, citrulline, and ammonia, which are released.

● The kidney takes up glutamine from the blood and releases ammonia into the urine and alanine and serine into the blood.

● The liver takes up alanine and other amino acids from the blood and converts the nitrogen to urea and the carbons to glucose and ketone bodies, which are released into the blood and oxidized by tissues for energy.

● Thus, amino acids from muscle protein serve as a source of energy for many other tissues.

A. Amino acid metabolism in muscle

–During **fasting,** amino acids are released from muscle protein. Some of the amino acids are partially oxidized in muscle and their remaining carbons are converted to other amino acids. Thus, although about **50%** of the amino acids released into the blood from muscle are **alanine** and **glutamine,** these two amino acids constitute much less than 50% of the total amino acid residues in muscle protein.

1. **Branched-chain amino acids** are oxidized in muscle. Some of their carbons are converted to glutamine and alanine before they are released into the blood (see Figure 7-13).

   a. **Valine** and **isoleucine** produce succinyl CoA, which feeds into the TCA cycle and forms malate, generating energy.

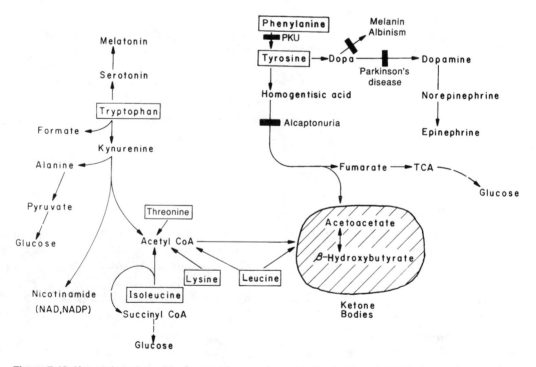

Figure 7-12. Ketogenic amino acids. Some of these amino acids (tryptophan, phenylalanine, and tyrosine) also contain carbons that can form glucose. Leucine and lysine are strictly ketogenic; they do not form glucose. A deficiency in various steps (■■) leads to the diseases indicated. *PKU* = phenylketonuria.

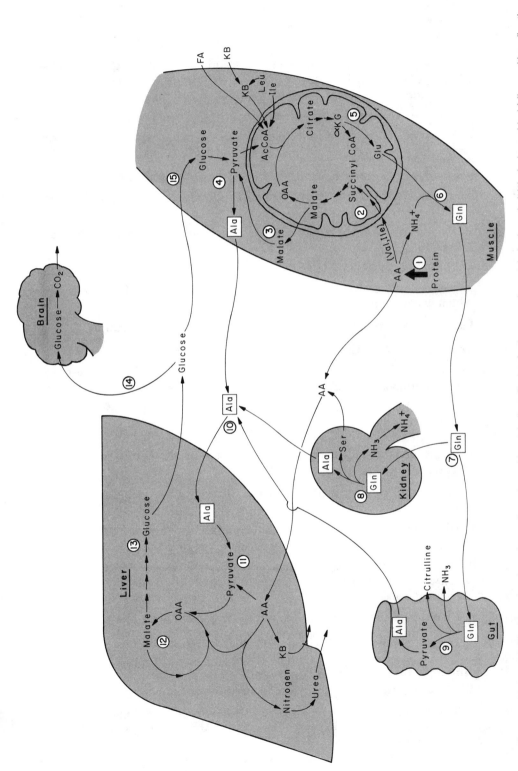

Figure 7-13. Interrelationships of various tissues in amino acid metabolism. During fasting, muscle protein is degraded (1). Amino acids (*AA*) are either directly released into the blood or converted to alanine (2, 3, 4) or glutamine (2, 5, 6) and then released into the blood. Glutamine is converted by the kidney to alanine and serine (8) or by the gut to alanine (9). Alanine is taken up by the liver (10). The liver converts the carbons of alanine and other amino acids to glucose (11, 12, 13) or to ketone bodies (*KB*) and the nitrogens to urea. Glucose is oxidized by tissues such as the brain (14) or muscle (15). Muscle can oxidize glucose, forming alanine, which is reconverted to glucose in the liver (the glucose–alanine cycle). *OAA* = oxaloacetate; *AcCoA* = acetyl CoA; *αKG* = α-ketoglutarate; *FA* = fatty acid.

      b. Malate may be converted by the malic enzyme to pyruvate, which is transaminated to alanine or oxidatively decarboxylated to acetyl CoA.

      c. Malate may also continue around the TCA cycle to α-ketoglutarate, generating additional energy.

      d. α-Ketoglutarate forms glutamate, which produces glutamine.

  2. The **alanine** released by muscle is also produced by the **glucose–alanine cycle,** which involves the transport of glucose from the liver to muscle and the return of carbon atoms to the liver as alanine.

      a. Glucose is oxidized in muscle to **pyruvate,** producing energy.

      b. Pyruvate can be transaminated to **alanine,** which travels to the liver carrying nitrogen for urea synthesis and carbon for gluconeogenesis.

B. Amino acid metabolism in the gut (see Figure 7-13)

  –The gut takes up glutamine and releases alanine, citrulline, and ammonia.

  1. In the gut, **glutamine** is converted to NH_4^+ and glutamate, which forms α-ketoglutarate.

  2. α-Ketoglutarate is converted to malate, generating energy. In the cytosol, malate is decarboxylated by the malic enzyme to form pyruvate, which is transaminated to alanine.

  3. Glutamate also is converted to ornithine, which forms citrulline.

C. Amino acid metabolism in the kidney (see Figure 7-13)

  –The kidney takes up **glutamine,** which is deaminated by glutaminase, forming ammonia and glutamate, which is converted to alanine and serine.

  1. Ammonia is released into the **urine,** where it buffers the hydrogen ions produced by phosphoric acid, sulfuric acid (produced from cysteine), and various metabolic acids (e.g., lactic acid and the ketone bodies, acetoacetic acid and β-hydroxybutyric acid).

  2. Alanine and **serine** produced from glutamate are released into the **blood**.

      a. Glutamate is deaminated or transaminated to form α-ketoglutarate, which enters the TCA cycle and is converted to malate.

      b. Malate enters the cytosol and is oxidized to oxaloacetate, which is converted by phosphoenolpyruvate carboxykinase to phosphoenolpyruvate (PEP).

      c. Phosphoenolpyruvate feeds into glycolysis and, along with glycolytic intermediates produced from glucose, forms alanine and serine which are released into the blood.

D. Amino acid metabolism in the liver (see Figure 7-13)

  –The liver takes up alanine, serine, and other amino acids from the blood and converts their nitrogen to urea and their carbons to glucose or ketone bodies, which are released into the blood and oxidized by other tissues.

VI. Tetrahydrofolate, Vitamin B_{12}, and S-Adenosylmethionine

- Groups containing a single carbon atom may be transferred from one compound to another.
- One-carbon groups at lower levels of oxidation than CO_2 (which is transferred by biotin) are transformed by tetrahydrofolate (FH_4), vitamin B_{12}, and S-adenosylmethionine.
- FH_4, which is produced from the vitamin folate, obtains one-carbon groups from serine, glycine, histidine, formaldehyde, and formate. The one-carbon groups may be oxidized and reduced while they are attached to FH_4.
- The one-carbon groups carried by FH_4 are transferred to dUMP to form dTMP; to glycine to form serine; to purine precursors to form C2 and C8 of the purine ring; and to vitamin B_{12}.
- Vitamin B_{12} is involved in two reactions in the body. It is used in the rearrangement of the methyl group of methylmalonyl CoA to form succinyl CoA, and it transfers a methyl group from 5-methyl FH_4 to homocysteine to form methionine.
- S-Adenosylmethionine (SAM), which is produced from methionine and ATP, is involved in the transfer of methyl groups to compounds such as creatine, phosphatidylcholine, epinephrine, melatonin, and methylated polynucleotides.

A. Tetrahydrofolate

1. The nature of tetrahydrofolate (FH_4) and its derivatives

    a. FH_4 cannot be synthesized in the body. It is produced from the vitamin folate (Figure 7-14A), which contains a pterin ring, *p*-aminobenzoic acid, and at least one glutamate residue. Additional glutamate residues are added in the body.

        –**NADPH and dihydrofolate reductase** convert folate to dihydrofolate (FH_2), which may undergo a second reduction by the same enzyme to form FH_4.

    b. The **one-carbon groups** of FH_4 may be oxidized and reduced (Figure 7-14B).

        (1) N^{10}-formyl-FH_4 releases H_2O to form the N^5,N^{10}-methenyl (or methylidyne) derivative.

        (2) N^5,N^{10}-methenyl-FH_4 may be reduced by NADPH to N^5,N^{10}-methylene-FH_4 in a reversible reaction.

        (3) The methylene derivative may be reduced to N^5-methyl-FH_4, which is not reoxidized under physiological conditions.

2. Sources of one-carbon groups carried by FH_4 (Table 7-1)

    –Serine, glycine, formaldehyde, histidine, and formate transfer one-carbon groups to FH_4 (Figure 7-15, *top*).

    a. Serine, glycine, and formaldehyde produce N^5,N^{10}-methylene-FH_4.

        (1) Serine transfers a one-carbon group to FH_4 and is converted to glycine in a reversible reaction. Because serine may be derived from glucose, this one-carbon group can be obtained from dietary carbohydrate.

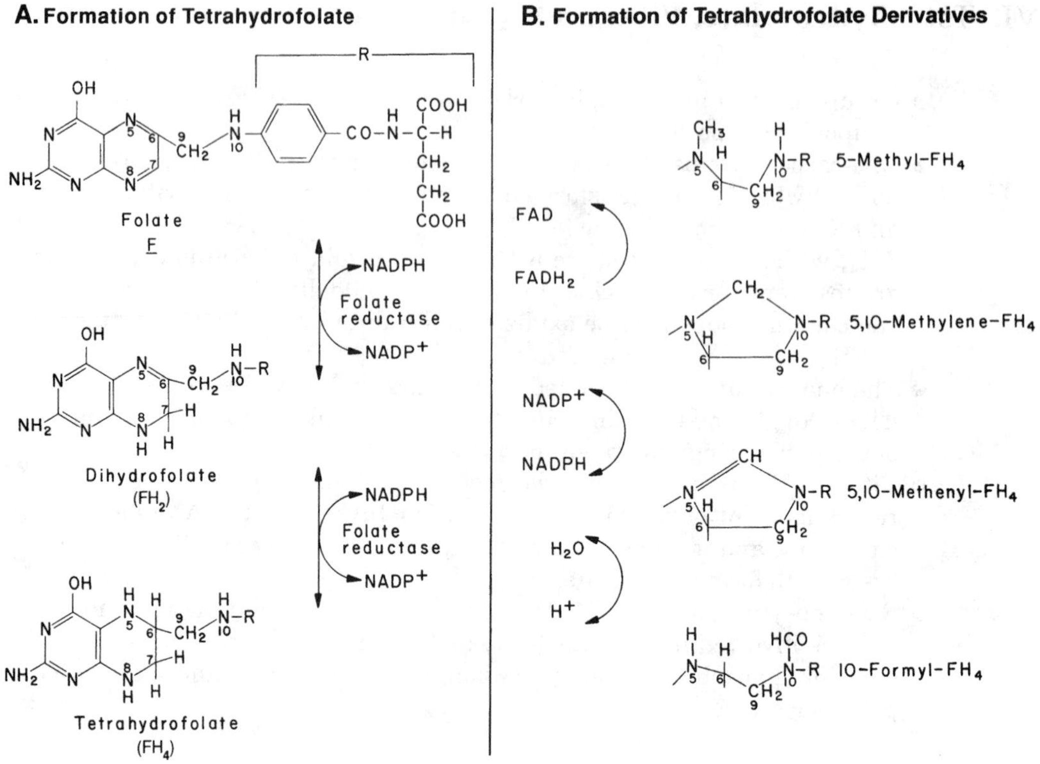

A. Formation of Tetrahydrofolate

B. Formation of Tetrahydrofolate Derivatives

Figure 7-14. Tetrahydrofolate (FH_4). (*A*) Reduction of folate by folate reductase. (*B*) The one-carbon groups carried by FH_4. Only atoms 5, 6, 9, and 10 of FH_4 are shown. The structure of R is given in the folate molecule shown in (*A*).

(2) When glycine transfers a one-carbon unit to FH_4, NH_4^+ and CO_2 are produced.

b. Histidine is degraded to formiminoglutamate (FIGLU). The formimino group reacts with FH_4, releasing NH_4^+ and producing glutamate and N^5,N^{10}-methenyl-FH_4.

c. Formate, which may be derived from tryptophan, produces N^{10}-formyl-FH_4.

3. Recipients of one-carbon groups (see Table 7-1)

–The one-carbon groups that tetrahydrofolate receives are transferred to various compounds (Figure 7-15, *bottom*).

Table 7-1. Sources and Recipients of Carbon for Forms of Tetrahydrofolate (FH_4)

Source	Form of FH_4	Recipient	Product
Formate	N^{10}-Formyl	Purine precursor	Purine (C2)
Histidine (via formiminoglutamate)	N^5,N^{10}-Methenyl (N^5,N^{10}-Methylidyne)	Purine precursor	Purine (C8)
Serine Glycine Formaldehyde	N^5,N^{10}-Methylene	dUMP Glycine	dTMP Serine
Reduction of N^5,N^{10}-Methylene-FH_4	N^5-Methyl	Vitamin B_{12}	Methyl-B_{12}

Sources of Carbon

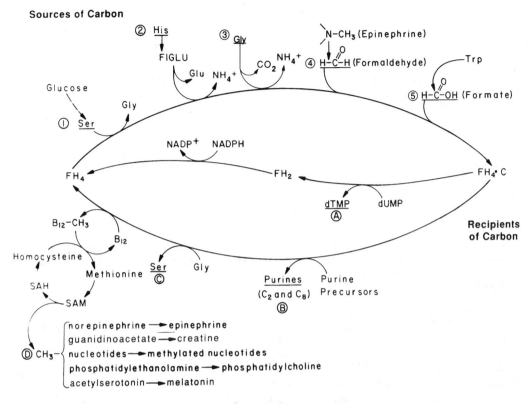

Figure 7-15. The sources of carbon (1–5) for the FH$_4$ pool and the recipients of carbon (A–D) from the pool. *FH$_4$•C* = FH$_4$ carrying a one-carbon unit. The FH$_4$ derivatives involved are listed in Table 7-1.

a. N^{10}-formyl-FH$_4$ provides C2 for synthesis of the **purines,** and N^5,N^{10}-methenyl-FH$_4$ provides C8. Purines are required for DNA and RNA synthesis.

b. N^5,N^{10}-methylene-FH$_4$ transfers a one-carbon group to deoxyuridine monophosphate (dUMP) to form deoxythymidine monophosphate (dTMP), which provides the **thymine** for DNA synthesis (Figure 7-16).

    (1) The methylene group is reduced to a methyl group in this reaction, and tetrahydrofolate is oxidized to dihydrofolate.

    (2) FH$_4$ is regenerated by the reduction of FH$_2$ in the NADPH-requiring reaction catalyzed by dihydrofolate reductase.

c. N^5,N^{10}-methylene-FH$_4$ provides the one-carbon group for the conversion of glycine to **serine.**

d. N^5-methyl-FH$_4$ transfers its methyl group to vitamin B$_{12}$ to form **methyl-B$_{12}$**. The methyl group is transferred from vitamin B$_{12}$ to homocysteine to form methionine (see Figure 7-15, *bottom*).

B. Vitamin B$_{12}$

1. Source of vitamin B$_{12}$

    a. Vitamin B$_{12}$ is produced by microorganisms, but it is not present in plants.

    b. Animals obtain vitamin B$_{12}$ from their intestinal flora or by consuming the tissues of other animals.

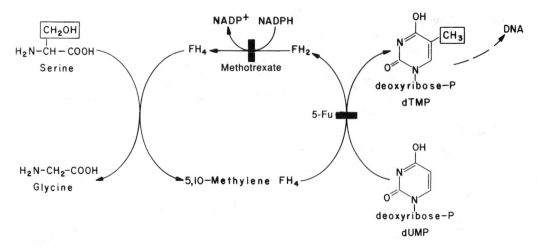

Figure 7-16. The transfer of a one-carbon unit from serine to deoxyuridine monophosphate (dUMP) to form deoxythymidine monophosphate (dTMP). FH_4 is oxidized to FH_2 (dihydrofolate) in this reaction. FH_2 is reduced to FH_4 by dihydrofolate reductase. ▬ indicate the steps at which the antimetabolites methotrexate and 5-fluorouracil (5-FU) act.

    c. Intrinsic factor, produced by gastric parietal cells, is required for absorption of vitamin B_{12} by the intestine.

    d. Vitamin B_{12} is stored and efficiently recycled in the body.

2. Functions of vitamin B_{12}

    –Vitamin B_{12} contains **cobalt** in a corrin ring that resembles a porphyrin. In the cofactor forms of the vitamin, an adenosyl moiety or methyl group is attached to the cobalt, forming adenosylcobalamin or methylcobalamin (Figure 7-17).

    a. Adenosylcobalamin is the cofactor for methylmalonyl CoA mutase, which catalyzes the rearrangement of methylmalonyl CoA to succinyl CoA (see Figure 7-10).

    –This reaction is involved in the production of succinyl CoA from valine, isoleucine, threonine, methionine, thymine, and the propionate formed by oxidation of fatty acids with an odd number of carbons.

    b. Methylcobalamin is involved in the transfer of methyl groups from FH_4 to **homocysteine** to form methionine (see Figure 7-10).

C. S-Adenosylmethionine

    1. *S*-Adenosylmethionine (SAM) is synthesized from methionine and ATP.

    2. Methyl groups are supplied by SAM for the following conversions (see Figure 7-15):

        a. Guanidinoacetate to **creatine**

        b. Phosphatidylethanolamine to **phosphatidylcholine**

        c. Norepinephrine to **epinephrine**

        d. Acetylserotonin to **melatonin**

        e. Polynucleotides to **methylated polynucleotides**

    3. When *S*-adenosylmethionine transfers its methyl group to an acceptor, *S*-adenosylhomocysteine is produced.

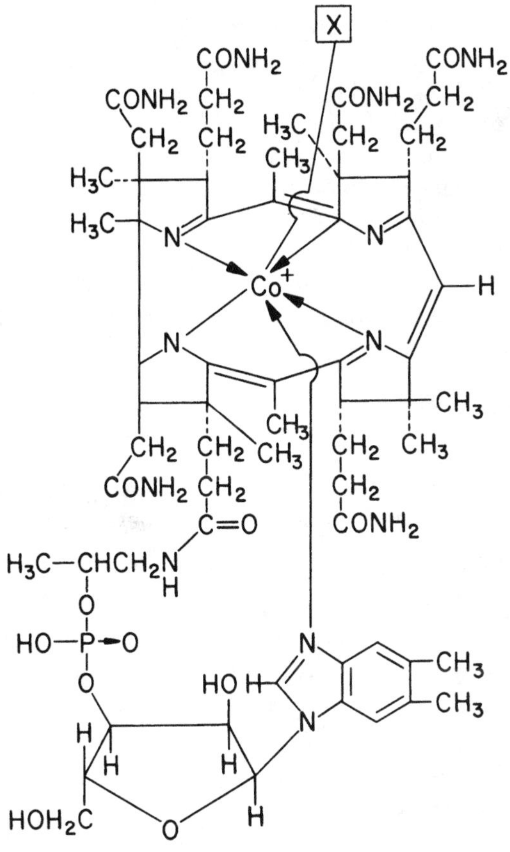

Figure 7-17. Vitamin B_{12}. *X* can be an adenosyl moiety or a methyl group in the coenzyme forms of the vitamin.

   4. *S*-Adenosylhomocysteine releases adenosine to form homocysteine, which can obtain a methyl group from vitamin B_{12} to form methionine. Methionine can react with ATP to regenerate SAM (see Figure 7-10).

VII. Special Products Derived from Amino Acids

- Amino acids are used to synthesize many nitrogen-containing compounds in the body.
- Creatine is produced from glycine, the guanidinium group of arginine, and the methyl group of *S*-adenosylmethionine.
  - Creatine phosphate, a storage form for high-energy phosphate, is produced from creatine and ATP. It spontaneously cyclizes to form creatinine, which is excreted in the urine.
- γ-Aminobutyric acid (GABA) is formed by decarboxylation of glutamate, histamine by decarboxylation of histidine, and putrescine (from which spermine and spermidine are formed) by decarboxylation of ornithine.
- Sphingosine, which is used for the synthesis of sphingolipids, is produced from serine and palmitoyl CoA.
- Serotonin and melatonin are derived from tryptophan, as is the nicotinamide portion of NAD^+ (which is also derived from the vitamin niacin).

- Thyroid hormone, 3,4-dihydroxyphenylalanine (dopa), melanin, dopamine, norepinephrine, and epinephrine are produced from tyrosine.
- During purine biosynthesis, the entire glycine molecule is incorporated into the growing ring structure, glutamine provides N3 and N9, and aspartate provides N1.
- Purines are degraded to form uric acid, a nitrogenous excretory product.
- During pyrimidine biosynthesis, the ring is formed by carbamoyl phosphate and aspartate.
- Heme is produced from glycine and succinyl CoA via a series of porphyrins.
- Heme is degraded to bilirubin, which is excreted in the bile.

A. Creatine (Figure 7-18)

1. Creatine is produced in the liver.

—Glycine combines with arginine to form ornithine and guanidinoacetate, which is methylated by *S*-adenosylmethionine to form creatine.

2. Creatine travels from the liver to other tissues where it is converted to creatine phosphate.

—ATP phosphorylates creatine to form creatine phosphate in a reaction catalyzed by creatine kinase (also known as creatine phosphokinase).

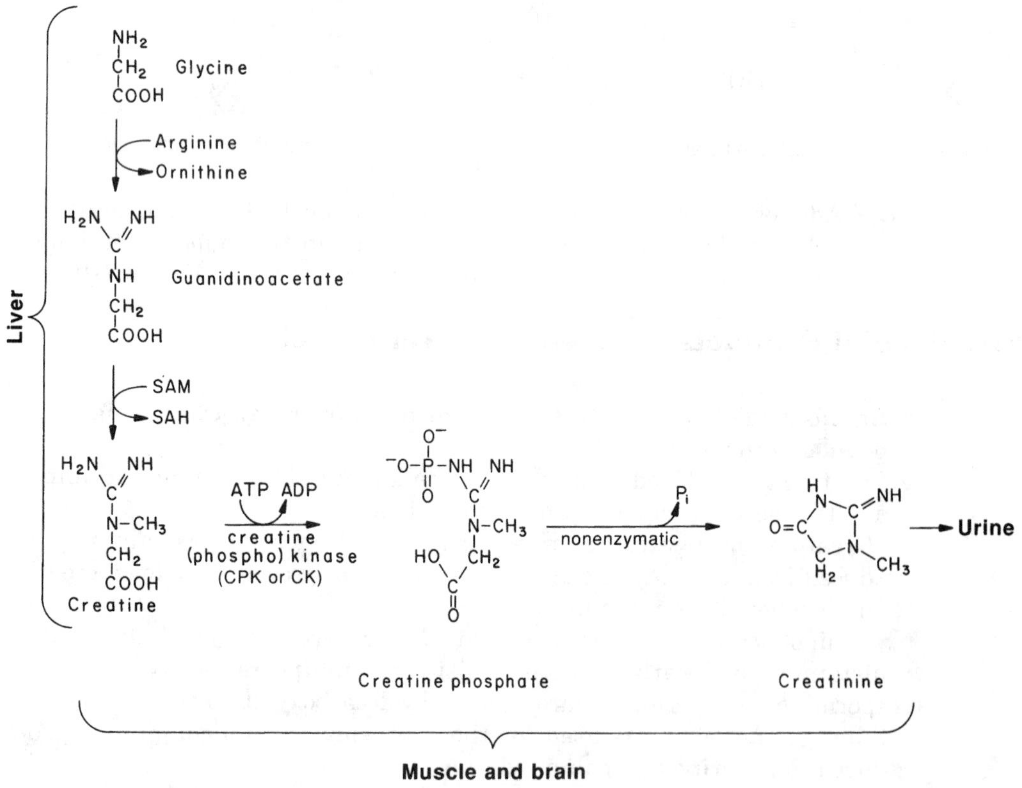

Figure 7-18. The synthesis of creatine phosphate and its spontaneous (nonenzymatic) conversion to creatinine. *SAM* = *S*-adenosylmethionine; *SAH* = *S*-adenosylhomocysteine.

    a. Muscle and brain contain large amounts of **creatine phosphate**.

    b. Creatine phosphate provides a small reservoir of high-energy phosphate that can readily regenerate ATP from ADP. It plays a particularly important role during the early stages of exercise in muscle, where the largest quantities of creatine phosphate are found.

    c. Creatine also transports high-energy phosphate from mitochondria to actomyosin fibers.

  3. Creatine phosphate spontaneously cyclizes, forming **creatinine,** which is **excreted by the kidney.**

    a. The amount of creatinine excreted per day depends on body muscle mass and kidney function and is constant at about 15 millimoles for the average man.

    b. The excretion of creatinine is a constant drain on the methyl pools of the body.

B. Products formed by amino acid decarboxylations

  –Amines are produced by decarboxylation of amino acids in reactions that utilize pyridoxal phosphate as a cofactor.

  1. γ-Aminobutyric acid (**GABA**), an inhibitory neurotransmitter, is produced by decarboxylation of **glutamate** (Figure 7-19).

  2. Histamine is produced by decarboxylation of **histidine**.

    –Histamine causes vasodilation and bronchoconstriction. In the stomach, it stimulates the secretion of HCl.

  3. Putrescine, produced by the decarboxylation of **ornithine,** is converted to spermine and spermidine.

    –Ornithine decarboxylase activity is associated with cell division.

  4. The initial step in **sphingosine** formation involves the condensation of **palmitoyl CoA with serine,** which undergoes a simultaneous decarboxylation.

    –Sphingosine is a precursor of ceramide from which sphingomyelin, cerebrosides, and gangliosides are produced (see Figure 6-15).

  5. Decarboxylations of amino acids are involved in the production of serotonin from tryptophan and of dopamine from tyrosine.

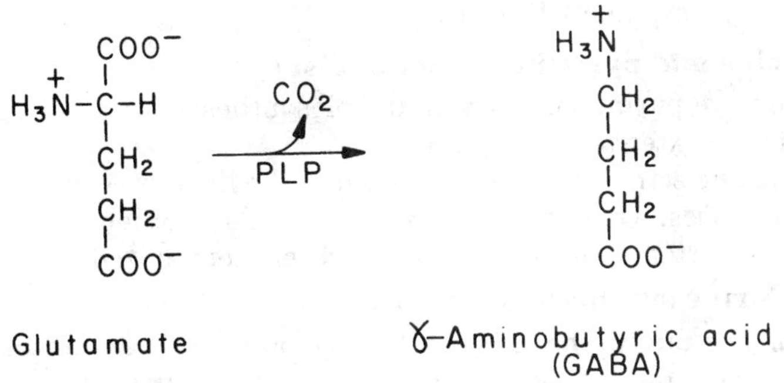

Figure 7-19. The decarboxylation of glutamate to form γ-aminobutyric acid (GABA). *PLP* = pyridoxal phosphate.

C. Products derived from tryptophan

–**Serotonin, melatonin,** and the nicotinamide moiety of **NAD** and **NADP** are formed from tryptophan.

1. Tryptophan is hydroxylated in a tetrahydrobiopterin-requiring reaction similar to the hydroxylation of phenylalanine. The product, 5-hydroxytryptophan, is decarboxylated to form serotonin.

2. **Serotonin** undergoes acetylation by acetyl CoA and methylation by *S*-adenosylmethionine to form melatonin in the pineal gland.

3. Tryptophan can be converted to the nicotinamide moiety of NAD and NADP (see Figure 7-12), although the major precursor of nicotinamide is the vitamin niacin (nicotinic acid). Thus, to a limited extent, tryptophan can spare the dietary requirement for niacin.

D. Products derived from phenylalanine and tyrosine

–Phenylalanine can be hydroxylated to form **tyrosine,** which subsequently can be hydroxylated to form **dopa** (3,4-dihydroxyphenylalanine). Both hydroxylation reactions are catalyzed by mixed-function oxidases, which require tetrahydrobiopterin as a cofactor.

1. The **thyroid hormones,** thyroxine (T_4) and triiodothyronine (T_3), are produced in the thyroid gland from tyrosine residues in thyroglobulin.

2. **Melanins,** which are pigments in skin and hair, are formed by polymerization of oxidation products (quinones) of dopa.

3. The **catecholamines** (dopamine, norepinephrine, and epinephrine) are derived from dopa in a series of reactions (see Figure 7-12).

   –Decarboxylation of dopa forms the neurotransmitter **dopamine.**

   –Hydroxylation of dopamine on its aliphatic chain by an enzyme that requires copper and vitamin C yields the neurotransmitter **norepinephrine.**

   –Methylation of norepinephrine in the adrenal medulla by *S*-adenosylmethionine forms the hormone **epinephrine.**

   a. The catecholamines are inactivated by monoamine oxidase (MAO), which produces NH_4^+ and H_2O_2 and converts the catecholamine to an aldehyde, and by catecholamine *O*-methyltransferase (COMT), which methylates the 3-hydroxy group.

   b. The major urinary excretory product of the deaminated, methylated catecholamines is **VMA** (vanillylmandelic acid, or 3-methoxy-4-hydroxymandelic acid).

E. Purine and pyrimidine metabolism

–De novo purine and pyrimidine biosynthesis occurs in the liver and, to a limited extent, in the brain.

–The nucleotides that are produced in the liver are converted to nucleosides and bases, which travel in red blood cells to other tissues where they are reconverted to nucleotides and further metabolized.

1. **Purine synthesis** (Figure 7-20, *left*)

   a. The **purine base** is synthesized on the **ribose moiety**.

      (1) 5'-Phosphoribosyl-1'-pyrophosphate (**PRPP**), which provides the ribose moiety, reacts with glutamine to form phosphoribosylamine.

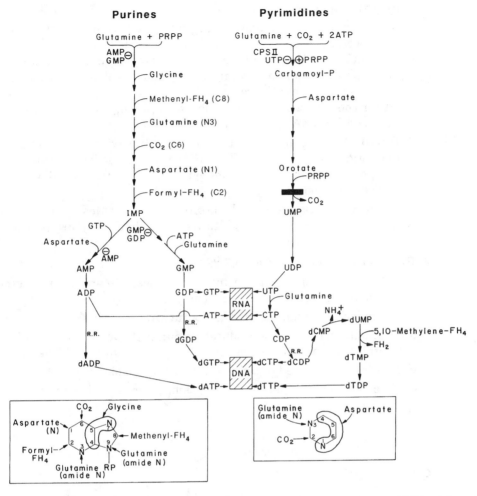

Figure 7-20. De novo synthesis of purines and pyrimidines. Ribonucleotide reductase (*R.R.*) catalyzes the reduction of the ribose moiety in ADP, GDP, and CDP to deoxyribose. The source of each of the atoms is indicated in the boxes at the bottom of the figure. In hereditary orotic aciduria, the enzymes converting orotate to UMP are defective (■■).

–This first step in purine biosynthesis produces N9 of the purine ring and is inhibited by AMP and GMP.

(2) The entire glycine molecule is added to the growing purine precursor. Then C8 is added by methenyl tetrahydrofolate, N3 by glutamine, C6 by CO_2, N1 by aspartate, and C2 by formyl tetrahydrofolate (see Figure 7-20, *bottom*).

(3) Inosine monophosphate (IMP), which contains the base hypoxanthine, is generated.

–IMP is cleaved in the liver. Its free base, or nucleoside, travels to various tissues where it is reconverted to the nucleotide.

b. IMP is the **precursor** of both **AMP** and **GMP**.

(1) Each product, by feedback inhibition, regulates its own synthesis from the IMP branch point as well as inhibits the initial step in the pathway.

(2) AMP and GMP can be phosphorylated to the triphosphate level.

(3) The nucleotide triphosphates can be used for energy-requiring processes or for **RNA synthesis**.

c. Reduction of the ribose moiety to deoxyribose can occur at the diphosphate level and is catalyzed by ribonucleotide reductase, which requires the protein thioredoxin.

—After the diphosphates are rephosphorylated, dATP and dGTP may be used for **DNA synthesis**.

d. Purine bases can be salvaged by **reacting with PRPP** to re-form nucleotides (Figure 7-21). The purine-salvage enzymes, located in tissues, are hypoxanthine–guanine phosphoribosyl transferase (**HGPRT**) and adenine phosphoribosyl transferase (**APRT**).

e. Pyrimidine bases travel mainly as nucleosides from the liver, where they are synthesized, to tissues, where they are phosphorylated to nucleotides.

2. Purine degradation (Figure 7-22*A*)

—In the degradation of the purine nucleotides, **phosphate and ribose** are removed first; then the nitrogenous base is oxidized.

a. Degradation of guanine produces **xanthine**.

b. Degradation of adenine produces **hypoxanthine,** which is oxidized to xanthine by xanthine oxidase; this enzyme requires molybdenum.

c. Xanthine is oxidized to **uric acid** by xanthine oxidase.

d. Uric acid, which is not very water soluble, is **excreted** by the **kidneys**.

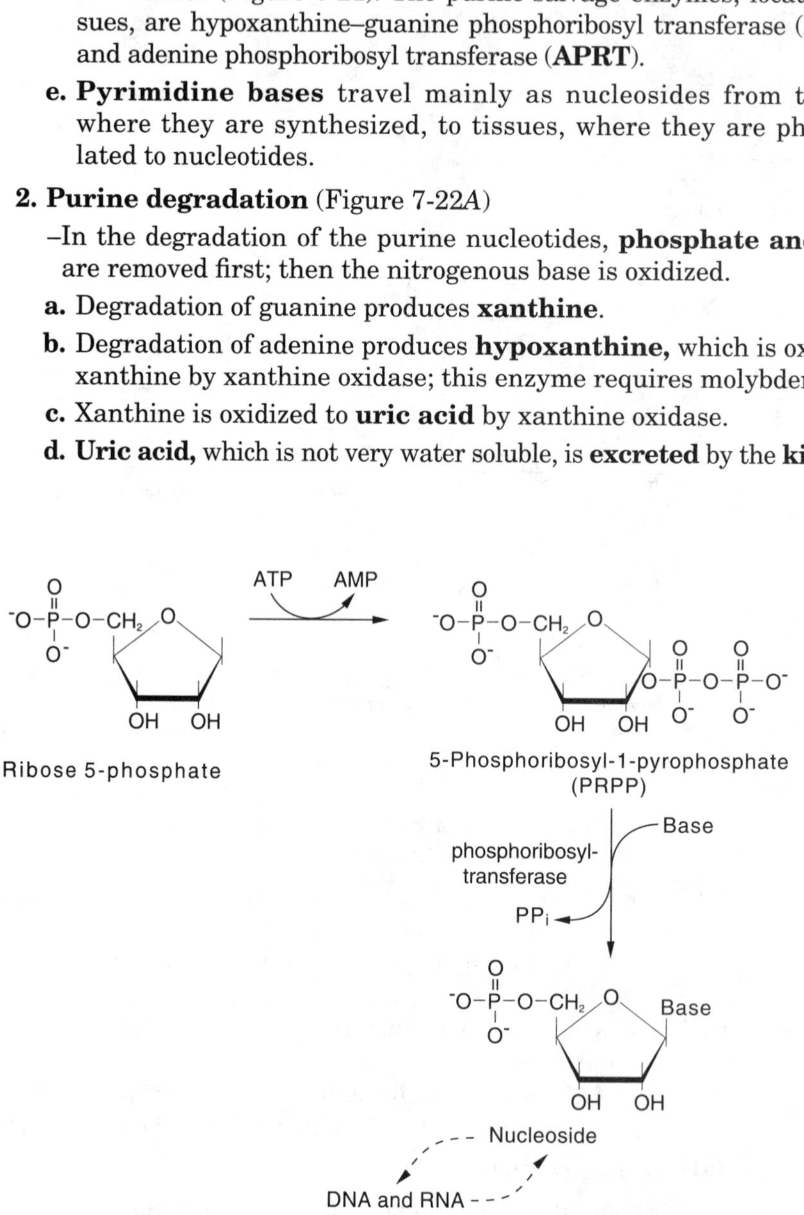

Figure 7-21. Salvage of the purine bases guanine, adenine, and hypoxanthine occurs in reactions catalyzed by phosphoribosyl transferases.

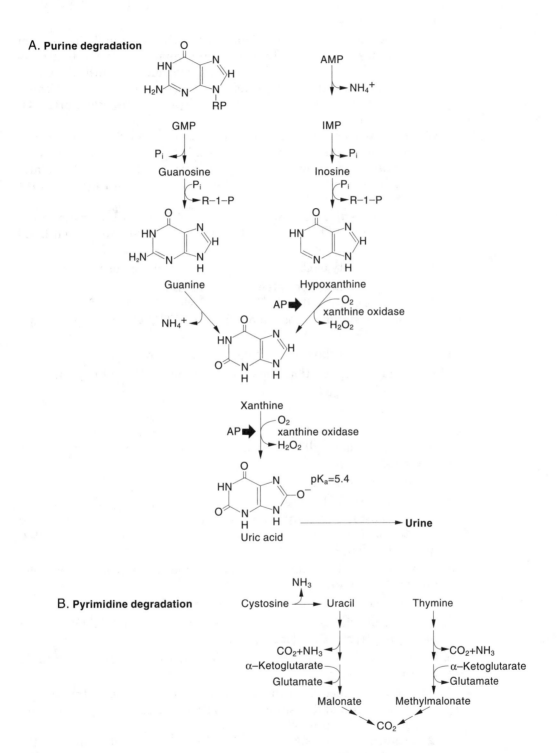

Figure 7-22. Purine and pyrimidine degradation. Allopurinol (*AP*), which inhibits xanthine oxidase, is used to treat gout.

3. Pyrimidine synthesis (see Figure 7-20, *right*)

  a. The **pyrimidine base** is synthesized prior to addition of the ribose moiety.

    (1) In the first reaction, glutamine reacts with CO_2 and 2 ATP to form carbamoyl phosphate. This reaction is analogous to the first reaction of the urea cycle. However, for pyrimidine synthesis, glutamine provides the nitrogen and the reaction occurs in the cytosol, where it is **catalyzed by carbamoyl phosphate synthetase II,** which is inhibited by UTP.

    (2) The entire aspartate molecule adds to carbamoyl phosphate. The molecule closes to yield a ring, which is oxidized, forming **orotate**.

    (3) **Orotate reacts with PRPP,** producing orotidine 5'-phosphate, which is decarboxylated to form **uridine monophosphate (UMP)**.

  b. **UMP is phosphorylated to UTP,** which obtains an amino group from glutamine to form CTP. UTP and CTP are used in the synthesis of RNA.

  c. The ribose moiety of CDP is reduced to deoxyribose, forming dCDP.

    (1) dCDP is dephosphorylated to form dUMP.

    (2) dUMP is converted to dTMP by methylene tetrahydrofolate.

    (3) Phosphorylations produce dCTP and dTTP, which are precursors of DNA.

4. Pyrimidine degradation (see Figure 7-22*B*)

  –In pyrimidine degradation, the carbons produce CO_2 and the nitrogens produce urea.

F. Heme metabolism

  –Heme, which consists of a **porphyrin ring** coordinated with iron, is found mainly in hemoglobin but is also present in myoglobin and the cytochromes.

1. Heme synthesis (Figure 7-23)

  a. In the first step of heme synthesis, glycine and succinyl CoA condense to form δ-**aminolevulinic acid** (δ-ALA). Pyridoxal phosphate is the cofactor for δ-aminolevulinate synthetase. Glycine is decarboxylated in this reaction.

  b. Two molecules of δ-aminolevulinate condense to form the pyrrole, **porphobilinogen**.

  c. Four pyrrole rings condense to form the first in a series of **porphyrins**.

  d. These porphyrins are altered by decarboxylation and oxidation, and **protoporphyrin IX** is formed.

  e. **Iron** (as Fe^{2+}) is introduced into the structure by **ferrochelatase, forming heme**.

    (1) Iron is obtained from the diet, travels in the blood in the protein **transferrin,** and is stored as **ferritin** in tissues such as liver and spleen (see Figure 7-23).

    (2) **Vitamin C** increases the uptake of iron from the intestinal tract.

    (3) **Ceruloplasmin,** a protein that contains copper, is involved in the oxidation of iron.

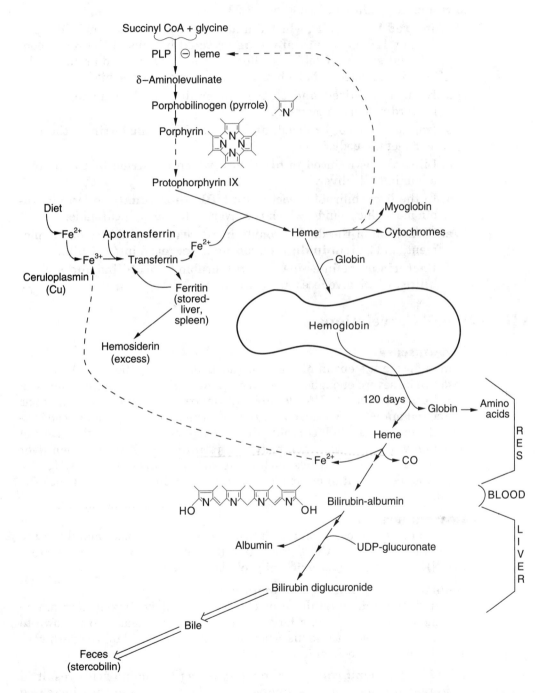

Figure 7-23. Hemoglobin synthesis and degradation. *RES* = reticuloendothelial system; *PLP* = pyridoxal phosphate.

(4) Excess iron is stored as **hemosiderin**.

f. Heme regulates its own **production** by repressing the synthesis of δ-aminolevulinate synthetase.

g. Heme stimulates synthesis of the protein globin by maintaining the translational initiation complex on the ribosome in its active state.

2. **Heme degradation** (see Figure 7-23)

   –After **red blood cells,** which contain hemoglobin, reach their life span of about 120 days, they are **phagocytized** by cells of the reticuloendothelial system. Globin is released and converted to amino acids. Heme is degraded to bilirubin, which is excreted in the bile.

   a. Heme is oxidized and cleaved to produce carbon monoxide and biliverdin, a green pigment.

   b. Iron is released, oxidized, and returned by transferrin to the iron stores of the body.

   c. Biliverdin is reduced to **bilirubin,** which is carried by the protein albumin to the liver.

   d. In the liver, bilirubin reacts with **UDP-glucuronate** to form bilirubin monoglucuronide, which is converted to the diglucuronide.

   e. Formation of the diglucuronide increases the solubility of the pigment, and **bilirubin diglucuronide** is **secreted** into the **bile.**

   f. Bacteria in the intestine convert bilirubin to **urobilins and stercobilins,** which give feces its brown color.

VIII. Clinical Correlations

A. Liver disease

Many important steps in nitrogen metabolism occur in the liver. Liver disease can be severe enough so that **urea production** may be compromised. Blood urea nitrogen **(BUN) levels** will **decrease,** and levels of the toxic compound ammonia will increase. Because the liver is involved in converting bilirubin to the diglucuronide that is excreted in the bile, the levels of bilirubin will increase in the body and **jaundice** will occur. When liver cells are damaged, enzymes such as aspartate transaminase (AST, also known as GOT, glutamate-oxaloacetate transaminase) will leak into the blood.

B. Renal failure

Nitrogenous excretory products are removed from the body mainly in the urine. In renal failure, these products will be retained. Blood urea nitrogen **(BUN), creatinine,** and **uric acid levels** will **rise.**

C. Inborn errors of amino acid metabolism

Although the individual diseases caused by defective enzymes in amino acid metabolism are rare, a large number of these diseases are known to occur, and studies of these diseases have helped to elucidate the pathways of amino acid metabolism in the human.

1. **Defective membrane-transport systems** for amino acids result in decreased resorption of amino acids by the kidney and thus increased excretion in the urine. In **cystinuria,** transport of cysteine is defective.

Cysteine is oxidized to cystine, which may crystallize, forming kidney stones. In **Hartnup's disease,** the transport of neutral amino acids is defective in both intestinal cells and kidney tubules, resulting in **deficiencies of essential amino acids**.

2. In **phenylketonuria (PKU),** the conversion of phenylalanine to tyrosine is defective. Phenylalanine accumulates and is converted to compounds such as the phenylketones, which give the urine a musty odor. Mental retardation occurs. PKU is treated by restriction of phenylalanine in the diet.

3. In **alcaptonuria,** homogentisate, a product of phenylalanine and tyrosine metabolism, accumulates because homogentisate oxidase is defective (see Figure 7-12). Homogentisate autooxidizes and the products polymerize, forming dark colored pigments, which accumulate in various tissues and cause a **degenerative arthritis**.

4. In **histidinemia,** histidase, which converts histidine to urocanate, is defective. Early cases were reported to be associated with mental retardation, but more recently, deleterious consequences have not been observed.

5. In **maple syrup urine disease,** the enzyme complex that decarboxylates the transamination products of the branched-chain amino acids is defective (see Figure 7-11). Valine, isoleucine, and leucine accumulate. Urine has the odor of maple syrup. **Mental retardation** occurs.

6. In **homocystinuria,** cystathionine synthase is defective. Therefore, homocysteine does not react with serine to form cysteine (see Figure 7-10). The homocysteine that accumulates is oxidized to homocystine and excreted in the urine. Some cases respond to increased doses of vitamin B_6, which forms pyridoxal phosphate, the cofactor for the synthase enzyme.

D. Disorders of purine and pyrimidine metabolism

Gout is a group of diseases caused by an increased conversion of purine bases to uric acid or a decreased excretion of uric acid by the kidney. Thus **uric acid,** which is very insoluble, accumulates, resulting in the precipitation of urate crystals in the joints, which causes **acute inflammatory arthritis**. Chronic cases are treated with allopurinol, a base that forms a nucleotide that inhibits xanthine oxidase and prevents hypoxanthine and xanthine from being converted to uric acid (see Figure 7-22A).

1. **Lesch-Nyhan syndrome** is caused by a defective hypoxanthine-guanine phosphoribosyl transferase (**HGPRT**). Purine bases cannot be salvaged (i.e., reconverted to nucleotides). The purines are converted to uric acid, which rises in the blood. Mental retardation and self-mutilation are characteristics of the disease.

2. In **hereditary orotic aciduria,** orotic acid is excreted in the urine because the enzymes that convert it to uridine monophosphate, orotate phosphoribosyl transferase and orotidine 5'-phosphate decarboxylase, are defective (see Figure 7-20). Pyrimidines cannot be synthesized, and therefore, normal growth does not occur. Oral administration of uridine bypasses the metabolic block and provides the body with a source of pyrimidines.

E. Diseases related to the metabolism of other nitrogen-containing compounds

1. In **Parkinson's disease,** dopamine levels are decreased because of a deficiency in conversion of dopa to dopamine (see Figure 7-12). The common characteristics are tremors, difficulty initiating voluntary movement, a masked face with staring expression, and a shuffling gait.

2. **Pheochromocytomas** are tumors that produce norepinephrine and epinephrine and cause **hypertension.**

3. In **albinism,** tyrosinase is defective and tyrosine cannot be converted to the skin pigment melanin.

F. Vitamin deficiencies that affect amino acid metabolism

1. **Pellegra** is caused by a deficiency of the vitamin niacin or of tryptophan. Niacin is required for the production of NAD and NADP. These compounds may also be generated from tryptophan. Pellegra results in the four D's: dermatitis, diarrhea, dementia, and death.

2. Vitamin B_6 is required for the formation of pyridoxal phosphate, an important cofactor in nitrogen metabolism. **Deficiencies of vitamin B_6** are caused by a lack of the vitamin in the diet or by the administration of drugs such as isoniazid, which interfere with its metabolism. Synthesis of neurotransmitters, NAD, and heme are decreased, resulting in neurologic and pellegra-like symptoms and anemia.

3. **Folate and vitamin B_{12} deficiencies** result in decreased DNA and RNA synthesis, which causes a megaloblastic anemia. Tetrahydrofolate (FH_4) is directly involved in providing carbon for thymine (and thus DNA) synthesis and for synthesis of the purines (see Figure 7-20). If vitamin B_{12}, which accepts methyl groups from 5-methyl-FH_4, is deficient, 5-methyl-FH_4 accumulates. Since this compound cannot be reconverted to other FH_4 derivatives, a **folate** deficiency is secondarily produced (the methyl trap theory). Vitamin B_{12} also is involved in converting methylmalonyl CoA to succinyl CoA (see Figure 7-10). In a vitamin B_{12} deficiency, methylmalonic acid is excreted in the urine. Neurologic symptoms, which cannot be alleviated by administration of folate, occur. The absence of intrinsic factor, which is produced by the stomach and necessary for vitamin B_{12} absorption in the intestine, causes **pernicious anemia**.

G. Chemotherapeutic drugs that affect nitrogen metabolism (see Figure 7-16)

1. **5-Fluorouracil** produces a nucleotide that binds to thymidylate synthetase and prevents the conversion of dUMP to dTMP. This inhibition of the synthesis of thymine prevents DNA synthesis and thus affects the proliferation of cells.

2. **Methotrexate** inhibits dihydrofolate reductase, which catalyzes reduction of FH_2 to FH_4. When dUMP is converted to dTMP, N^5,N^{10}-methylene tetrahydrofolate is converted to dihydrofolate, which must be reduced to tetrahydrofolate in order for the production of thymine to continue. If the reductase is inhibited by methotrexate, thymine synthesis also is inhibited, thus preventing DNA synthesis.

Review Test

Directions: Each of the numbered items or incomplete statements in this section is followed by answers or by completions of the statement. Select the **one** lettered answer or completion that is **best** in each case.

1. Which of the following proteolytic enzymes has the greatest effect on the activity of other proteolytic enzymes involved in digestion?

(A) Trypsin
(B) Chymotrypsin
(C) Carboxypeptidase A
(D) Pepsin
(E) Aminopeptidase

2. The compound shown below

$$\underset{H_2N - \overset{\overset{\displaystyle O}{\|}}{C} - CH_2 - CH_2 - \overset{\overset{\displaystyle NH_2}{|}}{CH} - COO^-}{}$$

(A) is an intermediate in the urea cycle
(B) is formed by transamination of oxaloacetate
(C) is the amino acid lysine
(D) contains two amino groups
(E) releases two moles of ammonia when converted to one mole of α-ketoglutarate

3. In the urea cycle

(A) carbamoyl phosphate is derived directly from glutamine and CO_2
(B) ornithine reacts with aspartate to generate argininosuccinate
(C) the α-amino group of arginine forms one of the nitrogens of urea
(D) ornithine directly reacts with carbamoyl phosphate to form citrulline
(E) N-acetylglutamate is a positive allosteric effector of ornithine transcarbamoylase

4. Each of the following enzymes is involved in the synthesis of serine from glucose EXCEPT

(A) phosphofructokinase
(B) aldolase
(C) phosphoserine phosphatase
(D) pyruvate kinase
(E) glyceraldehyde 3-phosphate dehydrogenase

5. A common intermediate in the synthesis of arginine, serine, and aspartate from glucose is

(A) oxaloacetate
(B) glyceraldehyde 3-phosphate
(C) pyruvate
(D) α-ketoglutarate
(E) ornithine

6. Via enzymes of the urea cycle, aspartate

(A) provides nitrogen for synthesis of arginine
(B) provides carbon for the synthesis of arginine
(C) is converted to malate
(D) is converted to oxaloacetate

7. Which of the following statements concerning glutamate is TRUE?

(A) It is produced in a transamination reaction in which aspartate reacts with oxaloacetate
(B) It undergoes a series of reactions in which it cyclizes to produce histidine
(C) It can be converted to arginine by a series of reactions, some of which require urea cycle enzymes
(D) It is produced by the action of glutamate dehydrogenase, an enzyme that requires NH_4^+ and FAD

8. Each of the following statements about serine is correct EXCEPT

(A) it is converted to pyruvate and ammonia by a dehydratase
(B) it may be synthesized from glucose via a glycolytic intermediate
(C) it is converted to glycine by a reaction requiring tetrahydrofolate
(D) it is the only amino acid that contains a hydroxyl group
(E) it is a nonessential amino acid

9. Pyridoxal phosphate is required for the enzymes catalyzing the reaction

(A) glutamate + NH_3 + ATP $\rightarrow$ glutamine + ADP + P_i
(B) glutamate + NAD^+ $\rightarrow$ α-ketoglutarate + NH_4^+ + NADH + H^+
(C) glutamine + H_2O $\rightarrow$ glutamate + NH_3
(D) pyruvate + glutamate $\rightarrow$ alanine + α-ketoglutarate

10. The carbons of cysteine are derived from

(A) threonine
(B) serine
(C) leucine
(D) methionine

11. Isocitrate dehydrogenase is required for the synthesis from glucose of the amino acid

(A) serine
(B) alanine
(C) aspartate
(D) glutamate
(E) cysteine

12. During the metabolism of the branched-chain amino acids

(A) valine is deaminated rather than transaminated
(B) none of the carbons of isoleucine is converted to succinyl CoA
(C) lipoic acid is not required
(D) leucine is converted to acetoacetate

13. The major amino acid that is released from muscle and converted to glucose in the liver is

(A) alanine
(B) glutamine
(C) valine
(D) aspartate
(E) glutamate

14. Each of the following statements about the kidney is correct EXCEPT

(A) it uses ammonia released from glutamine to buffer acids in the urine
(B) it converts glutamine to α-ketoglutarate
(C) it produces serine and alanine and releases them into the blood
(D) it synthesizes most of the urea that is excreted into the urine

15. De novo pyrimidine synthesis requires

(A) phosphoribosyl pyrophosphate (PRPP) for the initial step
(B) tetrahydrofolate for the incorporation of carbons 2 and 8
(C) both carbon and nitrogen of aspartate to form the ring
(D) NH_4^+ as a substrate for carbamoyl phosphate synthetase II
(E) glycine as the source of two nitrogens in the ring

16. The principal nitrogenous urinary excretion product in humans resulting from the catabolism of AMP is

(A) creatinine
(B) urea
(C) uric acid
(D) thiamine
(E) thymine

17. The conversion of propionyl CoA to succinyl CoA requires

(A) biotin
(B) vitamin B_{12}
(C) biotin and vitamin B_{12}
(D) biotin, vitamin B_{12}, and tetrahydrofolate

18. *S*-Adenosylmethionine (SAM) serves as the methylating agent for each of the following EXCEPT

(A) the conversion of norepinephrine to epinephrine
(B) the synthesis of creatine from guanidinoacetate
(C) the synthesis of phosphatidylcholine from phosphatidylethanolamine
(D) the conversion of dUMP to dTMP

19. Glycine is an important precursor in the pathway for the biosynthesis of each of the following EXCEPT

(A) heme
(B) creatine
(C) guanine
(D) valine

20. Each of the following statements about heme and iron metabolism is correct EXCEPT

(A) iron is stored in the liver as ferritin
(B) iron (as Fe^{2+}) is inserted into protoporphyrin IX in the last step of heme synthesis
(C) δ-aminolevulinate synthetase catalyzes the regulated and rate-limiting step in heme biosynthesis
(D) carbon from glycine is excreted as bilirubin
(E) the iron produced by heme degradation is excreted in the feces

21. Which of the following statements about bilirubin is TRUE?

(A) It is made more soluble in the liver by attachment of residues of glucose
(B) It is excreted mainly in the urine
(C) It is produced by oxidation of heme, with loss of carbon monoxide (CO)
(D) It contains iron in the Fe^{2+} state

22. Each of the following statements about nitrogen metabolism is correct EXCEPT

(A) cysteine "spares" methionine—ingestion of cysteine reduces the need for methionine in the diet
(B) the enzyme glutamate dehydrogenase catalyzes the fixation of ammonia—the addition of ammonia to an organic compound
(C) creatine requires glycine, arginine, and methionine for synthesis of its carbon skeleton
(D) creatine phosphate contains a high-energy bond
(E) formiminoglutamate (FIGLU) is an intermediate in glutamine degradation

23. Pregnant women frequently suffer from folate deficiencies. A deficiency of folate would decrease the production of

(A) creatine phosphate from creatine
(B) all of the pyrimidines required for RNA synthesis
(C) the thymine nucleotide required for DNA synthesis
(D) phosphatidyl choline from diacylglycerol and CDP-choline

24. Compared to a healthy person, a person with pernicious anemia

(A) produces less intrinsic factor
(B) excretes less methylmalonic acid in the urine
(C) requires less methionine in the diet
(D) has a higher rate of purine biosynthesis
(E) has lower blood levels of FIGLU

25. A 24-hour urine collection showed that an individual's excretion of creatinine was much lower than normal. Decreased excretion of creatinine could be caused by

(A) kidney failure
(B) decreased dietary intake of creatine
(C) a higher than normal muscle mass resulting from weight lifting
(D) a genetic defect in the enzyme that converts creatine phosphate to creatinine

26. A genetic defect in the ability to synthesize tetrahydrobiopterin would affect each of the following conversions EXCEPT

(A) phenylalanine to tyrosine
(B) tyrosine to dopamine
(C) dopa to melanin
(D) tryptophan to serotonin

27. Phenylketonuria, alcaptonuria, and albinism are caused by deficiencies in enzymes involved in the metabolism of

(A) tryptophan
(B) tyrosine
(C) histidine
(D) valine
(E) lysine

28. The plasma and urine of patients with maple syrup urine disease contain elevated levels of each of the following amino acids EXCEPT

(A) valine
(B) leucine
(C) lysine
(D) isoleucine

268 / *Biochemistry*

Directions: Each group of items in this section consists of lettered options followed by a set of numbered items. For each item, select the **one** lettered option that is most closely associated with it. Each lettered option may be selected once, more than once, or not at all.

Questions 29–32

Match each of the conditions below with the most likely clinical findings.

(A) Low blood urea nitrogen (BUN); high blood NH_4^+ and total bilirubin levels
(B) Dark brown stool; elevated total bilirubin in blood
(C) Low hematocrit; small, pale red blood cells
(D) Light colored stool; elevated conjugated bilirubin in blood

29. Bile duct obstruction
30. Hemolytic anemia
31. Iron deficiency anemia
32. Hepatitis

Questions 33–36

Match each condition below with the component that would most likely be elevated in the blood.

(A) Bilirubin
(B) Uric acid
(C) Creatine phosphokinase
(D) Blood urea nitrogen (BUN)

33. ~~Gout~~ Hepa.
34. ~~Myocardial infarction~~ Gout
35. ~~Hepatitis~~ Myo. inf.
36. Kidney disease

Questions 37–41

Match each property below with the appropriate enzyme.

(A) Pepsin
(B) Trypsin
(C) Carboxypeptidase A
(D) Enteropeptidase

37. Digests dietary proteins in the stomach
38. Is synthesized by intestinal cells
39. Cleaves bonds at the carboxyl end of the arginine and lysine residues within a polypeptide chain
40. Acts as an exopeptidase
41. Is produced by the action of HCl on its precursor

Questions 42–45

Match each description below with the appropriate compound.

(A)
$$OH$$
$$|$$
$$CH_2$$
$$|$$
$$H_3^+N-CH-COO^-$$

(B)
$$SH$$
$$|$$
$$CH_2$$
$$|$$
$$H_3^+N-CH-COO^-$$

(C)
$$CH_3$$
$$|$$
$$CH_2-OH$$
$$|$$
$$H_3^+N-CH-COO^-$$

(D)
$$CH_3$$
$$|$$
$$S$$
$$|$$
$$CH_2$$
$$|$$
$$CH_2$$
$$|$$
$$H_3^+N-CH-COO^-$$

42. Is produced by cleavage of cystathionine
43. Contains a carbon skeleton that can be converted to pyruvate by a single enzyme
44. Can be converted to glycine in a single reaction that requires tetrahydrofolate
45. Contains a carbon skeleton that can be converted to homocysteine

Questions 46–50

Match the descriptions below with the appropriate cofactor.

(A) Vitamin B_{12}
(B) Tetrahydrofolate (FH_4)
(C) Biotin
(D) Thiamine
(E) Pyridoxal phosphate

46. Required for the decarboxylation of the transamination product of valine
47. Required for the synthesis of deoxythymidylate from deoxyuridylate
48. Directly required for the synthesis of serine from glycine
49. Directly required for the conversion of methylmalonyl CoA to succinyl CoA
50. Required for the conversion of histidine to histamine

Questions 51–56

Match the descriptions below with the appropriate amino acid or pyrimidine.

(A) Tyrosine
(B) Tryptophan
(C) Threonine
(D) Thymine

51. Can be converted to epinephrine
52. Contains nonring carbons that can be cleaved from the ring structure to form alanine
53. Is synthesized by hydroxylation of an essential amino acid
54. May be converted to serotonin by reactions requiring tetrahydrobiopterin and molecular oxygen
55. May be converted to the moiety of NAD^+ that may also be derived from niacin
56. May be produced from uracil

Questions 57–60

Match each of the metabolites below with the compound to which it is related.

(A) Leucine
(B) Homocysteine
(C) Glutamate
(D) Tryptophan

57. NAD^+
58. HMG CoA
59. Proline
60. Methionine

Questions 61–64

Match each metabolic product below with the amino acid from which it is derived.

(A) Glutamate
(B) Tyrosine
(C) Histidine
(D) Tryptophan

61. Serotonin
62. γ-Aminobutyric acid (GABA)
63. Histamine
64. Epinephrine

Answers and Explanations

1–A. Trypsin activates the pancreatic zymogens, converting chymotrypsinogen to the active form, chymotrypsin, and the procarboxypeptidases to the active carboxypeptidases. It even autocatalyzes its own activation—the conversion of trypsinogen to trypsin (which is also catalyzed by enteropeptidase).

2–E. This compound is glutamine. It contains one amino and one amide group, both of which are converted to ammonia. The amide is removed by glutaminase, which forms ammonia and glutamate. Glutamate dehydrogenase releases ammonia from glutamate, forming α-ketoglutarate.

3–D. Carbamoyl phosphate is formed from NH_4^+, CO_2, and ATP. It reacts with ornithine to form citrulline, which reacts with aspartate to form argininosuccinate. Fumarate is released from argininosuccinate, and arginine is formed. Urea is produced from the guanidinium group on the side chain of arginine, not from the amino group. Ornithine is regenerated. *N*-acetylglutamate is an allosteric activator of carbamoyl phosphate synthetase I.

4–D. Serine is synthesized from glucose. The pathway branches from glycolysis at phosphoglyceric acid, which is reduced, transaminated, and dephosphorylated by phosphoserine phosphatase. Pyruvate kinase is a glycolytic enzyme that functions beyond the branch point for serine synthesis.

5–B. In the synthesis of these three amino acids from glucose, serine is produced from the glycolytic intermediate phosphoglyceric acid. Arginine is produced from the TCA cycle intermediate α-ketoglutarate, and aspartate by transamination of oxaloacetate. Therefore, glyceraldehyde 3-phosphate is the only common intermediate.

6–A. When argininosuccinate is cleaved to form arginine, the carbons that were derived from aspartate are released as fumarate and the nitrogen of aspartate is incorporated into arginine.

7–C. Glutamate can be reduced to glutamate semialdehyde and then transaminated to form ornithine, which can be converted to arginine via enzymes of the urea cycle. Glutamate semi-aldehyde cyclizes to form proline. (Histidine cannot be synthesized in the human.) Aspartate and α-ketoglutarate undergo a transamination reaction that produces oxaloacetate and gluta-mate. The reaction catalyzed by glutamate dehydrogenase requires NADH or NADPH; it pro-duces glutamate from α-ketoglutarate and NH_4^+.

8–D. Both serine and threonine contain a hydroxyl group. Serine dehydratase produces pyru-vate and ammonia from serine. Serine can be produced from glucose via the phosphoglyceric acid intermediates of glycolysis; thus it is nonessential. Tetrahydrofolate (FH_4) reacts with ser-ine to form glycine and N^5,N^{10}-methylene-FH_4.

9–D. The transamination of pyruvate and glutamate requires pyridoxal phosphate.

10–B. Serine reacts with homocysteine to form cystathionine, which is cleaved to form cys-teine, NH_4^+, and α-ketobutyrate. Methionine provides the sulfur via homocysteine.

11–D. Formation of glutamate from glucose involves the TCA cycle intermediate α-keto-glutarate, which is formed from isocitrate in a reaction catalyzed by isocitrate dehydrogenase. α-Ketoglutarate is converted to glutamate either by glutamate dehydrogenase or a transami-nase. Formation of serine, alanine, aspartate, and cysteine from glucose do not require isoci-trate dehydrogenase.

12–D. Valine, isoleucine, and leucine (the branched-chain amino acids) are transaminated and then oxidized by an α-keto acid dehydrogenase that requires lipoic acid as well as thiamine pyrophosphate, coenzyme A, FAD, and NAD^+. Four of the carbons of valine and isoleucine are converted to succinyl CoA. Isoleucine also produces acetyl CoA. Leucine is converted to HMG CoA, which is cleaved to acetoacetate and acetyl CoA.

13–A. Alanine and glutamine are the major amino acids released from muscle. Glutamine is further metabolized in the gut and the kidney. Alanine is the major amino acid that is con-verted to glucose in the liver.

14–D. Most of the urea that is excreted in the kidney is produced in the liver.

15–C. Options A and B are true for purine but not pyrimidine biosynthesis. During pyrimidine synthesis, the entire aspartate molecule is incorporated into the ring. Glutamine is the sub-strate for carbamoyl phosphate synthetase II, the enzyme involved in pyrimidine biosynthesis. (NH_4^+ is the substrate for synthetase I used in urea synthesis.) Glycine supplies one nitrogen for purine synthesis.

16–C. The purine bases, including adenine, are oxidized to uric acid, which is excreted in the urine.

17–C. The conversion of propionyl CoA to methylmalonyl CoA requires biotin, and the conver-sion of methylmalonyl CoA to succinyl CoA requires vitamin B_{12}. FH_4 is not involved.

18–D. Tetrahydrofolate is involved in the conversion of dUMP to dTMP.

19–D. Valine is an essential amino acid and is not synthesized in the human. Glycine reacts with succinyl CoA in the first step of heme synthesis and with arginine in the first step of crea-tine synthesis. The entire glycine molecule is incorporated into the growing purine ring.

20–E. Iron is returned to the body's iron stores and is not excreted. Bleeding is the only signifi-cant means by which iron is lost from the body.

21–C. Bilirubin is produced by oxidation of heme after its iron is released; CO is released in this reaction. Bilirubin diglucuronide is excreted into the bile by the liver.

22–E. FIGLU is produced during the degradation of histidine.

23–C. The only pyrimidine that requires folate for its synthesis is thymine (dUMP → dTMP). Folate is required for incorporation of carbons 2 and 8 into all purine molecules.

24–A. Pernicious anemia occurs when the stomach does not produce adequate intrinsic factor for absorption of vitamin B_{12}, which is required for the conversion of methylmalonyl CoA to succinyl CoA and homocysteine to methionine. A vitamin B_{12} deficiency results in the excretion of methylmalonic acid and an increased dietary requirement for methionine. The methyl group transferred from vitamin B_{12} to homocysteine to form methionine comes from 5'-methyl tetrahydrofolate, which accumulates in a vitamin B_{12} deficiency, causing a decrease in folate levels and symptoms of folate deficiency, including increased levels of FIGLU and decreased purine biosynthesis.

25–A. The amount of creatine in liver cells determines its rate of synthesis from glycine, arginine, and SAM. In muscle, creatine is converted to creatine phosphate, which is nonenzymatically cyclized to form creatinine. The amount of creatinine excreted by the kidneys each day depends on body muscle mass. In kidney failure, the excretion of creatinine into the urine will be low.

26–C. Tetrahydrobiopterin is involved in hydroxylation reactions that occur in phenylalanine to tyrosine, tyrosine to dopamine, and tryptophan to serotonin, but not in the conversion of dopa to melanin.

27–B. PKU is caused by a deficiency of phenylalanine hydroxylase, which converts phenylalanine to tyrosine. A defect in tyrosine degradation causes homogentisic acid to accumulate and produce dark pigments (alcaptonuria). A defect in the conversion of tyrosine to the skin pigment melanin causes albinism.

28–C. In maple syrup urine disease, the branched-chain amino acids (valine, leucine, and isoleucine) can be transaminated but not oxidatively decarboxylated because the α-keto acid dehydrogenase is defective.

29–D. If the bile duct is obstructed, conjugated bilirubin from the liver does not enter the gut, but instead backs up into the blood. The dark brown color of stool is caused by stercobilin, a pigment produced by bacterial oxidation of bilirubin in the gut.

30–B. Increased lysis of red blood cells causes bilirubin to be elevated in the blood. The liver conjugates the bilirubin and excretes it into the gut where it is converted to stercobilin.

31–C. A lower than normal amount of heme is produced from protoporphyrin IX in an iron-deficiency anemia. Red blood cells are pale and small.

32–A. In hepatitis, the liver's ability to produce urea from NH_4^+ and to conjugate and excrete bilirubin is decreased.

33–C. Bilirubin is conjugated and excreted by the liver. If the liver is sick, bilirubin will accumulate in the blood and in tissues, producing jaundice.

34–A. In gout, uric acid crystals precipitate in joints, causing severe pain.

35–B. When heart muscle is damaged, cellular enzymes leak into the blood. Creatine phosphokinase (CPK or CK) is found mainly in muscle cells and to a lesser extent in the brain. The presence of the MB fraction of CPK in blood is indicative of a heart attack.

36–D. The kidney excretes urea. If the kidney is diseased, urea will accumulate in the blood.

37–A. Pepsin acts in the stomach.

38–D. Enteropeptidase, which is produced by intestinal cells, cleaves trypsinogen to trypsin.

39–B. Trypsin is an endopeptidase that cleaves polypeptide chains at arginine and lysine residues.

40–C. Carboxypeptidases A and B are pancreatic exopeptidases, which cleave one amino acid at a time from the C-terminal end of a polypeptide chain.

41–A. Pepsin is produced as pepsinogen (a zymogen), which is initially cleaved by HCl to form the active protease. Subsequently, pepsin autocatalyses its own activation.

42–B. Cysteine is produced by cleavage of cystathionine.

43–A. Serine dehydratase converts serine to pyruvate and NH_4^+.

44–A. Serine can transfer a 1-carbon group to tetrahydrofolate, forming glycine and 5,10-methylene-FH_4.

45–D. The carbon skeleton of methionine can be converted to homocysteine.

46–D. The branched-chain amino acids (valine, isoleucine, and leucine) are transaminated and then oxidatively decarboxylated by an enzyme that requires thiamine, lipoic acid, coenzyme A, FAD, and NAD.

47–B. N^5,N^{10}-methylene tetrahydrofolate reacts with dUMP to form dTMP. Dihydrofolate is produced and reduced to FH_4 by dihydrofolate reductase.

48–B. N^5,N^{10}-methylene-FH_4 is involved in the conversion of glycine to serine.

49–A. Vitamin B_{12} is required for the conversion of methylmalonyl CoA to succinyl CoA.

50–E. Decarboxylations of amino acids require pyridoxal phosphate.

51–A. Tyrosine can be converted to dopa, dopamine, norepinephrine, and then to epinephrine.

52–B. The nonring carbons of tryptophan are cleaved from the ring to form alanine.

53–A. Tyrosine is synthesized by hydroxylation of the essential amino acid phenylalanine.

54–B. Tryptophan is hydroxylated by a mixed-function oxidase that requires tetrahydrobiopterin and O_2. 5-Hydroxytryptophan is decarboxylated to form serotonin.

55–B. The ring portion of tryptophan may be converted to the nicotinamide moiety of NAD, which may also be derived from the vitamin niacin.

56–D. Uracil as dUMP is converted to thymine (dTMP).

57–D. The nicotinamide moiety of NAD may be derived from tryptophan.

58–A. Leucine forms HMG CoA, which is cleaved to form acetyl CoA and acetoacetate.

59–C. Proline is derived from glutamate and may be converted to glutamate.

60–B. Methionine reacts with ATP to form S-adenosylmethionine (SAM), which donates a methyl group and forms S-adenosylhomocysteine. After release of adenosine, homocysteine is produced. Homocysteine is converted to methionine when it receives a methyl group from vitamin B_{12} (which obtains it from methyl-FH_4).

61–D. Serotonin is derived from tryptophan by a hydroxylation and then a decarboxylation reaction.

62–A. GABA is produced by decarboxylation of glutamate.

63–C. Histamine is produced by decarboxylation of histidine.

64–B. Epinephrine is produced from tyrosine.

8

Molecular Endocrinology

Overview

- Communication between cells is essential for human survival and is provided by the nervous and endocrine systems.
- Endocrine glands produce hormones that travel through the blood to other tissues where they elicit a response (Table 8-1).
- The action of hormones at the molecular level involves receptors.
  - Polypeptide hormones and epinephrine react with receptors in the cell membrane, causing the production of intracellular compounds (second messengers) that allow the external signal from the hormone to produce its intracellular effects.
  - Steroid hormones (as well as the thyroid hormones, 1,25-dihydroxycholecalciferol, and retinoic acid cross) the cell membrane and bind to intracellular receptors. Ultimately, the hormone-receptor complex binds to DNA and activates or inactivates genes, which produce proteins that have physiologic effects.
- Hormones often act as a chain of chemical messengers. For example, a hormone produced by the hypothalamus may stimulate the anterior pituitary to produce another hormone that subsequently causes an endocrine gland to produce still another hormone that ultimately acts on its target cells.

I. Synthesis of Hormones

- An amino acid may be converted to a compound that serves as a hormone (e.g., epinephrine, the thyroid hormones), or a series of amino acids may be joined by peptide bonds to produce a polypeptide hormone (e.g., insulin, prolactin).
- The steroid nucleus may be biochemically modified to produce a number of hormones (e.g., progesterone, cortisol).

A. Epinephrine

1. **Tyrosine** is produced by hydroxylation of the essential amino acid phenylalanine and is hydroxylated to form **dihydroxyphenylalanine** (dopa), which is subsequently decarboxylated to form **dopamine**.

Table 8-1. Abbreviations for Hormones and Related Compounds

ACTH	Adrenocorticotropic hormone
ABP	Androgen binding protein
ADH	Antidiuretic hormone (also known as VP)
ANF	Atrionatriuretic factor (or atriopeptin)
CCK	Cholecystokinin
CG	Chorionic gonadotropin (hCG, human CG)
CLIP	Corticotropin-like intermediate lobe peptide
CRH	Corticotropin releasing hormone
DHEA	Dehydroepiandrosterone
DHT	Dihydrotestosterone
1,25-DHC	1,25-Dihydroxycholecalciferol
E_2	Estradiol
FSH	Follicle stimulating hormone
GH	Growth hormone
GIP	Gastric inhibitory polypeptide
GnRH	Gonadotropin releasing hormone
GRH	Growth hormone releasing hormone (somatocrinin) (GHRH)
GRIH	Growth hormone release inhibiting hormone (somatostatin)
hCG	Human chorionic gonadotropin
IGF	Insulin-like growth factor
LH	Luteinizing hormone
LPH	Lipotropin
MSH	Melanocyte stimulating hormone
POMC	Proopiomelanocortin
PRF	Prolactin releasing factor
PRIH	Prolactin release inhibiting hormone (PIH)
PRL	Prolactin
PTH	Parathyroid hormone
T_3	Triiodothyronine
T_4	Thyroxine (tetraiodothyronine)
TRH	Thyrotropin releasing hormone
TSH	Thyroid stimulating hormone
VIP	Vasoactive intestinal polypeptide
VP	Vasopressin (ADH)

    2. Another hydroxylation reaction produces **norepinephrine,** which is methylated (mainly in the adrenal medulla) to produce **epinephrine**.

B. Thyroid Hormones (Figure 8-1)

    1. The follicular cells of the thyroid gland produce the protein **thyroglobulin,** which is secreted into the colloid.

    2. Iodine, concentrated in the follicular cells by a pump in the cell membrane, is oxidized by a peroxidase. Iodination of **tyrosine residues** in thyroglobulin produces monoiodotyrosine (MIT) and diiodotyrosine (DIT), which undergo **coupling reactions** to produce 3,5,3'-triiodothyronine (T_3) and 3,5,3',5'-tetraiodothyronine (T_4).

    3. Thyroid stimulating hormone (**TSH**) stimulates **pinocytosis** of thyroglobulin, and **lysosomal proteases** cleave peptide bonds, releasing free T_3 and T_4 into the blood.

C. Polypeptide hormones

    1. Polypeptide hormones are **gene products**.

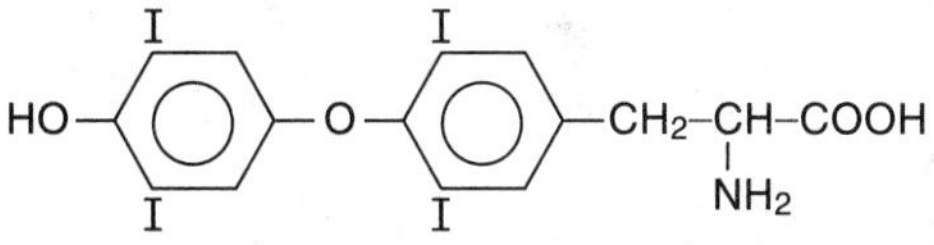

3, 5, 3', 5'-Tetraiodothyronine (T_4)

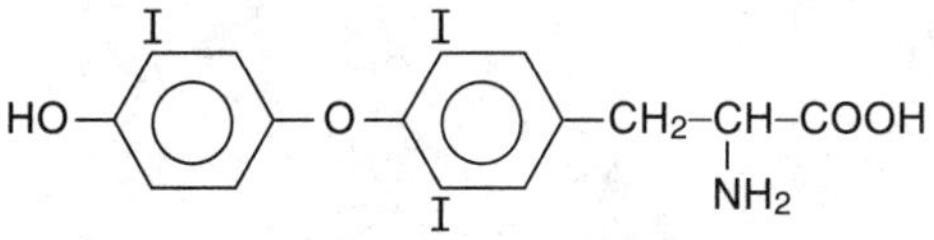

3, 5, 3'-Triiodothyronine (T_3)   **Figure 8-1.** The thyroid hormones. T_4 = thyroxine.

2. mRNA is transcribed from the gene and translated on ribosomes attached to the rough endoplasmic reticulum (RER). A polypeptide **precursor** (or prehormone) larger than the active hormone is formed.

3. The signal peptide is removed in the RER.

4. The polypeptide may be further modified in the Golgi complex and secreted from the cell by the process of exocytosis.

D. Steroid hormones

–Most steroid hormones are derived from cholesterol (Figure 8-2) by cleavage of the side chain to form **pregnenolone,** which subsequently forms **progesterone.**

1. Progesterone

–The A ring of pregnenolone is oxidized to form progesterone.

2. Testosterone

–The side chain of the D ring of progesterone is removed to form testosterone, or testosterone is produced from pregnenolone via dehydroepiandrosterone (DHEA).

3. Estradiol

–17β-Estradiol (E_2) is produced from testosterone by aromatization of the A ring.

4. Adrenal steroids

–**Cortisol** and **aldosterone** are produced from progesterone.

5. 1,25-Dihydroxycholecalciferol (1,25-DHC)

a. 1,25-DHC, the **active form of vitamin D** (see Figure 4-10), may be produced by two hydroxylations of dietary vitamin D_3 (cholecalciferol), first at position 25 (in the liver) and then at position 1 (in the kidney).

b. 7-Dehydrocholesterol, a precursor of cholesterol produced from acetyl CoA, may be converted by ultraviolet light to cholecalciferol and then hydroxylated to form 1,25-DHC (also called calcitriol).

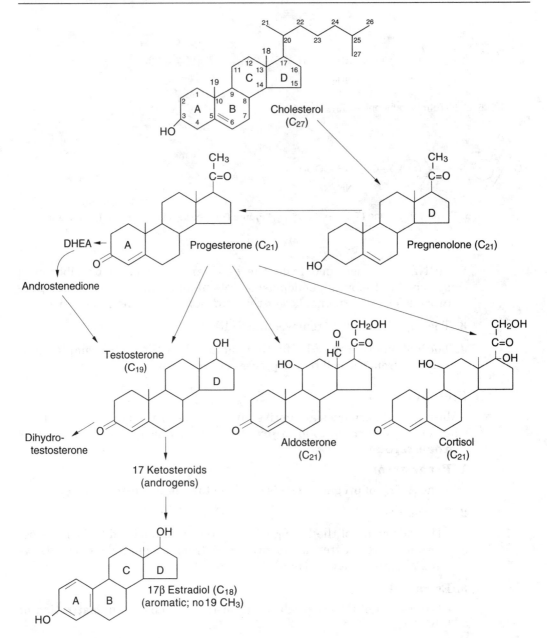

Figure 8-2. Synthesis of the steroid hormones. The rings of the precursor cholesterol are *lettered*. Dihydrotestosterone is produced from testosterone by reduction of the C – C double bond in ring A. DHEA = dehydroepiandrosterone.

II. General Mechanisms of Hormone Action

- Hormones interact with receptors that are located either inside the cell or within the cell membrane.
- Cells are exposed to many hormones. Whether a given hormone will elicit a response in a particular cell depends on the complement of receptors that the cell contains.

- In general, polypeptide hormones and epinephrine (the first messengers) react with receptors in the cell membrane, altering the intracellular concentration of compounds known as second messengers (e.g., cyclic AMP [cAMP], cyclic GMP [cGMP], inositol trisphosphate [IP$_3$], diacylglycerol [DAG], and Ca^{2+}). These second messengers permit an external signal from a hormone to produce intracellular effects.
- The steroid and thyroid hormones, 1,25-DHC, and retinoic acid cross the cell membrane and bind to intracellular receptors, forming complexes that activate or inactivate genes.

A. Hormone action mediated by second messengers

1. Production and effect of second messengers

  a. Hormones bind to receptors on the external surface of the cell membrane, and the hormone-receptor complexes interact with **G proteins** (so-called because they bind guanine nucleotides).

  b. The G proteins may activate or inhibit the process that **increases** the intracellular concentration of the **second messenger** within the cell.

  c. Most of the **second messengers activate protein kinases** that phosphorylate specific cellular proteins altering their activity. For example, glucagon stimulates the degradation of liver glycogen via the second messenger cAMP (see Chapter 5 IV E).

  d. Phosphatases may **dephosphorylate** these proteins.

  –The activity of the phosphatases may be controlled by hormones such as insulin, which opposes the action of glucagon (see Chapter 5 IV F).

  e. Many details of these systems (e.g., the specific proteins that are activated and their cellular effects) are currently unknown.

2. Hormones that act through cyclic nucleotides: the polypeptide hormones and epinephrine (Figure 8-3)

  a. These hormones form hormone-receptor complexes that interact with G proteins (Figure 8-4), activating **adenyl cyclase,** which **converts ATP to cAMP**.

  –cAMP activates protein kinase A, which subsequently **phosphorylates** certain intracellular proteins, altering their activity.

  b. Some of these hormone-receptor complexes lower cAMP levels, either by inhibiting adenyl cyclase or by activating the **phosphodiesterase** that cleaves cAMP to AMP.

  c. At least one hormone, **atrionatriuretic factor (ANF),** activates guanyl cyclase, which produces cGMP.

  –cGMP activates protein kinase G.

  –ANF is released from atrial cells of the heart and produces effects that include increased urine volume, excretion of sodium ions, and vasodilation.

3. Hormones that act through calcium and the phosphatidylinositol bisphosphate (PIP$_2$) system (Figure 8-5)

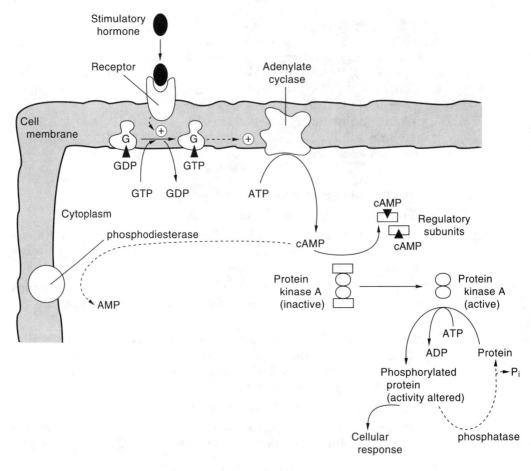

Figure 8-3. The production and action of cyclic AMP (cAMP). G = G protein. G proteins function when GTP is bound; 0 = free catalytic subunits; ☐ = regulatory subunits of protein kinase A; ⊕ = stimulates.

–Some hormones (e.g., thyrotropin releasing hormone [TRH] and oxytocin), after interacting with G proteins, alter the amount and distribution of calcium ions within the cell and activate **protein kinase C.**

a. The hormone-G protein complexes may **open calcium channels** within the cell membrane, allowing extracellular calcium to move into the cell.

b. The complexes may **activate phospholipase C,** which cleaves PIP_2 in the cell membrane to produce two messengers, IP_3 and DAG (Figure 8-6).

(1) **IP_3 causes Ca^{2+} to be released** from intracellular stores, such as those in the endoplasmic reticulum.

–Ca^{2+}, either directly or complexed with calmodulin, may interact with proteins, altering their activity.

(2) **DAG activates protein kinase C,** which phosphorylates certain proteins, altering their activity.

A. Before hormone binds

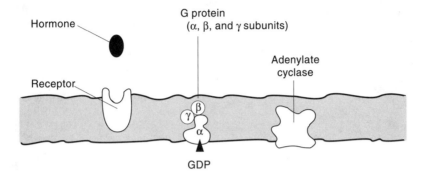

B. After hormone binds

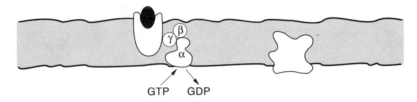

C. G proteins dissociate

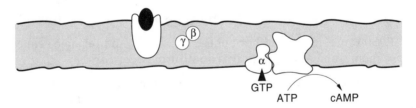

D. GTPase cleaves GTP to GDP and P_i

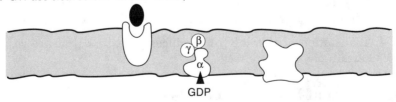

Figure 8-4. The G proteins. (*A*) Before hormone binds. (*B*) After hormone binds. GDP on G protein α is exchanged for GTP. (*C*) G proteins dissociate. α-Subunit with GDP activates adenylate cyclase, which converts ATP to cAMP. (*D*) GTPase cleaves GTP to GDP and P_i. Adenylate cyclase is no longer active. α, β, and γ subunits of G protein reassociate. Hormone dissociates from the receptor.

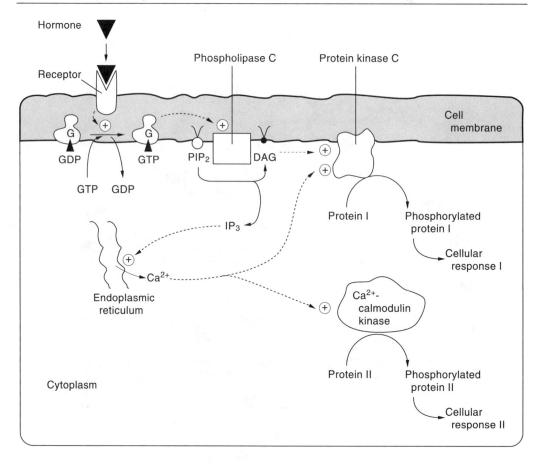

Figure 8-5. Signal transduction involving Ca^{2+} and the phosphatidylinositol bisphosphate (PIP_2) system. DAG = diacylglycerol; IP_3 = inositol 1,4,5-trisphosphate; G = G proteins; $\oplus$ = stimulates.

B. Hormone action mediated by gene activation (Figure 8-7)

1. Steroid and thyroid hormones, 1,25-DHC, and retinoic acid cross the cell membrane and bind to **intracellular receptors**.

2. These **receptors** have amino acid sequences that are very similar and contain domains that bind the hormone and domains that bind to DNA.

3. When the hormone-receptor complex is activated, it **binds to a regulatory element** (i.e., hormone response element [HRE]) **on DNA** that stimulates or inhibits the synthesis of mRNA (see Figure 8-7).

4. Translation of this mRNA produces **proteins** that are responsible for certain physiologic effects.

III. Regulation of Hormone Levels

● In order to maintain homeostasis or to repeat physiologic processes such as the menstrual cycle, hormone levels must be regulated.

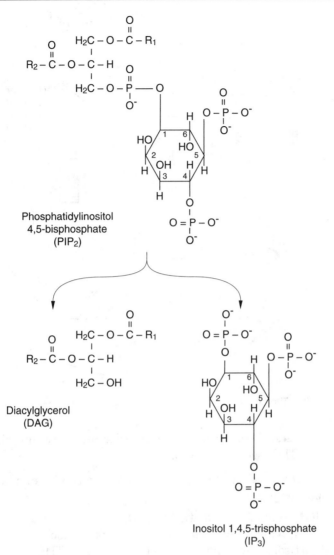

Phosphatidylinositol
4,5-bisphosphate
(PIP$_2$)

Diacylglycerol
(DAG)

Inositol 1,4,5-trisphosphate
(IP$_3$)

Figure 8-6. Structures of phosphatidylinositol bisphosphate, diacylglycerol, and inositol trisphosphate.

A. Regulation of hormone synthesis and secretion

1. The release of hormones is stimulated either by **changes** in the **environment** or physiologic state or by a stimulatory hormone from another tissue that acts on the cells that release the hormone. For example:

    a. A decrease in blood pressure causes a sequence of events that ultimately cause the adrenal gland to release **aldosterone**.

    b. The anterior pituitary releases ACTH, which stimulates the adrenal gland to release cortisol (Figure 8-8).

2. The physiologic effect of the hormone or the hormone itself causes a **decrease in the signal** that initially promoted the synthesis and release of the hormone. For example:

    a. Aldosterone causes an increased resorption from the kidney tubule of Na^+, and consequently of water, increasing blood pressure.

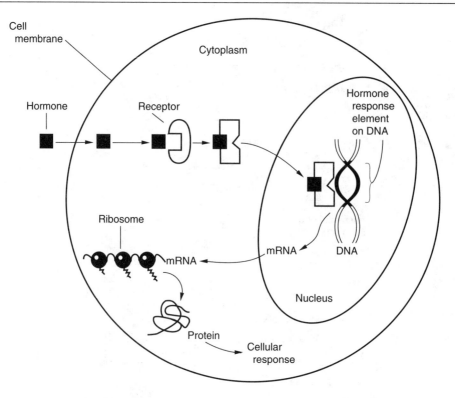

Figure 8-7. The mechanism of action of hormones of the steroid–thyroid family.

  b. Cortisol feeds back on the anterior pituitary, inhibiting the release of ACTH (see Figure 8-8).

 3. Complex feedback loops may **regulate** the **level** of a hormone in the blood (see Figure 8-8).

  –**CRH,** released from the hypothalamus, stimulates the release of ACTH, which stimulates the release of cortisol. Increased blood levels of cortisol inhibit the release of CRH and ACTH.

B. Hormone inactivation

 1. After hormones exert their physiologic effects, they are inactivated and excreted or degraded.

 2. Some hormones are converted to compounds that are no longer active and may be readily excreted from the body.

  –Cortisol, a steroid hormone, is reduced and conjugated with glucuronide or sulfate and excreted in the urine and the feces.

 3. Some hormones, particularly the polypeptides, are taken up by cells by the process of endocytosis and subsequently **degraded by lysosomal enzymes**.

  –The **receptor,** which is internalized along with the hormone, may either be **degraded** by lysosomal proteases, or it may be **recycled** to the cell membrane.

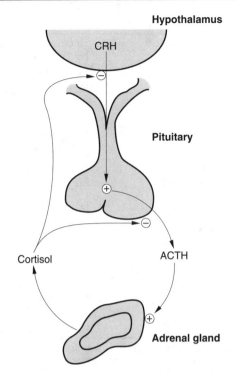

Figure 8-8. Hormone feedback regulation. CRH = corticotropin releasing hormone; ACTH = adrenocorticotropic hormone; ⊕ = activates; ⊖ = inhibits.

IV. The Actions of Specific Hormones

● Tissues that produce hormones include the hypothalamus, anterior and posterior pituitary, adrenal cortex and medulla, gonads, thyroid and parathyroid glands, heart, brain, cells of the gastrointestinal tract, and the pancreas.

A. Hypothalamic hormones

–The hypothalamus produces **vasopressin (VP)**, **oxytocin,** and other hormones (mainly peptides and polypeptides) that regulate the synthesis and release of hormones from the anterior pituitary (Table 8-2).

B. Hormones of the posterior pituitary

1. VP (also called antidiuretic hormone [ADH]) and **oxytocin** travel through nerve axons to the posterior pituitary where they are stored, each complexed with a neurophysin.

2. They are released into the blood in response to the appropriate stimulation.

3. VP stimulates the **resorption of water** by kidney tubules.

4. Oxytocin promotes the **ejection of milk** from the mammary gland and the **contraction of the uterus** during childbirth.

C. Hormones of the anterior pituitary (see Table 8-2)

1. Prolactin (PRL)

–Prolactin, produced in response to suckling of an infant, stimulates the **synthesis of milk proteins** during lactation.

Table 8-2. Actions of Hypothalamic and Anterior Pituitary Hormones

Hypothalamic Hormone	Anterior Pituitary Hormone	Target Gland Hormone	Process Affected by the Target Gland Hormone
CRH	ACTH and other hormones from the POMC gene	Adrenal Cortisol	Response to stress
TRH	TSH	Thyroid T_3, T_4	Energy production (fuel metabolism)
GnRH	LH, FSH	Gonads Estradiol Progesterone Testosterone	Reproduction Menstrual cycle Pregnancy Production of sperm
GRH (Somatocrinin)	GH	Various cells IGF (Somatomedin)	Cell growth
GRIH (Somatostatin)	GH$\ominus$	IGF$\ominus$	
PRF(?)	PRL		Lactation (mammary gland)
PRIH (Dopamine?)	PRL$\ominus$		Lactation (inhibited)

The hypothalamic hormones in column 1 stimulate the release of the corresponding anterior pituitary hormones in column 2 unless inhibition is indicated by $\ominus$. The anterior pituitary hormones act on the target glands in column 3 (or in column 4, in the case of PRL), which release hormones that have the effect listed in column 4. *Question marks* (?) indicate hormones that have not been fully established.

2. Growth hormone (GH)

–Although growth hormone (GH) stimulates the **release of insulin-like growth factors** (IGFs) primarily from the liver, it **antagonizes the effects of insulin on carbohydrate and fat metabolism**.

3. TSH stimulates the **release of T_3 and T_4** from the thyroid gland.

4. Luteinizing hormone (**LH**) and follicle stimulating hormone (**FSH**) **stimulate** the **gonads** to release hormones that are involved in reproduction.

5. The protein product of the **proopiomelanocortin (POMC)** gene may be cleaved to generate a number of polypeptides.

a. Adrenocorticotropic hormone (ACTH) affects the production of cortisol and aldosterone by the adrenal cortex.

b. Lipotropin (LPH) may be cleaved to form melanocyte stimulating hormone and endorphins.

c. Melanocyte stimulating hormone (MSH) stimulates the production of the pigment melanin by the melanocytes in the skin.

d. Endorphins produce analgesic effects.

D. Thyroid hormone

1. T_3 may be the only metabolically active thyroid hormone. T_3 binds to the thyroid receptor with 10 times the affinity of T_4.

a. Although the thyroid secretes some T_3, the majority is produced by deiodination of T_4, a process that occurs in nonthyroidal tissue.

b. During starvation, T_4 may be converted to reverse T_3 (rT_3), which is not active.

2. Thyroid hormone binds to nuclear receptors and **regulates gene expression**.

–It may stimulate production of the Na^+-K^+ ATPase.

3. Thyroid hormone is necessary for **growth, development, and maintenance** of almost all tissues of the body.

4. It **stimulates** oxidative metabolism and causes the **basal metabolic rate** (BMR) to increase.

E. Hormones that stimulate growth

1. Insulin and **GH** stimulate growth and promote protein synthesis.

2. GH antagonizes the action of insulin, stimulating gluconeogenesis and promoting lipolysis.

F. Hormones that mediate the response to stress

–**Glucocorticoids** (particularly cortisol) and **epinephrine** act in concert to supply fuels to the blood so that energy can be produced to combat stressful situations (Figure 8-9).

1. Glucocorticoids

–In response to ACTH, the adrenal cortex produces glucocorticoids. Cortisol is the major glucocorticoid in humans.

a. Glucocorticoids have **anti-inflammatory effects**.

–They induce the synthesis of **lipocortin**, a protein that inhibits phospholipase A_2, the rate-limiting enzyme in prostaglandin, thromboxane, and leukotriene synthesis (see Figure 6-16).

b. Glucocorticoids **suppress the immune response** by causing the lysis of lymphocytes.

c. Glucocorticoids **influence metabolism** by causing the movement of fuels from peripheral tissues to the liver, where gluconeogenesis and glycogen synthesis are stimulated (see Figure 8-9).

(1) Amino acids are released from muscle protein.

(2) Lipolysis occurs in adipose tissue.

(3) In addition to providing amino acids and glycerol as carbon sources, **glucocorticoids promote gluconeogenesis** by inducing synthesis of the enzyme phosphoenolpyruvate carboxykinase (PEPCK).

(4) Glucose, produced by gluconeogenesis promoted by glucocorticoids, is **stored as glycogen** in the liver.

(5) Glucocorticoids prepare the body during stressful conditions so that fuel stores are ready for the "alarm" reaction mediated by epinephrine.

2. Epinephrine

a. Among its effects, epinephrine increases blood glucose by **stimulating liver glycogenolysis** (see Figure 8-9).

b. Epinephrine also **stimulates lipolysis** in adipose tissue.

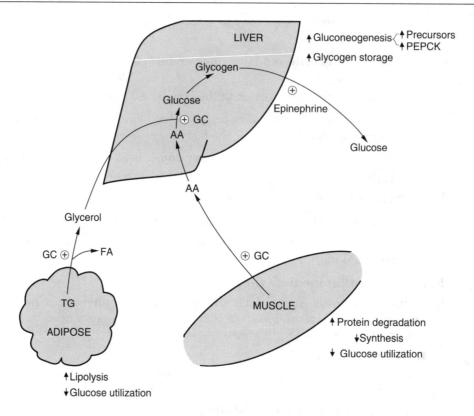

Figure 8-9. The effects of glucocorticoids (GC) and epinephrine on fuel metabolism.

G. Hormones that regulate salt and water balance

–In addition to **VP** (see IV B) and **ANF** (see II V A 2 c), **aldosterone** is involved in regulating salt and water balance.

1. Synthesis of aldosterone

   a. Renin (produced by the juxtaglomerular cells of the kidney in response to decreased blood pressure, blood volume, or sodium ion concentration) **cleaves angiotensinogen to angiotensin I.**

   b. Angiotensin I is **cleaved** to the 8 amino acid peptide hormone **angiotensin II** by angiotensin converting enzyme (ACE), which is made in the lung.

   –Further cleavage to angiotensin III may occur.

   c. Angiotensin II acts directly on vascular smooth muscle cells, causing **vasoconstriction,** which increases blood pressure.

   d. Angiotensin II and III (and also decreased serum [Na^+] and increased serum [K^+]) **stimulate** the glomerulosa cells of the adrenal cortex to produce and secrete **aldosterone.**

   –ACTH has a permissive effect (i.e., it maintains cells so that they can respond to angiotensin II).

2. Action of aldosterone

   a. Aldosterone causes the production of proteins in cells of the distal tubule and the collecting ducts of the kidney.

    (1) A **permease** is **produced** that allows Na^+ to enter cells from the lumen.

    (2) **Citrate synthase** is **induced,** which stimulates tricarboxylic acid (TCA) cycle activity and results in the generation of ATP.

    (3) Energy is thus provided to drive the **Na^+-K^+ ATPase,** which also may be induced.

  b. Overall, K^+ and H^+ are lost, Na^+ is retained, water is resorbed, and blood volume and pressure are increased.

  c. These changes cause a decrease in the release of renin by the juxtaglomerular cells.

  d. This homeostatic system maintains blood pressure and volume.

H. Hormones that control reproduction

–The hypothalamus produces gonadotropin releasing hormone (**GnRH**), which causes the anterior pituitary to release **FSH** and **LH,** which **act on** both the **ovary** and the **testis.**

1. The action of FSH and LH on the ovary

a. The menstrual cycle

    (1) Initially, **FSH acts on** the **follicles** to promote maturation of the ovum and to stimulate estradiol production and secretion.

    (2) **Estradiol acts on** the uterine **endometrium,** causing it to thicken and vascularize in preparation for implantation of a fertilized egg.

    (3) A **surge of LH** at the midpoint of the menstrual cycle **stimulates** the ripe **follicle to ovulate,** leaving the residual follicle, which forms the **corpus luteum** and **secretes** both **progesterone and estradiol.**

    (4) **Progesterone** causes the endometrium to continue to thicken and vascularize and increase its secretory capacity.

b. Events in the absence of fertilization

    (1) The **corpus luteum regresses** due to declining LH levels. It produces diminishing amounts of progesterone and estradiol.

    (2) Because of the low steroid hormone levels, the cells die and the degenerating **endometrium is sloughed** into the uterine cavity and excreted (**menstruation**).

    (3) The low levels of estradiol and progesterone cause feedback inhibition to be relieved, and the hypothalamus releases GnRH, initiating a new menstrual cycle.

c. Events following fertilization

    (1) The **corpus luteum** is **maintained** initially by **human chorionic gonadotropin (hCG)** produced by the cells of the developing embryo (trophoblast).

    (2) Subsequently, the **placenta produces hCG and progesterone.**

    (3) After the corpus luteum dies, the placenta continues to produce large amounts of progesterone.

    (4) Near **term,** hCG and, subsequently, **progesterone levels fall.**
       –Fetal cortisol may cause the decline in progesterone.

    (5) **Prostaglandin $F_{2\alpha}$ ($PGF_{2\alpha}$) and oxytocin** (released from both

maternal and fetal pituitaries) **stimulate uterine contractions,** and the infant is delivered.

2. The action of FSH and LH on the testis

   a. LH stimulates Leydig cells to produce and secrete **testosterone**.

   b. FSH acts on Sertoli cells of the seminiferous tubule to promote the **synthesis of androgen binding protein (ABP)**.

   c. ABP binds testosterone and transports it to the site of spermatogenesis, where **testosterone is reduced** to the more potent androgen dihydrotestosterone (DHT).

   d. Testosterone plays a role in **spermatogenesis** in the adult male.

   (1) Testosterone is responsible for masculinization during early development.

   (2) At puberty, testosterone promotes sexual maturation of the male.

I. Hormones that promote lactation

   –Many hormones are necessary for development of the mammary glands during adolescence.

1. Preparation of the mammary gland for lactation

   a. During pregnancy, **prolactin, glucocorticoids, and insulin** are the major hormones responsible for differentiation of mammary alveolar cells into secretory cells capable of producing milk.

   b. PRL stimulates the **synthesis of the milk proteins,** particularly casein and α-lactalbumin.

   (1) α-Lactalbumin, the major protein in human milk, serves as a **nutrient**.

   (2) α-Lactalbumin binds to galactosyl transferase, decreasing its K_m for glucose and, thus, **stimulating synthesis of** the milk sugar **lactose** (see Chapter 5 VII B 2 b).

   c. Progesterone inhibits milk protein production and secretion during pregnancy.

   d. At term, when progesterone levels fall, the inhibition of milk protein synthesis is relieved.

2. Regulation of milk secretion during lactation

   a. PRL causes milk proteins to be produced and secreted into the alveolar lumen.

   b. Oxytocin causes contraction of the myoepithelial cells surrounding the alveolar cells and the lumen, and **milk is ejected** through the nipple.

   c. The secretion of both PRL and oxytocin by the pituitary is stimulated by suckling of the infant and by other factors.

J. Hormones involved in growth and differentiation

1. Retinoids are produced in the body **from** dietary **vitamin A** (see Figure 4-10).

   –Dietary β-carotene is cleaved to 2 molecules of retinal.

2. Retinal is a functional component of the reactions of the **visual cycle**.

3. Retinol (an alcohol) and retinal (an aldehyde) may be interconverted by oxidation and reduction reactions.

4. Retinoic acid may be produced by oxidation of retinal and cannot be reduced.

5. **Retinol and retinoic acid** are involved in **growth** and also in **differentiation** and **maintenance** of **epithelial tissue.**

6. **Retinol** also is involved in **reproduction.**

7. Some of the functions of **retinoic acid** result from its ability to **activate genes.**

K. Hormones that regulate Ca^{2+} metabolism

–Calcium has many important functions. It is involved in blood coagulation, activation of muscle phosphorylase, and secretory processes. It combines with phosphate to form the hydroxyapatite of bone.

–Parathyroid hormone (PTH) and 1,25-DHC are the major regulators of Ca^{2+} metabolism.

–Calcitonin also may be involved, but its role in humans is unclear.

1. Parathyroid hormone (PTH)

–**PTH,** produced in response to low calcium levels, **acts to increase Ca^{2+} levels** in the extracellular fluid.

a. PTH promotes Ca^{2+} and phosphate mobilization from bone.

b. PTH acts on renal tubules to resorb Ca^{2+} and excrete phosphate.

c. PTH stimulates the synthesis of 1,25-DHC.

2. 1,25-Dihydroxycholecalciferol

a. The major action of 1,25-DHC is to **stimulate** the synthesis of a protein involved in Ca^{2+} **absorption by intestinal cells.**

b. 1,25-DHC acts synergistically with PTH in bone resorption and promotes resorption of Ca^{2+} by renal tubular cells.

L. Hormones that regulate the utilization of nutrients

1. Gut hormones

a. Gastrin from the gastric antrum and the duodenum stimulates gastric acid and pepsin secretion.

b. Cholecystokinin (CCK) from the duodenum and jejunum stimulates contraction of the gallbladder and the secretion of pancreatic enzymes.

c. Secretin from the duodenum and jejunum stimulates the secretion of bicarbonate by the pancreas.

d. Gastric inhibitory polypeptide (GIP) from the small bowel enhances insulin release and inhibits secretion of gastric acid.

e. Vasoactive intestinal polypeptide (VIP) from the pancreas relaxes smooth muscles and stimulates bicarbonate secretion by the pancreas

2. Insulin and glucagon

–The two major hormones that **regulate fuel metabolism,** insulin and glucagon, are produced by the pancreas (discussed extensively in Chapters 5, 6, and 7).

V. Clinical Correlations

A. Hypothyroidism

In patients with hypothyroidism, the stimulatory effect of thyroid hormone on the oxidation of fuels is diminished. As a consequence, the generation of ATP is reduced, causing a sense of **weakness, fatigue,** and **hypokinesis**. The reduced metabolic rate is associated with diminished heat production, causing **cold intolerance** and **decreased sweating**. With less demand for the delivery of fuels and oxygen to peripheral tissues, the circulation is slowed, causing a reduction in heart rate and, when far advanced, a reduction in blood pressure.

B. Hyperthyroidism

When the thyroid gland secretes excessive quantities of thyroid hormone, the rate of oxidation of fuels by muscle and other tissues is increased. With enhanced oxidative metabolism, heat production is increased, leading to a sense of **heat intolerance** and the need to dissipate heat through **increased sweating**. Thyroid hormone excess raises the tone of the sympathetic (adrenergic) nervous system, **raising the heart rate and systolic blood pressure**. In addition, tremulousness, a sense of restlessness, and insomnia often occur. Since stored fuels in muscle and fat tissue are being utilized at an excessive rate, **weight loss** occurs in spite of increased caloric intake.

C. Growth hormone excess

Excessive secretion of growth hormone (GH) occurs as a result of a **benign tumor** of the **anterior pituitary gland**. If the hypersecretion begins prior to closure of the growth centers in the long bones, **excessive height** (gigantism) occurs. If hypersecretion begins after the growth centers have closed, the bones grow in bulk and width, leading to a condition called **acromegaly**. Soft tissue overgrowth occurs as well, leading to **organomegaly,** thickness of the skin, and coarseness of the facial features.

Chronic GH excess may lead to **glucose intolerance** because of its diabetogenic actions. If the pituitary tumor grows beyond the confines of the sella turcica, the tumor may encroach on the optic nerves, causing visual difficulties, or may lead to other cranial nerve dysfunction, progressive headaches, and eventually, symptoms of increased intracranial pressure.

D. Prolactin excess

The most common secretory neoplasm of the anterior pituitary gland is a **prolactin-secreting adenoma** (prolactinoma). An early symptom of prolactin excess is a **milky discharge from the breasts**. Because hyperprolactinemia may suppress the secretion of gonadotropic hormones (e.g., LH, FSH), **menstrual irregularity, amenorrhea,** and **infertility** may occur in women. In men, **gonadotropic hormone dysfunction** may lead to reduced libido, sexual impotence, and infertility.

E. Hypoglycemia

Whenever insulin levels in the blood are chronically elevated as occurs, for example, in patients with insulin-secreting pancreatic tumors or in patients inadvertently given excessive quantities of exogenous insulin, the transport of glucose from the blood into tissues such as skeletal muscle and fat cells is enhanced, leading to hypoglycemia. The clinical manifestations of a **reduction in blood glucose levels** are those related to stimulation of the sympathetic nervous system by **hypoglycemia** (e.g., sweating, palpitations of the heart, tremulousness) and those due to inadequate delivery of glucose to the brain, also known as **neuroglycopenic sequelae** (e.g., irritability, slurring of speech, confusion, drowsiness, and eventually, coma).

F. Hyperparathyroidism

When the secretion of parathyroid hormone is excessive, the physiologic effects of this peptide on the gut, the skeleton, and the kidney tubules are enhanced. The percentage of dietary calcium absorbed into the circulation is increased, calcium ions are released from bone and enter the blood more rapidly, and the renal tubules reabsorb more calcium than usual from the luminal urine, all leading to **hypercalcemia**. Chronic hypercalcemia is associated with vague generalized musculoskeletal pain, fatigue, and eventually, slowed mentation.

The osteolytic effect of excessive parathyroid hormone action may lead to demineralization of the skeleton (**osteoporosis**) and fracture. Chronic renal filtration of blood rich in calcium leads to saturation of the tubular fluid with calcium salts; as a consequence, **renal calculi** (kidney stones) may occur.

G. Menopause

Clinically, menopause is defined as the last physiologic menstrual cycle in which pituitary gonadotropins (e.g., LH, FSH) have stimulated the maturation of a primordial follicle and caused ovulation. The transition from full reproductive potential to final ovarian failure is known as the **climacteric,** a period that begins about age 40. As estradiol levels in the blood gradually fall below normal, the early symptoms of the climacteric occur, including anxiety, mood swings, irritability, and in some instances, depression. Vasomotor instability with early morning paresthesias in the extremities and ascending bouts of sweating and flushing (**"hot flash"**) also occur.

Late symptoms of the climacteric are those related to chronic exposure of the urogenital tract to less than normal physiologic levels of estradiol. These include vaginal dryness and laxity of the supportive tissues of the urogenital tract, causing a predisposition to **urinary tract infections** and **painful intercourse**.

H. Glucocorticoid excess

Hypercortisolemia has an adverse effect on virtually every tissue of the body. Central nervous system effects range from **hyperirritability** to **depression**. The catabolic effect on protein-containing tissues leads to a reduction in the ground substance of bone and eventual **osteoporosis,** loss of muscle protein causing weakness, and thinning and tearing of dermal and epidermal structures, which is manifest as reddish stripes, or striae, over the lower abdomen, the lateral thorax, and other areas where skin tension is increased. Similar catabolic effects on the elastin of vessel walls leads to vascular fragility with easy bruising and hemorrhaging of the skin. A suppressive effect on immunocompetence may increase the likelihood of infection.

The diabetogenic actions of cortisol may lead to **glucose intolerance** or overt **diabetes mellitus**. A peculiar tendency for the disposition of fat in the face (moon facies), posterior neck (buffalo hump), thorax, and abdomen, while sparing the distal extremities, causes a distinct "central obesity."

This constellation of clinical signs and symptoms resulting from chronic hypercortisolemia is referred to as **Cushing's syndrome**.

I. Mineralocorticoid deficiency

A deficiency of adrenocortical secretion of aldosterone is usually accompanied by a reduction in the secretion of other adrenal steroid hormones as well. The loss of adrenocortical steroids is known as **Addison's disease**.

The mineralocorticoid deficiency leads to a net loss of sodium ions and water into the urine with a reciprocal retention of potassium ions (**hyperkalemia**) and hydrogen ions (**mild metabolic acidosis**). The subsequent contraction of the effective plasma volume may lead to a reduction in blood pressure. If volume loss is profound, perfusion of vital tissues such as the brain could lead to lightheadedness and possible loss of consciousness.

Review Test

Directions: Each of the numbered items or incomplete statements in this section is followed by answers or by completions of the statement. Select the **one** lettered answer or completion that is **best** in each case.

1. Which of the following acts to increase the release of Ca^{2+} from the endoplasmic reticulum?

(A) Diacylglycerol (DAG)
(B) Inositol trisphosphate (IP_3)
(C) Parathyroid hormone (PTH)
(D) 1,25-Dihydroxycholecalciferol (1,25-DHC)

2. A dietary deficiency of iodine would

(A) directly affect the synthesis of thyroglobulin on ribosomes
(B) result in increased secretion of thyroid stimulating hormone (TSH)
(C) result in decreased production of thyrotropin releasing hormone (TRH)
(D) result in increased heat production

3. Which of the following is true of testosterone?

(A) May be converted to a more active androgen in its target cells
(B) Acts by binding to receptors on the cell surface
(C) Is produced from estradiol (E_2)
(D) Stimulates the synthesis of gonadotropin releasing hormone (GnRH) by the hypothalamus

4. Which of the following is true of epinephrine?

(A) Acts only through the phosphatidylinositol bisphosphate system
(B) Is synthesized from tyrosine
(C) Causes the level of cAMP in liver cells to decrease
(D) Functions like a steroid hormone

5. GnRH stimulates the release of

(A) GH
(B) T_3 and T_4
(C) PRL
(D) IGF
(E) LH and FSH

6. A key intermediate for the synthesis of both testosterone and cortisol from cholesterol is

(A) 7-hydroxycholesterol
(B) pregnenolone
(C) aldosterone
(D) retinoic acid

7. In the synthesis of 1,25-DHC from 7-dehydrocholesterol

(A) the steroid ring structure remains intact
(B) cholesterol is an intermediate
(C) ultraviolet light is required
(D) three hydroxylations occur

8. A patient with central obesity, thin limbs, and purple striae on the abdomen complained of muscle weakness, depression, and blurred vision. The patient's blood glucose was 280 mg/dl (reference range = 70–100 mg/dl). No ketone bodies were present in the urine. Plasma cortisol levels were 56 µg/ml (reference range = 3–31 µg/ml), and plasma ACTH levels were 106 pg/ml (reference range = 0–100 pg/ml). A low dose (1 mg) of dexamethasone (synthetic glucocorticoid) was administered in the evening. This dose failed to suppress the plasma cortisol level by the next morning. After a high-dose (8 mg) dexamethasone suppression test, the plasma cortisol level was 21 µg/ml. Based on this information, if the patient's problem is due to a single cause, the most likely diagnosis is

(A) non–insulin-dependent diabetes mellitus
(B) insulin-dependent diabetes mellitus
(C) a secretory tumor of the anterior pituitary
(D) a secretory tumor of the posterior pituitary
(E) a secretory tumor of the adrenal cortex

Questions 9–11

A patient complains of nervousness, palpitations, sweating, and weight loss without loss of appetite, and has a goiter. Suspecting a defect in thyroid function, the physician orders a total serum T_4. The test is performed by radioimmunoassay. The standard curve for the assay, which measures T_4 in 0.1 ml of serum, is shown below. Normal levels of $T_4 = 4 - 10$ µg/dl. In an assay of 0.1 ml of the patient's serum, 15% of the radioactive T_4 was bound to the antibody.

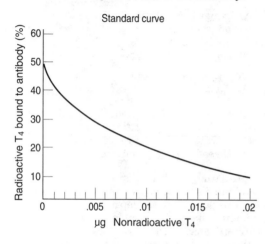

Standard curve

9. According to the radioimmunoassay, the approximate blood levels of T_4 are

(A) 0.015 µg/dl
(B) 0.15 µg/dl
(C) 15 µg/dl
(D) 20 µg/dl
(E) 30 µg/dl

10. The patient is

(A) hypothyroid
(B) normal
(C) hyperthyroid

11. TSH levels in the patient's blood were also measured by radioimmunoassay. If pituitary function is normal, the patient's TSH levels will most likely be

(A) higher than normal
(B) normal
(C) lower than normal

Directions: Each group of items in this section consists of lettered options followed by a set of numbered items. For each item, select the **one** lettered option that is most closely associated with it. Each lettered option may be selected once, more than once, or not at all.

Questions 12–17

Match the characteristic below with the appropriate hormone.

(A) Luteinizing hormone (LH)
(B) Prolactin (PRL)
(C) Thyroid stimulating hormone (TSH)
(D) Growth hormone (GH)
(E) Follicle stimulating hormone (FSH)

12. Has its release inhibited by thyroxine
13. Binds to receptors on Leydig cells
14. Stimulates production of insulin-like growth factor (IGF)
15. Stimulates the synthesis of milk proteins
16. Stimulates the production of progesterone by the corpus luteum
17. Stimulates the production of estradiol by the immature ovarian follicle

Questions 18–22

Match the characteristic with the appropriate steroid hormone.

(A) Cortisol
(B) Aldosterone
(C) Both cortisol and aldosterone
(D) Neither cortisol nor aldosterone

18. Action mediated by a second messenger
19. Synthesized from cholesterol by cells of the adrenal cortex
20. Receptors that have a DNA binding domain
21. Associated with induction of phosphoenolpyruvate carboxykinase (PEPCK)
22. Secreted in response to angiotensin II

Questions 23–26

Match the characteristic below with the appropriate hormone.

(A) Oxytocin
(B) Vasopressin
(C) Both oxytocin and vasopressin
(D) Neither oxytocin nor vasopressin

23. Produced by the anterior pituitary
24. Found associated with neurophysin in secretory granules
25. Associated with diuresis
26. Produced from the proopiomelanocortin (POMC) gene

Answers and Explanations

1–B. Phosphatidylinositol bisphosphate is cleaved to IP_3 and DAG. IP_3 causes the release of Ca^{2+} from the endoplasmic reticulum, while DAG activates protein kinase C. PTH stimulates the release of Ca^{2+} from bone, and 1,25-DHC stimulates the absorption of Ca^{2+} from the intestine.

2–B. When iodine is deficient in the diet, the thyroid does not make normal amounts of thyroid hormone. Consequently, there is less feedback inhibition of TSH production and release. Low levels of thyroid hormone result in decreased heat production.

3–A. Testosterone is reduced to dihydrotestosterone (DHT), the more active hormone. Testosterone is a steroid hormone, so it activates genes. It is a precursor of E_2. Testosterone inhibits the synthesis of GnRH.

4–B. Epinephrine is synthesized from tyrosine. It functions like a polypeptide hormone, binding to receptors on the cell membrane. cAMP levels rise in response to epinephrine.

5–E. GnRH stimulates the release of two pituitary hormones, LH and FSH. GRH stimulates the release of GH; TSH, the release of T_3 and T_4; and PRF, the release of prolactin. IGF stimulates growth.

6–B. Both pregnenolone and progesterone are intermediates in the synthesis of steroid hormones from cholesterol. 7-Hydroxycholesterol is an intermediate in bile salt synthesis, and aldosterone is a mineralocorticoid produced well beyond the branch point for the synthesis of the adrenal and gonadal steroids. Retinoic acid is derived from vitamin A.

7–C. 1,25-DHC is formed from an intermediate in cholesterol synthesis, 7-dehydrocholesterol. The B ring of this compound is cleaved by a reaction requiring ultraviolet light and, subsequently, hydroxyl groups are added at position 25 in the liver and at position 1 in the kidney.

8–C. The fact that ACTH levels initially were elevated and that administration of a glucocorticoid caused the plasma cortisol level to decrease indicates that the problem is in the anterior pituitary. It is overproducing ACTH, causing cortisol to be overproduced by the adrenal gland. The patient's hyperglycemia was caused by the elevated cortisol.

9–C. If 15% of the radioactive T_4 is bound to antibody, the amount of T_4 in 0.1 ml of the patient's serum is 0.015 μg/0.1 ml or 15 μg/dl. (1.0 dl = 100 ml.)

10–C. The patient's T_4 level is above the normal range—the patient is hyperthyroid.

11–C. Thyroid hormone suppresses TSH secretion by the anterior pituitary. If thyroid hormone levels are elevated, TSH levels will be lower than normal.

12–C. Thyroxine inhibits the release of TSH by the anterior pituitary.

13–A. LH binds to receptors on Leydig cells and stimulates the release of testosterone.

14–D. GH stimulates the release of IGF by the liver.

15–B. Prolactin stimulates the synthesis of milk proteins.

16–A. LH stimulates the corpus luteum to produce progesterone.

17–E. FSH stimulates maturation of the ovarian follicle, which produces estradiol.

18–D. Steroid hormones do not act through second messengers. They enter the cell and activate genes.

19–C. Steroid hormones are synthesized from cholesterol. Cortisol and aldosterone are made in the adrenal cortex.

20–C. Steroid hormones bind to receptors that subsequently bind to DNA, as do 1,25-DHC, thyroid hormone, and retinoic acid (from vitamin A).

21–A. Glucocorticoids, such as cortisol, activate the gene for PEPCK.

22–B. Angiotensin II stimulates the synthesis and secretion of aldosterone.

23–D. Oxytocin and vasopressin are produced by the hypothalamus and stored in and secreted from the posterior pituitary.

24–C. Both oxytocin and vasopressin are bound to neurophysins.

25–D. Vasopressin has an antidiuretic action. Neither acts as a diuretic.

26–D. POMC is produced by the anterior pituitary.

Comprehensive Examination

Directions: Each of the numbered items or incomplete statements in this section is followed by answers or by completions of the statement. Select the **one** lettered answer or completion that is **best** in each case.

1. A patient who needed to lose weight began eating at fast-food restaurants. He did not change his exercise level. However, the composition of his diet was altered in that his carbohydrate intake decreased by 50 g/day and his fat intake increased by 50 g/day. Otherwise, his diet remained the same. On this diet

(A) he gained weight
(B) he lost weight
(C) his weight remained the same

2. A patient who is obese and has hypertension requires a weight reduction diet. She weighs 176 pounds and has a sedentary lifestyle. What is the approximate number of calories the patient burns each day at this weight?

(A) 1920
(B) 2500
(C) 4220
(D) 5490

3. Which of the following compounds can be synthesized in humans?

(A) Riboflavin
(B) Linoleic acid
(C) Leucine
(D) Thiamine
(E) Niacin

4. Each of the following metabolites provides carbon for glucose synthesis by the process of gluconeogenesis EXCEPT

(A) amino acids from muscle protein
(B) lactate from red blood cells and exercising muscle
(C) glycerol from adipose triacylglycerols
(D) even-chain fatty acids from adipose triacylglycerols

5. A person who accidentally ingested a compound that completely inhibited fructose 1,6-bisphosphatase could still form substantial amounts of blood glucose from

(A) muscle glycogen stores
(B) lactate produced by red blood cells
(C) ingested fructose
(D) ingested galactose
(E) ingested fructose and galactose

6. A person who accidentally ingested a compound that completely inhibited phosphoenolpyruvate carboxykinase could still form substantial amounts of blood glucose from

(A) muscle glycogen stores
(B) lactate produced by red blood cells
(C) ingested fructose
(D) ingested galactose
(E) ingested fructose and galactose

7. A solution contains 2×10^{-3} moles per liter of a weak acid (pK = 3.5) and 2×10^{-3} moles per liter of its conjugate base. Its pH is

(A) 4.1
(B) 3.9
(C) 3.5
(D) 3.1
(E) 2.7

8. Hydroxymethylglutaryl CoA

(A) is formed by catabolism of valine
(B) gives rise to ketone bodies by cleavage to acetyl CoA and acetoacetyl CoA
(C) serves as a precursor of cholesterol
(D) is formed from glutamic acid by the direct action of HMG CoA synthetase

9. The cytochrome P_{450} system of liver in normal individuals has a capacity (V_m) to oxidize approximately 10 nmol of drug X per minute per gram of liver. When the concentration of drug X in the liver is 2 μM, oxidation products are formed at the rate of 4 nmol/min/g of liver. What is the K_m of cytochrome P_{450} for this drug?

(A) 4 nM
(B) 5 nM
(C) 10 nM
(D) 2 μM
(E) 3 μM

10. Enzyme Y was purified from a tissue sample obtained from a patient. The kinetic properties of this enzyme and those of the same enzyme isolated from a normal individual are shown in the graph below. Which of the following statements is TRUE?

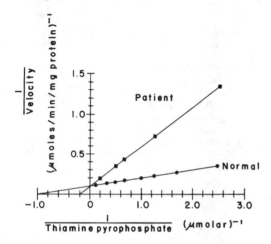

(A) The enzymes do not have the same V_m
(B) A lower concentration of thiamine pyrophosphate (TPP) is required to saturate the patient's enzyme
(C) The patient's enzyme has a K_m for thiamine pyrophosphate that is less than that for the normal enzyme
(D) Administration of thiamine to the patient should result in a greater proportion of the enzyme in the active enzyme-TPP complex

11. A deficiency of pantothenic acid would directly affect the reaction catalyzed by

(A) citrate synthase
(B) isocitrate dehydrogenase
(C) succinate dehydrogenase
(D) fumarase
(E) malate dehydrogenase

12. The oxidation of acetyl CoA by the citric acid cycle plays a major role in providing energy in each of the following tissues EXCEPT

(A) muscle
(B) brain
(C) liver
(D) red blood cells
(E) kidney

13. Which of the following statements about adult hemoglobin (HbA) is TRUE?

(A) HbA is composed of two β and two γ subunits
(B) Four subunits combine to form the primary structure of HbA
(C) Each subunit of HbA contains one heme
(D) HbA binds 1 mole of O_2 per mole of protein
(E) The β chain of HbA is more hydrophobic than the β chain of sickle cell hemoglobin (HbS)

14. In scurvy, defective collagen is produced because hydroxylation of procollagen does not occur at the normal rate. Hydroxylation of lysine residues on procollagen

(A) is not required for polymerization of collagen
(B) requires vitamin C
(C) occurs before incorporation of lysine into the polypeptide chain
(D) occurs after the mature collagen molecule is secreted from the cell

15. After an overnight fast

(A) fatty acids are the primary fuel for the brain
(B) glucose is synthesized from even-chain fatty acids
(C) ketone bodies are formed in the liver
(D) ketone bodies are a major fuel for red blood cells

16. Which of the following possible characteristics of protein synthesis in eukaryotes is TRUE?

(A) Is initiated by formyl-methionine
(B) Begins with the binding of mRNA to the 30S ribosomal subunit
(C) Can occur on 80S ribosomes attached to the rough endoplasmic reticulum
(D) Does not require peptidyl transferase for synthesis of peptide bonds
(E) Occurs on mRNA that is in the process of being synthesized

17. Which of the following statements concerning genes and transcription in eukaryotes is TRUE?

(A) The conversion of a UAG codon to UAA in mRNA would result in the incorporation of an incorrect amino acid into the polypeptide chain
(B) Genes always occur in one or a small number of copies in the genome
(C) The order of gene sequences in chromosomes is always the same in highly differentiated cells as in germ cells
(D) The primary product of transcription (hnRNA) often contains nucleotide sequences that do not code for the amino acid sequence of the protein encoded by a gene

18. The sequence for a portion of a gene responsible for a lysosomal storage disease (Tay-Sachs) has been determined. The normal gene sequence and the mutant gene sequence are given below. (There is a dot above every fifth base and a number above every tenth base.)

```
            •      10       •      20        •
Normal CGTATATCCTATGGCCCTGACCCAG
Mutant CGTATATCTATCCTATGGCCCTGAC
```

The amino acid sequence in this region of the normal protein is Arg - Ile - Ser - Tyr - Gly - Pro - Asp. Which of the following statements about this portion of the gene sequence is TRUE?

(A) The messenger RNA produced from this region of the mutant gene codes for the amino acid sequence given above
(B) The codon used for arginine in this sequence is AGA
(C) The mutant protein will be shorter than the normal protein
(D) The mutant gene contains a deletion that causes a frameshift mutation
(E) The mutant gene has a point mutation

19. During synthesis of eukaryotic mRNA

(A) RNA polymerase II binds to a promoter that contains a TATA box
(B) exons are transcribed from DNA and then cleaved from hnRNA
(C) a cap site serves as the signal for cleavage and addition of the poly(A) tail
(D) the template strand of DNA is covalently bound to histones

20. Bacterial cells growing in a medium containing glucose and all 20 amino acids are transferred to a medium in which the only sugar is lactose and NH_4^+ is the only source of nitrogen. In cells growing in the second medium, compared with those growing in the first medium

(A) cAMP levels will be lower
(B) CAP protein (cAMP-binding protein) will be bound to the *lac* promoter
(C) the *lac* repressor will be bound to the *lac* operator
(D) RNA polymerase will bind with lower affinity to the *trp* promoter
(E) attenuation of transcription of the *trp* operon will increase

21. Which of the following enzymes is NOT involved in DNA repair?

(A) Ligase
(B) DNA polymerase
(C) Reverse transcriptase
(D) An endonuclease
(E) A glycosylase

22. Use the figures below to answer question 22.

Normal

G   C   A   T

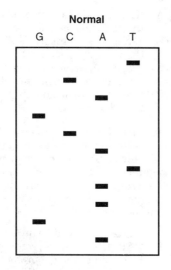

Mutant

G   C   A   T

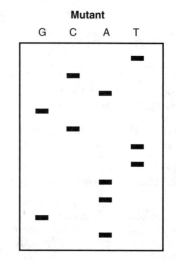

The gene responsible for a recessive genetic disorder was isolated and sequenced by the dideoxynucleotide method. The sequencing gel patterns obtained for a segment of the normal and mutant genes are shown above. A screening test was developed to identify individuals likely to have this disorder. This test used two restriction enzymes: *Kpn*I, which cleaves the sequence 5'-GGTACC-3', and *Eco*RI, which cleaves 5'-GAATTC-3'. One of these enzymes produced a DNA fragment from the normal and mutant genes that contained the sequences shown above. The other enzyme cleaved within one of these sequences. Which of the following statements about the sequences shown on the gels is TRUE?

(A) The normal sequence is cleaved by *Kpn*I
(B) The mutant sequence is cleaved by *Kpn*I
(C) The normal sequence is cleaved by *Eco*RI
(D) The mutant sequence is cleaved by *Eco*RI

23. Which of the following statements concerning Okazaki fragments is TRUE?

(A) They are produced by the action of an endonuclease on DNA
(B) They are produced from the parental DNA strand that is oriented 3' to 5' in the direction that the replication fork is moving
(C) They are joined by a phosphorylase
(D) They contain a very short sequence of RNA joined by a phosphodiester bond to a short sequence of DNA
(E) They require only DNA polymerase for their synthesis

24. Use the following figure to answer question 24.

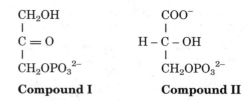

Compound I   **Compound II**

During glycolysis, compound I is converted to compound II shown above. In this conversion

(A) inorganic phosphate is produced
(B) NADH is consumed
(C) ATP is produced
(D) only one enzyme is required

25. After 2 days of fasting, the major process by which blood glucose is produced is

(A) glycolysis
(B) gluconeogenesis
(C) glycogenolysis
(D) the pentose phosphate pathway

26. An individual accidentally ingests a compound that inhibits glucose 6-phosphatase. After an overnight fast, this individual, compared with a healthy person, would have a higher

(A) rate of gluconeogenesis
(B) rate of glycogenolysis
(C) level of liver glycogen
(D) level of blood glucose

27. To determine the genetic father of a child (C), a laboratory test was performed that depended on a restriction fragment length polymorphism (RFLP) in a region upstream from a known gene. This region contains a variable number of tandem repeats (VNTR). Consequently, *Hpa*I produces restriction fragments that differ in size. The restriction fragment map is shown below.

DNA, extracted from a child (C), the child's mother (M), and three men (F1, F2, and F3) was digested with *Hpa*I and subjected to electrophoresis. A probe that hybridized to the known gene was used, so that only bands containing this gene were visualized. The gel is shown below. Numbers on the left refer to the size in kilobase (kb) pairs of the *Hpa*I restriction fragments that bound the probe.

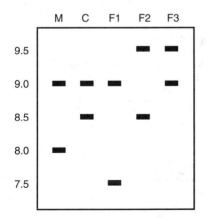

The child's father is most likely to be

(A) F1
(B) F2
(C) F3
(D) none of these men

28. Which of the following characteristics is TRUE of fatty acid oxidation?

(A) Is used to produce energy in all cells of the body
(B) Involves a C3-C4 enoyl intermediate
(C) Always begins with palmitoyl CoA
(D) Can result in ketone body formation

29. Which of the following statements concerning digestion of dietary lipids is TRUE?

(A) A deficiency of pancreatic lipase would eventually lead to a prostaglandin deficiency
(B) 2-Monoacylglycerols and free fatty acids are produced by lipoprotein lipase
(C) Phospholipids act as detergents for micelle formation
(D) Bicarbonate ions (HCO_3^-) lower the pH of the small intestine so that lipase and bile salts are more active

30. A muscle-building nutritional supplement that contains succinate was advertised with the statement, "Succinate will provide your muscles with energy even when oxygen is depleted." In the absence of oxygen, the oxidation of 1 mole of succinate in muscle would yield how many moles of ATP?

(A) 0
(B) 1
(C) 2
(D) 3
(E) 5

Questions 31 and 32

Malate + NAD$^+$ → oxaloacetate + NADH + H$^+$
$\Delta G^{\circ'}$ = +7.1 kcal/mol

Oxaloacetate + acetyl CoA → citrate + CoASH
$\Delta G^{\circ'}$ = –7.7 kcal/mol

Citrate → isocitrate
$\Delta G^{\circ'}$ = +1.5 kcal/mol

31. The standard free energy ($\Delta G^{\circ'}$) for each of the reactions in the conversion of malate to isocitrate is shown above. What is the $\Delta G^{\circ'}$ (in kcal/mol) for the net conversion of malate to isocitrate?

(A) –13.3
(B) –2.1
(C) –0.9
(D) +0.9
(E) +1.5

32. The equilibrium constant for the net conversion of malate to isocitrate is

(A) 0
(B) greater than 1
(C) less than 1

33. The direct conversion of A to B (shown below)

```
    COO⁻           COO⁻
     |              |
    C = O          C = O
     |              |
    CH₃            CH₂
                    |
                   COO⁻

     A              B
```

(A) occurs in the brain, red blood cells, and liver
(B) occurs during gluconeogenesis and lipogenesis
(C) is catalyzed by an enzyme that is allosterically inhibited by acetyl CoA
(D) requires tetrahydrofolate

34. In skeletal muscle, increased hydrolysis of ATP during muscular contraction

(A) decreases the rate of palmitate oxidation to acetyl CoA
(B) decreases the rate of NADH oxidation by the electron transport chain
(C) results in activation of phosphofructokinase 1
(D) results in an increased proton gradient across the inner mitochondrial membrane

35. If a person consumed a chemical that irreversibly inhibited all cytochromes in the electron transport chain

(A) the amount of heat generated from NADH oxidation would increase
(B) the rate of succinate oxidation would not change
(C) ATP could still be generated from the oxidation of NADH by transfer of electrons to O$_2$
(D) death would result from a lack of ATP
(E) an electrochemical potential could still be generated

36. In red blood cells, a pyruvate kinase deficiency would be expected to increase

(A) the life span of the cells
(B) the rate of fatty acid oxidation
(C) ATP production
(D) the NADH/NAD$^+$ ratio
(E) the activity of hexokinase

37. A person with a deficiency of muscle phosphorylase would

(A) produce a higher than normal amount of lactate during a brief period of intense exercise (push-ups)
(B) be incapable of performing mild exercise of long duration (a 10-mile walk)
(C) have lower than normal amounts of glycogen in muscle tissue
(D) be less dependent than normal on blood glucose to supply energy for exercise
(E) produce normal amounts of blood glucose in response to increased glucagon

38. In the presence of a drug that inhibits cAMP phosphodiesterase, which of the following enzymes would be phosphorylated and inactive in liver?

(A) Phosphorylase kinase
(B) Pyruvate kinase
(C) Phosphorylase
(D) Protein kinase A

39. In a fasting individual, each of the following stimulates the production of blood glucose by gluconeogenesis EXCEPT

(A) an increased supply of substrates
(B) induction of phosphoenolpyruvate carboxykinase
(C) a decrease of the portal blood glucose level below the K$_m$ of glucokinase
(D) a cAMP-mediated activation of pyruvate kinase

40. A woman with a lactase deficiency who eats no dairy products

(A) cannot produce mucopolysaccharides that contain galactose
(B) cannot produce lactose during lactation
(C) is likely to suffer from a calcium deficiency
(D) is likely to have high cellular levels of galactose 1-phosphate

41. Which of the following liver enzymes becomes less active when a diabetic person in ketoacidosis is treated with insulin?

(A) Fructose 1,6-bisphosphatase
(B) Pyruvate kinase
(C) Pyruvate dehydrogenase
(D) Phosphofructokinase 1 (PFK1)

42. Which of the following statements about proteoglycans is TRUE?

(A) The polysaccharide chain is not covalently linked to the protein
(B) Short chains of disaccharide repeating units are usually present
(C) Sulfation occurs before the monosaccharides are incorporated into the mucopolysaccharide chain
(D) Deficiencies of lysosomal enzymes cause accumulation of partially degraded products

43. Each of the following statements about the carbohydrate structures of glycoproteins is correct EXCEPT

(A) they do not contain *N*-acetylneuraminic acid (NANA)
(B) they are degraded in vivo by lysosomal enzymes
(C) they require nucleoside diphosphate sugars for their synthesis
(D) they are synthesized in the endoplasmic reticulum and the Golgi complex
(E) they are found on the outer surface of the cell membrane

44. In cystic fibrosis, the pancreatic ducts become obstructed by viscous mucus. Consequently, digestion of which of the following substances would be most impaired?

(A) Starch
(B) Fat
(C) Lactose
(D) Sucrose

45. A gallstone that blocked the upper part of the bile duct would cause an increase in

(A) the formation of chylomicrons
(B) the recycling of bile salts
(C) the excretion of bile salts
(D) the excretion of fat in the feces

46. A 20-year-old woman with diabetes mellitus was admitted to the hospital in a semiconscious state with fever, nausea, and vomiting. Her breath smelled of acetone. A urine sample was strongly positive for ketone bodies. Which of the following statements about this woman is TRUE?

(A) A blood glucose test would probably show that her blood glucose level was well below 80 mg/dl
(B) An insulin injection would decrease her ketone body production
(C) She should be given a glucose infusion to regain consciousness
(D) Glucagon should be administered to stimulate glycogenolysis and gluconeogenesis in the liver
(E) The acetone was produced by decarboxylation of the ketone body β-hydroxybutyrate

47. De novo fatty acid biosynthesis from cytosolic citrate

(A) requires an enzyme that contains thiamine
(B) is at a maximum when the blood glucagon: insulin ratio is high
(C) is regulated by an enzyme that contains biotin
(D) yields stearic acid as its major product

48. Each of the following statements about the conversion of glucose to triacylglycerol in the liver is correct EXCEPT

(A) citrate serves to transport acetyl units across the mitochondrial membrane
(B) reducing equivalents are provided by the reactions of the pentose phosphate pathway
(C) reducing equivalents are provided by the malic enzyme (a decarboxylating malate dehydrogenase)
(D) the glycerol moiety may be derived from dihydroxyacetone phosphate (DHAP) but not from blood glycerol
(E) phosphatidic acid is an intermediate

49. Each of the following statements concerning liver and adipose cells is correct EXCEPT

(A) adipose cells lack glycerol kinase
(B) liver cells contain a hormone-sensitive lipase
(C) adipose cells have a transport system for glucose that is regulated by insulin
(D) liver cells secrete lipoproteins when blood insulin levels are high
(E) adipose cells secrete lipoprotein lipase when blood insulin levels are high

50. A woman was told by her physician to go on a low-fat diet. She decided to continue to consume the same number of calories by increasing her carbohydrate intake, while decreasing her fat intake. Which of the following blood lipoprotein levels would be decreased as a consequence of this diet?

(A) VLDL
(B) IDL
(C) LDL
(D) Chylomicrons

51. Hormone-sensitive lipase is activated by elevated levels of

(A) cAMP
(B) ADP
(C) insulin
(D) apoprotein CII

52. A person with insulin-dependent diabetes mellitus failed to take insulin regularly and was found to have high VLDL levels. As a consequence, which of the following compounds in the blood would be elevated?

(A) Triacylglycerols
(B) Cholesterol
(C) Both triacylglycerols and cholesterol
(D) Lipoprotein lipase

53. Each of the following statements about phosphatidylcholine (PC) is correct EXCEPT

(A) it can transfer a fatty acyl group to cholesterol
(B) it can be synthesized by methylation of phosphatidylethanolamine
(C) it can be synthesized from CDP-choline and 1,2-diacylglycerol
(D) it is derived exclusively from dietary choline

54. Which of the following statements concerning metabolism of arachidonic acid is TRUE?

(A) It is converted to prostaglandins by a process that is stimulated by aspirin
(B) It is converted to prostaglandins by a process that is stimulated by glucocorticoids
(C) It is produced from thromboxanes and leukotrienes
(D) It is derived from palmitate
(E) It is cleaved from membrane phospholipids by a phospholipase

55. A person with a familial hyperlipidemia caused by a deficiency of LDL receptors was treated with an HMG CoA reductase inhibitor. This drug would cause

(A) cellular levels of squalene to increase
(B) cellular levels of HMG CoA to decrease
(C) blood cholesterol levels to decrease
(D) blood triacylglycerol levels to increase
(E) cellular acyl cholesterol acyl transferase (ACAT) activity to increase

56. Which of the following statements concerning bile salts is TRUE?

(A) They are derived from cholesterol in all tissues of the body
(B) They contain ionic groups with a positive charge
(C) They can contain glycine or serine residues
(D) They are secreted in the bile and resorbed in the intestine
(E) They are most effective as detergents below pH 3

57. Which of the following statements about reactions of the urea cycle is TRUE?

(A) Aspartate reacts with ornithine to form citrulline
(B) A total of six high-energy phosphate bonds are cleaved during production of one molecule of urea
(C) N-acetylglutamate is a positive allosteric effector of carbamoyl phosphate synthetase I
(D) The enzyme arginase releases fumarate from argininosuccinate
(E) Glutamine is the substrate that directly provides the nitrogen for carbamoyl phosphate synthesis

58. Decreased activity of which one of the following enzymes could cause an increase in urinary excretion of citrulline?

(A) Argininosuccinate synthetase
(B) Carbamoyl phosphate synthetase I (CPSI)
(C) Formiminotransferase (FIGLU + FH$_4$ → Glutamate + Formimino-FH$_4$)
(D) Glutamate dehydrogenase

59. After an overnight fast, alanine is converted to glucose in the liver. As a result of this conversion

(A) some alanine nitrogen appears in aspartate
(B) none of the alanine nitrogen appears in NH$_4^+$
(C) one alanine carbon is lost as CO$_2$
(D) blood urea nitrogen (BUN) levels decrease

60. De novo creatine synthesis requires

(A) alanine
(B) arginine
(C) AMP
(D) NAD$^+$
(E) urea

61. The cofactor required for the reaction that produces ornithine from glutamate semialdehyde is

(A) tetrahydrofolate
(B) pyridoxal phosphate
(C) thiamine pyrophosphate
(D) NAD$^+$
(E) vitamin B$_{12}$

62. A deficiency of each of the following substances could cause increased blood levels of homocystine EXCEPT

(A) cystathionine synthetase
(B) S-adenosylhomocysteine (SAH) cleaving enzyme (SAH → adenosine + homocysteine)
(C) vitamin B$_{12}$
(D) folate

63. In a person with phenylketonuria (a deficiency of phenylalanine hydroxylase)

(A) phenylalanine can be replaced by tyrosine in the diet
(B) tyrosine is an essential amino acid, but phenylalanine is not
(C) phenylalanine is an essential amino acid, but tyrosine is not
(D) phenylalanine and tyrosine are both essential amino acids

64. A person consuming a diet deficient in methionine is likely to have decreased synthesis of

(A) creatine from glycine
(B) dTMP from dUMP
(C) alanine from glucose
(D) methylmalonyl CoA from propionyl CoA
(E) hydroxymethylglutaryl CoA from acetoacetyl CoA

65. Which of the following cofactors is required for the synthesis of γ-amino butyric acid, serotonin, epinephrine, dopamine, and histamine from their respective amino acid precursors?

(A) Tetrahydrobiopterin
(B) Tetrahydrofolate
(C) Pyridoxal phosphate
(D) Thiamine pyrophosphate
(E) Vitamin B$_{12}$

66. Which of the following statements is TRUE of de novo pyrimidine synthesis but not of de novo purine synthesis?

(A) The base is synthesized while attached to ribose 5-phosphate
(B) One-carbon fragments are donated by folic acid derivatives
(C) Carbamoyl phosphate donates a carbamoyl group
(D) The entire glycine molecule is incorporated into a precursor of the base
(E) Glutamine is a nitrogen donor

67. Allopurinol prevents the conversion of

(A) IMP to GMP
(B) adenosine to inosine
(C) xanthine to uric acid
(D) dUMP to dTMP
(E) cytosine to uracil

68. Each of the following changes occurs in hemolytic anemia EXCEPT

(A) a decrease in the rate of formation of bilirubin diglucuronides in the liver
(B) an increase in the rate of secretion of bilirubin diglucuronides into the gallbladder
(C) an increase in the rate of conversion of bile pigments to stercobilins in the intestine
(D) a decrease in the amount of hemoglobin in the blood

69. When excessive amounts of iron are present in the diet, the excess iron is stored as

(A) hemoglobin
(B) transferrin
(C) hemosiderin
(D) ferritin

70. The symptoms of a dietary deficiency of niacin (which results in pellagra) will be less severe if the diet has a high content of

(A) tyrosine
(B) tryptophan
(C) thiamine
(D) thymine

71. Each of the following statements about nitrogen metabolism is correct EXCEPT

(A) glutamate is produced from α-ketoglutarate by fixation of ammonia or by transamination
(B) in the degradation of histidine, the intermediate formiminoglutamate (FIGLU) is cleaved to form glutamate
(C) alanine may be produced from serine by the action of a dehydratase followed by the action of a transaminase
(D) vitamin B_{12} can transfer a methyl group to propionyl CoA to form methylmalonyl CoA

72. A 52-year-old patient with a round face, acne, and a large hump on the back of his neck complains that he is too weak to mow his lawn. His fasting blood glucose level is 170 mg/dl (reference range = 70–100 mg/dl); plasma cortisol level, 62 μg/ml (reference range = 3–31 μg/ml); and plasma ACTH level, 0 pg/ml (reference range = 0–100 pg/ml). If the patient's condition is due to a single cause, the most likely diagnosis is

(A) non–insulin-dependent diabetes mellitus
(B) insulin-dependent diabetes mellitus
(C) a secretory tumor of the anterior pituitary
(D) a secretory tumor of the posterior pituitary
(E) a secretory tumor of the adrenal cortex

73. Women who take oral contraceptives (estrogen for 25 days, accompanied by progesterone for the last 10 days)

(A) produce a normal corpus luteum
(B) ovulate on the 14th day of the cycle
(C) have a peak of estradiol on the 13th day of the cycle
(D) do not produce normal amounts of FSH or LH

74. As part of the treatment for hypopituitarism caused by damage to the pituitary during surgery

(A) TSH and ACTH should be given orally
(B) water intake should be restricted to compensate for low vasopressin levels
(C) thyroxine tablets should be prescribed and taken regularly by the patient
(D) cortisol should be administered daily except during periods of increased stress
(E) estrogen and progesterone are the only hormones needed by women who wish to remain fertile

75. A woman whose thyroid gland was surgically removed was treated with 0.10 mg of thyroxine/daily (tablet form). After 3 months of treatment, her serum TSH levels were constant at 6.0 MIU/ml (reference range = 0.3–5.0 MIU/ml). She complained of fatigue, weight gain, and hoarseness. Her dose of thyroid hormone should

(A) be increased
(B) be decreased
(C) remain the same

76. In heart cells deprived of oxygen during a myocardial infarction

(A) the mitochondrial proton pump slows down, preventing ATP synthesis by oxidative phosphorylation
(B) the citric acid cycle will accelerate to provide more electrons for ATP synthesis
(C) the electron transport chain will accelerate to provide more protons for ATP synthesis
(D) anaerobic glycolysis will decrease, and the conversion of glucose to CO_2 will increase
(E) product inhibition by $NADH^+$ and ADP will slow many of the reactions in the citric acid cycle

Directions: Each group of items in this section consists of lettered options followed by a set of numbered items. For each item, select the **one** lettered option that is most closely associated with it. Each lettered option may be selected once, more than once, or not at all.

Questions 77–80

Match the circled letters in the structure shown below with the appropriate type of bond or interaction.

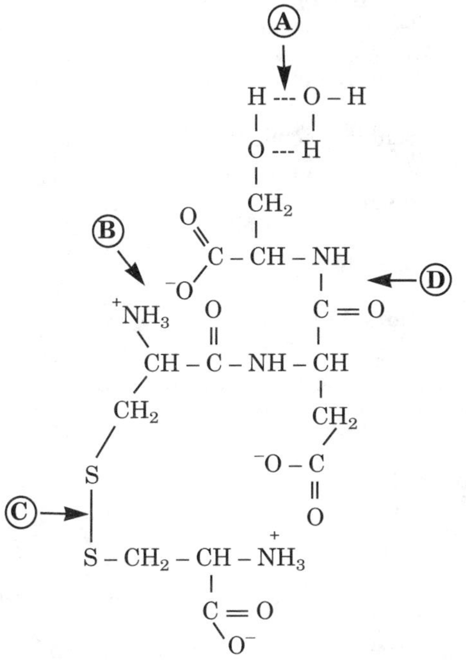

77. Hydrogen bond
78. Electrostatic interaction
79. Disulfide bond
80. Peptide bond

Questions 81–84

Indicate whether the blood levels of the compounds below would be higher, lower, or the same in a person with insulin-dependent diabetes mellitus (IDDM) who fails to take insulin for 2 days compared with a normal person who has just finished dinner.

(A) Higher
(B) Lower
(C) The same

81. Glucose
82. Glucagon
83. Urea
84. Ketone bodies

Questions 85–90

Match each property below with the appropriate type of nucleic acid.

(A) rRNA
(B) tRNA
(C) mRNA
(D) hnRNA
(E) DNA

85. Contains a CCA sequence to which an amino acid is attached
86. Is synthesized only during the S phase of the cell cycle
87. Is synthesized mainly in the nucleolus of the cell
88. Interacts with histones to form nucleosomes
89. Is found only in the nucleus
90. Contains A-T base pairs

Questions 91–93

Match each description with the appropriate nucleotide.

(A) UTP
(B) dTTP
(C) ATP
(D) GTP
(E) dUTP

91. A precursor that adds nucleotides to growing DNA chains
92. Substrate from which the tail at the 3' end of mRNA is produced
93. Substrate for the cap structure of hnRNA

Questions 94–96

Match each description below with the appropriate component of the translation process in eukaryotes.

(A) eIF-2
(B) EF-2
(C) EF-1
(D) Peptidyl transferase

94. Required for translocation during protein synthesis
95. Required during the initiation of protein synthesis
96. Required for binding of methionyl-tRNA to codon 36 of an mRNA

Questions 97–99

Match each antibiotic below with the appropriate step in translation that it inhibits in prokaryotes.

(A) Initiation
(B) Binding of aminoacyl-tRNA to the "A" site on the ribosome
(C) Peptide bond formation
(D) Translocation

97. Tetracycline
98. Streptomycin
99. Erythromycin

Questions 100–102

The blood levels of glucose, galactose, and fructose were measured in normal persons and in persons with various enzyme deficiencies soon after they drank tea with milk and sugar. Match each comparison below with the appropriate sugar(s).

(A) Glucose
(B) Galactose
(C) Fructose
(D) Glucose and galactose
(E) Glucose, galactose, and fructose

100. Lower in the blood of a person with a lactase deficiency than in the normal person
101. Higher in the blood of a person with a galactose 1-phosphate uridyl transferase deficiency than in the normal person
102. Higher in the blood of a person with a fructokinase deficiency than in the normal person

Questions 103–105

Match each enzyme below with its required cofactor.

(A) Thiamine pyrophosphate
(B) NAD⁺
(C) NADP⁺
(D) Biotin

103. Malate dehydrogenase
104. Transketolase
105. Glucose 6-phosphate dehydrogenase

Questions 106–110

For each untreated condition below, select the blood or urine value that best distinguishes the condition from the others. All values are measured after an overnight fast and are compared with those of a normal individual.

(A) Increased MB fraction of serum creatine phosphokinase (CPK)
(B) Increased blood ketone bodies
(C) Decreased creatinine in the urine
(D) Decreased blood lactate
(E) Decreased blood urea nitrogen (BUN)

106. Insulin-dependent diabetes mellitus
107. Myocardial infarction
108. Hepatitis
109. Renal failure
110. Alcoholism

Questions 111–113

Match each description below with the most appropriate hormone.

(A) LH
(B) FSH
(C) GH
(D) MSH

111. Is encoded by the proopiomelanocortin (POMC) gene
112. Stimulates testosterone production
113. Causes acromegaly when secreted in excessive amounts in adults

Questions 114–116

Match each biological activity below with the appropriate hormone.

(A) Oxytocin
(B) Aldosterone
(C) Vasopressin
(D) Prolactin

114. Increases sodium and water resorption by renal tubule cells
115. Causes synthesis of α-lactalbumin
116. Causes uterine contractions

Questions 117–119

Indicate whether the blood levels of the compounds below (following an overnight fast) would be higher, lower, or the same in a person with a carnitine deficiency compared with a normal person.

(A) Higher
(B) Lower
(C) The same

117. Fatty acids
118. Ketone bodies
119. Glucose

Questions 120–122

Each reaction below is the first step in a biosynthetic pathway. In each case, select the final end product of the pathway.

(A) Heme
(B) Purines
(C) Pyrimidines
(D) Urea

120. Glycine + succinyl CoA $\rightarrow$ δ-aminolevulinic acid + CoASH + CO_2
121. CO_2 + NH_4^+ + 2 ATP $\rightarrow$ carbamoyl phosphate + 2 ADP + P_i
122. Glutamine + phosphoribosyl pyrophosphate $\rightarrow$ phosphoribosyl amine + glutamate + PP_i

Questions 123–126

Match each description below with the appropriate lipid–protein complex.

(A) Chylomicrons
(B) VLDL
(C) LDL
(D) HDL
(E) Fatty acid–albumin complexes

123. The major donor of cholesterol to peripheral tissues
124. The first lipoprotein to increase in concentration in the blood after ingestion of 400 g of jelly beans (carbohydrate)
125. The site of the LCAT reaction
126. Composed mainly of triacylglycerols synthesized in intestinal epithelial cells

Questions 127–130

Each condition below can be caused by a problem with the metabolism of a particular compound. Match each condition with the appropriate compound.

(A) Glycogen
(B) Collagen
(C) Dopamine
(D) Galactose
(E) A sphingolipid

127. Ehlers-Danlos syndrome
128. Parkinson's disease
129. Tay-Sachs disease
130. McArdle's disease

Questions 131–134

A dietary deficiency of a vitamin can cause each of the conditions below. Match each condition with the appropriate vitamin.

(A) Vitamin C
(B) Niacin
(C) Vitamin D
(D) Biotin
(E) Thiamine

131. Pellagra
132. Scurvy
133. Beriberi
134. Rickets

Answers and Explanations

1–A. Fat contains 9 calories per gram, whereas carbohydrate contains 4 calories per gram. Therefore, when the patient substituted 50 g of fat for 50 g of carbohydrate, he took in more calories per day and gained weight.

2–B. The patient's weight (176 lb × 0.454 lb per kg) is 80 kg. Since the basal metabolic rate (BMR) is approximately 24 Cal/kg/day, her BMR (24 Cal/kg/day × 80 kg) is 1920 Cal/day. She requires 30% more calories for her activity (sedentary) or 1920 × 1.3 = 2500 Cal.

3–E. Although niacin is a vitamin, it may be synthesized to a limited extent from tryptophan. The other compounds are required in the diet.

4–D. All of these compounds except even-chain fatty acids serve as carbon sources for gluconeogenesis.

5–D. Galactose is converted to glucose 6-phosphate, which is cleaved by glucose 6-phosphatase to form blood glucose. Lactate and fructose require fructose 1,6-bisphosphatase to form blood glucose. Muscle glycogen does not produce blood glucose.

6–E. Both fructose and galactose can be converted to blood glucose in the absence of phosphoenolpyruvate carboxykinase, but this enzyme is required for conversion of lactate to blood glucose. Muscle glycogen is not converted to blood glucose.

7–C. The pH and pK are related as follows: pH = pK + log ([A⁻]/[HA]). Thus, when the concentrations of a weak acid and its base are equal, the pH equals the pK.

8–C. HMG CoA is not formed from glutamic acid, but from acetyl CoA and acetoacetyl CoA. It is also formed by degradation of leucine in muscle. It is cleaved to form acetyl CoA and the ketone body acetoacetate. It is reduced to mevalonic acid in cholesterol biosynthesis.

9–E. According to the Michaelis-Menten equation, the velocity (v) is related to the concentration of substrate, [X], as follows: $v = V_m[X]/(K_m + [X])$. Thus, $4 = (10 \times 2)/(K_m + 2)$. $4 K_m + 8 = 20$. $4 K_m = 20 - 8 = 12$. $K_m = 12/4 = 3 \mu M$.

10–D. The Y intercept ($1/V_m$) is the same for the normal enzyme and the patient's enzyme, so the enzymes have the same V_m. However, the X intercepts ($-1/K_m$) differ. The normal enzyme has a K_m of 1 μM, whereas the patient's enzyme has a K_m of 5 μM. Therefore, more thiamine pyrophosphate is required to saturate the patient's enzyme, and raising the body levels of thiamine should cause more of the patient's enzyme to be in the active complex.

11–A. Acetyl CoA, which contains pantothenic acid as part of its coenzyme A moiety, is a substrate for citrate synthase.

12–D. Of these tissues, only the red blood cell does not have an active TCA cycle because it lacks mitochondria.

13–C. Hemoglobin contains two α chains and two β chains, which combine to form the quaternary structure. Each mole of subunit binds 1 mole of heme, which binds 1 mole of O_2. Therefore, 1 mole of HbA binds 4 moles of O_2. In the β chain of HbS, valine replaces glutamate at position 6, so HbS is more hydrophobic than HbA.

14–B. Hydroxylation of lysine residues requires vitamin C and occurs after lysine has been incorporated into the polypeptide chain but before secretion from the cell. Hydroxylysine residues form cross links between collagen molecules, causing polymerization.

15–C. After an overnight fast, fatty acids are released from adipose tissue, oxidized by muscle (but not the brain), and converted to ketone bodies in the liver. Glucose is not synthesized from even-chain fatty acids, and ketone bodies are not oxidized by red blood cells because they lack mitochondria.

16–C. A, B, and E occur in prokaryotes, not eukaryotes. 80S ribosomes are utilized in eukaryotes; during synthesis of proteins that are secreted, the ribosomes are attached to the RER. Peptidyl transferase is required.

17–D. hnRNA contains sequences (introns) that are removed to produce mRNA. UAG and UAA are both stop codons. Some genes occur in multiple copies within the genome. Gene sequences may undergo rearrangement in germ cells as they differentiate (e.g., during development of immunoglobulin-producing cells).

18–C. The mutant gene has a four-base insertion (TATC) starting at position 9. Consequently, a frameshift occurs, and the mutant gene encodes a protein with a different amino acid sequence beyond this point. The insertion causes the sequence TGA at position 22 to come into frame. This corresponds to UGA, termination codon in the mRNA. Therefore, the mutant protein will be shorter than the normal protein. Although AGA is a codon for arginine, another codon for arginine (CGU) is used in this sequence.

19–A. Transcription is catalyzed by RNA polymerase II, which binds to promoter regions, including a TATA box. The DNA template strand is not covalently bound to histones. The primary transcript (hnRNA) is capped at the 5' end and polyadenylated at the 3' end; introns are removed by splicing to form mRNA.

20–B. In the absence of glucose and the presence of lactose, the *lac* repressor will be inactive, cAMP levels will rise, and CAP protein will bind to the *lac* promoter stimulating transcription of the operon. Tryptophan levels in the cell will be low, so the repressor for the *trp* operon will be inactive and the operon will be transcribed by RNA polymerase. Attenuation of transcription of this operon will decrease.

21–C. Reverse transcriptase uses an RNA template to synthesize DNA. It is not involved in repair.

22–D. The gene sequence is read 5' to 3' from the bottom to the top of the gel. The normal and mutant sequences are the same except for a point mutation that converted an A to a T. Thus, the mutant gene contains the sequence 5'-GAATTC-3', which is cleaved by *Eco*RI, but the normal gene does not. *Kpn*I would not cleave within the sequences shown on the gels.

23–D. Okazaki fragments are synthesized in the 5' to 3' direction, starting with an RNA primer to which DNA precursors are attached. The RNA is subsequently removed, replaced with DNA, and the fragments are joined by a ligase. A group of enzymes, not just DNA polymerase, is involved in this process.

24–C. Dihydroxyacetone phosphate (Compound I) is isomerized to glyceraldehyde 3-phosphate, which is oxidized to 1,3-bisphosphoglycerate in a reaction in which inorganic phosphate and NAD^+ are utilized. Then 3-phosphoglycerate (Compound II) is formed and ATP is generated.

25–B. Blood glucose is maintained after about 2 hours of fasting by glycogenolysis, which is subsequently supplemented by gluconeogenesis. However, after about 1 day of fasting, liver glycogen is depleted, so thereafter gluconeogenesis is solely responsible for maintaining blood glucose.

26–C. After an overnight fast, glycogenolysis and gluconeogenesis act to maintain blood glucose levels in a normal person. Both pathways produce glucose 6-phosphate and require glucose 6-phosphatase to produce free glucose. If the phosphatase is inhibited, blood glucose levels will be lower and liver glycogen stores higher than normal.

27–B. Because two fragments containing this gene are produced from each person, there are two copies (alleles) of this gene in the genome. The restriction fragments have a different number of tandem repeats; one fragment is inherited from the mother and the other from the father. The child received the 9.0-kb fragment from the mother. An 8.5-kb fragment could only have come from F2, so he is most likely to be the father.

28–D. Almost all fatty acids can undergo β-oxidation, resulting in release of acetyl CoA, which can be converted to ketone bodies in the liver. The oxidation pathway contains a C2-C3 enoyl intermediate. In most tissues, but not the brain and red blood cells, oxidation of fatty acids produces energy.

29–A. Pancreatic lipase catalyzes the breakdown of dietary triacylglycerols into free fatty acids and 2-monoacylglycerols, an essential step in the digestion of dietary lipids. Since prostaglandins are produced from linoleate, an essential fatty acid, a deficiency of pancreatic lipase would eventually cause a prostaglandin deficiency.

30–A. When oxygen is depleted, the electron transport chain stops, NADH builds up, and the TCA cycle is inhibited. Therefore succinate, an intermediate of the cycle, will not be oxidized. In the absence of oxygen, no ATP is produced.

31–D. For a series of coupled reactions, the individual $\Delta G^{\circ\prime}$ values may be added to give the value of $\Delta G^{\circ\prime}$ for the overall reaction.

32–C. $\Delta G^{\circ\prime} = -2.303\ RT \log K_{eq}$. For the conversion of isocitrate to malate, $\Delta G^{\circ\prime} = +0.9$ and log K_{eq} is negative. Therefore, K_{eq} is less than 1.

33–B. Pyruvate carboxylase, which converts pyruvate (A) to oxaloacetate (B), is found in brain and liver but not in red blood cells or muscle. It is involved in the synthesis of glucose and fatty acids. It is activated by acetyl CoA and requires biotin and ATP.

34–C. A decrease in the concentration of ATP stimulates processes that generate ATP. The proton gradient decreases, NADH oxidation increases, and fuel utilization increases. Palmitate is oxidized, and glycolysis increases because of activation of phosphofructokinase 1 by AMP. As ATP decreases, AMP rises (ATP + AMP ↔ 2 ADP).

35–D. If the cytochromes are inhibited, ATP production, the electrochemical potential, heat production from NADH oxidation, and succinate oxidation all decrease.

36–D. A pyruvate kinase deficiency would slow glycolysis in red blood cells and cause ATP production to decrease and NADH to increase. Intermediates of glycolysis before the blocked step would accumulate, and glucose 6-phosphate would inhibit hexokinase. The life span of the cells would decrease. Fatty acids cannot serve as a source of energy because these cells lack mitochondria.

37–E. A muscle phosphorylase deficiency would prevent muscle glycogen from being oxidized during exercise. Therefore, lactate levels would be low, and the person could not tolerate intense exercise of brief duration and would rely on fuels from the blood (glucose, fatty acids, and ketone bodies) for energy. The person could engage in mild exercise of long duration, using these blood fuels. The liver would not be affected (a different phosphorylase isozyme is present) and could respond to glucagon by breaking down glycogen.

38–B. Under these conditions, cAMP levels would remain elevated. Phosphorylation of pyruvate kinase causes its inactivation. Phosphorylase kinase and phosphorylase are activated by phosphorylation. Protein kinase A is not regulated by phosphorylation.

39–D. cAMP causes inactivation of pyruvate kinase and, thereby, promotes glucose production.

40–C. Lactase is a digestive enzyme that cleaves lactose to galactose and glucose. Galactose can be produced from glucose and would be metabolized normally in this woman. Because of her low intake of dairy products, she might develop a calcium deficiency and would have normal or low levels of galactose 1-phosphate.

41–A. Insulin stimulates activation of pyruvate kinase, pyruvate dehydrogenase, and phosphofructokinase 2 (PFK2). PFK2 then catalyzes formation of fructose 2,6-bisphosphate, which is an activator of PFK1 and an inhibitor of fructose 1,6-bisphosphatase, a gluconeogenic enzyme.

42–D. The mucopolysaccharides of proteoglycans contain long chains of repeating disaccharide units that are covalently linked to a protein. Sulfation occurs after the monosaccharides are incorporated into the mucopolysaccharide chain. Proteoglycans are degraded by lysosomal enzymes.

43–A. The carbohydrates present in glycoproteins are synthesized from UDP-sugars and CMP-NANA.

44–B. Lactose and sucrose are digested by disaccharidases on the brush border of intestinal epithelial cells. Starch is digested by salivary and pancreatic amylase. Therefore, its digestion would be less affected by a lack of pancreatic juice than fat, which is digested mainly by pancreatic lipase. A common finding in cystic fibrosis is steatorrhea (fatty stools).

45–D. In this situation, bile salts from the pancreas could not enter the digestive tract. Therefore, recycling and excretion of bile salts, digestion of fats, and formation of chylomicrons would all decrease. As a consequence, fat in the feces (steatorrhea) would increase.

46–B. The acetone on her breath (produced from decarboxylation of acetoacetate) and the ketone bodies in her urine indicate that she is in diabetic ketoacidosis. Her blood glucose levels would be high because her insulin levels are too low to stimulate glucose transport into muscle and adipose tissue and to stimulate glycogen and triacylglycerol synthesis in liver. An insulin injection would reduce her blood glucose levels and decrease the release of fatty acids from adipose triacylglycerols. Consequently, ketone body production would decrease.

47–C. Fatty acid synthesis is maximal in the fed state when insulin is elevated. Glucose is converted to citrate, which is cleaved to oxaloacetate and acetyl CoA. The regulatory enzyme acetyl CoA carboxylase requires biotin and converts acetyl CoA to malonyl CoA, which provides the two-carbon units for elongation of the fatty acyl chain by the fatty acid synthase complex. The major product is palmitate. Thiamine is not required for this process.

48–D. The liver has glycerol kinase, so blood glycerol can be used.

49–B. Hormone-sensitive lipase is found in adipose tissue.

50–D. Chylomicrons are blood lipoproteins produced from dietary fat. VLDL are produced mainly from dietary carbohydrate. IDL and LDL are produced from VLDL.

51–A. The hormone-sensitive lipase of adipose tissue is activated by glucagon via a cAMP-mediated process. Apoprotein CII is the activator of lipoprotein lipase.

52–C. VLDL levels are elevated because the decreased insulin and increased glucagon cause lipolysis of adipose triacylglycerols. The fatty acids and glycerol are repackaged in VLDL, which are secreted by the liver. Therefore, both triacylglycerols and cholesterol are elevated in the blood. Lipoprotein lipase is decreased because its synthesis and secretion by adipose tissue are stimulated by insulin.

53–D. In the absence of dietary choline, PC can be synthesized de novo from glucose. The last step in this pathway involves methylation of phosphatidylethanolamine.

54–E. Arachidonic acid is cleaved from membrane phospholipids by phospholipase A_2, which is inhibited by glucocorticoids. It requires essential fatty acids for its synthesis. It can be converted to leukotrienes, or be oxidized by a cyclooxygenase, which is inhibited by aspirin, and converted to prostaglandins and thromboxanes.

55–C. Reduction of HMG CoA to mevalonic acid is an early step in cholesterol synthesis. Inhibition of this step would lead to an increase in cellular levels of HMG CoA and a decrease in squalene, an intermediate beyond this step, and cholesterol. The decreased cholesterol levels in cells cause ACAT activity to decrease and synthesis of LDL receptors to increase. Because the receptors function (but at a less than normal rate), more receptors cause more LDL to be taken up from the blood. Consequently, blood cholesterol levels decrease, but blood triacylglycerol levels do not change much, since LDL does not contain much triacylglycerol.

56–D. Bile salts are synthesized from cholesterol only in the liver. Glycine or taurine is conjugated to the carboxyl group on the side chain. Ionization occurs with a pK of about 4 for the glycoconjugates and 2 for the tauroconjugates; these carry a negative charge. Bile salts are secreted in the bile, participate in lipid digestion, are resorbed in the ileum, and are recycled by the liver.

57–C. In the urea cycle, ammonia provides the nitrogen for synthesis of carbamoyl phosphate, which reacts with ornithine to form citrulline. Aspartate reacts with citrulline to form argininosuccinate, which releases fumarate. The product arginine is cleaved by arginase to form urea and regenerate ornithine. Overall, four high-energy phosphate bonds are cleaved. Carbamoyl phosphate synthetase I is activated by *N*-acetylglutamate.

58–A. If argininosuccinate synthetase activity is low, less citrulline will be converted to argininosuccinate than normal. Citrulline levels will rise, and citrulline will be excreted in the urine.

59–A. During fasting, all the carbons of alanine are converted to glucose in the liver, and the nitrogens of alanine are converted to urea.

60–B. Creatine is synthesized from glycine, the guanidinium group of arginine, and the methyl group of methionine.

61–B. Glutamate semialdehyde is transaminated to form ornithine. Glutamate provides the nitrogen and is converted to α-ketoglutarate. The cofactor for transamination is pyridoxal phosphate.

62–B. Homocystine is produced when two homocysteine molecules form a disulfide bond. Homocysteine is produced when *S*-adenosylhomocysteine (SAH) releases adenosine. It is used in the synthesis of cystathionine. Homocysteine accepts a methyl group (from FH_4 via vitamin B_{12}) to form methionine. If the enzyme that cleaves adenosine from SAH is deficient, homocysteine levels will decrease. The other deficiencies would lead to increased levels of homocysteine.

63–D. Tyrosine is produced by hydroxylation of the essential amino acid phenylalanine. In PKU, tyrosine cannot be synthesized in adequate amounts and is required in the diet.

64–A. *S*-Adenosylmethionine (SAM) provides a methyl group for the synthesis of creatine. It is not involved in the other reactions.

65–C. In the synthesis of each of these compounds, a decarboxylation of an amino acid occurs. Amino acid decarboxylation reactions, as well as transaminations, require pyridoxal phosphate.

66–C. In pyrimidine biosynthesis, carbamoyl phosphate, produced from glutamine, CO_2, and ATP, reacts with aspartate to form a base which, after oxidation, reacts with phosphoribosyl pyrophosphate (PRPP) to form a nucleotide. This nucleotide is decarboxylated to form UMP. In purine biosynthesis, the base is produced on ribose 5-phosphate. Glycine is incorporated into the precursor, and tetrahydrofolate derivatives donate carbons 2 and 8. Glutamine is a nitrogen donor for both purine and pyrimidine biosynthesis.

67–C. Xanthine oxidase is involved in the conversion of the purine bases to uric acid. It catalyzes the oxidation of hypoxanthine to xanthine and of xanthine to uric acid. Allopurinol is used in the treatment of gout, which is caused by the precipitation of uric acid crystals in the joints.

68–A. When red blood cells lyse, hemoglobin is degraded. Bilirubin is produced from heme at an increased rate. The liver converts bilirubin to the diglucuronide at a rapid rate and excretes it into the bile. In the intestine, bacteria convert bilirubin to stercobilins, which give stool its brown color.

69–C. Iron is normally carried in the blood on transferrin and stored as ferritin. If an abnormally large amount of iron is present, hemosiderin is produced. Hemoglobin contains iron complexed with heme.

70–B. Although dietary niacin is the major source of the nicotinamide ring of NAD, it may also be produced from tryptophan.

71–D. Vitamin B_{12} is involved in the rearrangement of methylmalonyl CoA to form succinyl CoA.

72–E. ACTH levels were low, so the elevated cortisol levels were most likely caused by a secretory tumor of the adrenal cortex. The elevated cortisol levels caused the elevation of blood glucose.

73–D. The constant levels of estrogen provided by use of oral contraceptives prevent the release of FSH, which causes the egg to mature, and LH, which causes ovulation and maintains the corpus luteum.

74–C. Thyroxine can be taken orally to compensate for low TSH levels. TSH and ACTH are polypeptide hormones; therefore, they would be digested by pancreatic proteases in the gut. Low vasopressin causes water to be lost in the urine (diabetes insipidus). Cortisol is required particularly during stress. GnRH, FSH, and LH are required for production of a mature egg. Estrogen and progesterone, taken alone, will suppress ovulation.

75–A. Thyroid hormone feeds back on the anterior pituitary and inhibits the release of TSH. This patient's TSH levels are elevated, so her thyroid hormone levels are too low, and her dose should be increased.

76–A. A lack of O_2 causes a slowing of the proton pump and ATP synthesis decreases. The electron transport chain, the citric acid cycle, and the conversion of glucose to CO_2 all decrease. NADH slows the citric acid cycle, but ADP allosterically activates isocitrate dehydrogenase, stimulating the cycle.

77–A. The hydroxyl group of serine forms hydrogen bonds with water.

78–B. A positively charged amino group (on a cysteine residue) and a negatively charged carboxyl group (on a serine residue) form an electrostatic interaction.

79–C. Two cysteine residues are covalently joined by a disulfide bond.

80–D. In peptides, adjacent amino acids are joined covalently by peptide bonds.

81–A. A person with IDDM who is not taking insulin behaves metabolically like a person who is fasting except that glucose levels are elevated. The low insulin levels cause decreased transport of glucose into muscle and adipose cells and decreased conversion of glucose to glycogen and triacylglycerols in liver.

82–A. As insulin decreases, glucagon rises and stimulates glycogenolysis and gluconeogenesis.

83–A. When insulin is low and glucagon is high, the carbon skeletons of amino acids derived from muscle protein are converted to glucose in the liver by gluconeogenesis. The amino acid nitrogen is converted to urea.

84–A. When the insulin:glucagon ratio is low, fatty acids are released from adipose tissue and converted to ketone bodies by the liver.

85–B. A CCA sequence is present at the 3' end of tRNA.

86–E. DNA is replicated during S phase.

87–A. All species of rRNA are produced in the nucleolus except 5S rRNA, which is produced in the nucleoplasm.

88–E. Approximately 140 base pairs of DNA are wrapped around a core of histones to form a nucleosome.

89–D. hnRNA is produced in the nucleus and converted to mRNA, which migrates to the cytoplasm. DNA is found in the nucleus and in mitochondria.

90–E. A-T pairs are found in DNA. RNA contains A-U base pairs. Both DNA and RNA contain G-C base pairs.

91–B. The deoxyribonucleoside triphosphates dTTP, dATP, dGTP, and dCTP are the precursors for elongation of DNA chains.

92–C. The poly(A) tail of mRNA is produced from ATP.

93–D. GTP is the substrate for the cap of hnRNA (and, therefore, mRNA).

94–B. Elongation factor 2 (EF-2) is required for translocation.

95–A. Eukaryotic initiation factors (such as eIF-2) are required for initiation.

96–C. Although initiation factors are required for binding of the methionyl-tRNA that initiates synthesis of a polypeptide chain, internal methionine residues require the elongation factors for their incorporation into growing polypeptide chains. EF-1 is required for binding of aminoacyl-tRNA to the appropriate codon on mRNA.

97–B. Tetracycline prevents aminoacyl-tRNA from binding.

98–A. Streptomycin prevents formation of the initiation complex.

99–D. Erythromycin prevents translocation.

100–D. Galactose and glucose would be lower in a person with a lactase deficiency, because lactose in the milk would not be cleaved to produce galactose and glucose.

101–B. Galactose would be higher, because in classical galactosemia (the uridyl transferase deficiency), galactose can be phosphorylated but it cannot be metabolized further. Galactose 1-phosphate and its precursor, galactose, increase.

102–C. Fructose derived from the table sugar (sucrose) would not be converted to fructose 1-phosphate, so it accumulates in the blood and is excreted in the urine, producing a benign fructosuria.

103–B. NAD^+ is required for the conversion of malate to oxaloacetate.

104–A. The transketolase of the pentose phosphate pathway requires thiamine pyrophosphate.

105–C. Glucose 6-phosphate dehydrogenase, the first enzyme in the pentose phosphate pathway, requires $NADP^+$.

106–B. In the absence of insulin, a person with IDDM will behave metabolically like a person undergoing prolonged starvation except that blood glucose levels will be elevated. Lipolysis in adipose tissue will produce fatty acids, which will be converted to ketone bodies in the liver.

107–A. CPK is found in large amounts in muscle cells. When a tissue is damaged, cellular enzymes leak into the blood. Heart muscle has more of the MB isozyme of CPK than skeletal muscle. (Brain has the BB isozyme.)

108–E. Urea is made in the liver. If the liver is infected, less urea will be produced and the BUN will decrease. Consequently, ammonia levels will increase.

109–C. The kidney excretes nitrogenous waste products, including urea, ammonia, creatinine, and uric acid. If the kidneys fail, these waste products will not be excreted into the urine.

110–E. Long-term exposure of the liver to alcohol can cause cirrhosis. Because the amount of functional liver tissue decreases, urea production decreases.

111–D. The POMC gene produces a protein that is cleaved to form ACTH, lipotropin, MSH, and endorphins.

112–A. LH binds to plasma membrane receptors on Leydig cells and stimulates the production of testosterone.

113–C. An excess of growth hormone (GH) after epiphyseal closure causes acromegaly, a condition characterized by enlargement of the jaw, nose, hands, feet, and skull.

114–B. Aldosterone causes resorption of sodium ions and, consequently, the resorption of water. Vasopressin (antidiuretic hormone) causes water resorption.

115–D. Prolactin causes synthesis of the milk protein α-lactalbumin, which stimulates lactose synthesis.

116–A. Oxytocin causes uterine contractions during labor and contraction of mammary myoepithelial cells during lactation.

117–A. Carnitine is required for the transport of fatty acids into mitochondria where they undergo β-oxidation. If carnitine is deficient, fatty acids accumulate.

118–B. Ketone bodies are produced in the liver from acetyl CoA generated by fatty acid oxidation. In a carnitine deficiency, fatty acids are not oxidized, so ketone bodies are not produced.

119–B. More blood glucose will be oxidized for energy because fatty acids cannot be used.

120–A. Heme.

121–D. Urea.

122–B. Purines.

123–C. Cells take up LDL by endocytosis, and cholesterol is released from cholesterol esters by a lysosomal enzyme.

124–B. VLDL are synthesized by the liver from dietary sugar (present in jelly beans).

125–D. HDL removes cholesterol from cell membranes. The lecithin cholesterol acyl transferase (LCAT) reaction, which converts the cholesterol to cholesterol esters, occurs on HDL.

126–A. Chylomicrons contain triacylglycerols synthesized from dietary lipid in intestinal epithelial cells.

127–B. A defect in the synthesis or processing of collagen causes Ehlers-Danlos syndrome.

128–C. Dopamine levels are low in Parkinson's disease.

129–E. Because a hexosaminidase is deficient in Tay-Sachs disease, partially degraded sphingolipids (gangliosides) accumulate in lysosomes.

130–A. Glycogen accumulates because muscle phosphorylase is deficient in McArdle's disease (a glycogen storage disease).

131–B. A dietary deficiency of niacin causes pellagra.

132–A. Scurvy is caused by lack of vitamin C.

133–E. Lack of thiamine in the diet causes beriberi.

134–C. A dietary deficiency of vitamin D causes rickets.

Index

Note: Page numbers in *italic* denote illustrations; those followed by t denote tables; those followed by Q indicate questions; and those followed by E indicate explanations.

Urea
cycle, 237–239, *238*, 265Q,
269E, 306Q, 311Q, 316E,
320E
liver disease, 262
starvation state, 10, *11*, 309Q,
318E
Uric acid
gout, 263, 268Q, 271E
purine degradation, 258,
266Q, 270E
Uridine, pyrimidine
metabolism, 263
Uridine monophosphate (UMP),
pyrimidine synthesis, 260
Uridine triphosphate (UTP),
pyrimidine synthesis, 260
Uridyl transferase deficiency, 174
Urobilins, heme degradation, 262

V

Valine
branched-chain amino acid
metabolism, 246
degradation, 243–244, 244,
244, 266Q, 270E
Vanillylmandelic acid (VMA),
from catecholamines, 256
Vasoactive intestinal polypeptide (VIP), 291
Vasopressin, actions of, 285,
286t, 297Q, 298E

Vectors, DNA cloning, 80
Very-long-chain fatty acids,
ω-oxidation, 206
Very-low-density lipoproteins
(VLDL)
composition, 200, 200t
fed (absorptive) state, 4–7, *5*
metabolism, 200–201, *201*,
223Q, 228E
triacylglycerol synthesis, 195,
306Q, 315E
adipose tissue, regulation,
197–199
Viral infections, DNA synthesis,
85
Vitamin A
action of, 290
deficiencies and excesses, 121
structure of, 109, *110,* 126Q,
129E
Vitamin B$_6$, amino acid metabolism and, 264
Vitamin B$_{12}$
amino acid metabolism and,
264, 266Q, 268Q, 270E,
272E, 308Q, 317E
metabolism, 251–252, *253*
Vitamin C. *See* Ascorbic acid
Vitamin D, 110, *110,* 126Q, 129E
deficiency, 121, 311Q, 320E
as hormone metabolism, 217
Vitamin E, 109, *110*
Vitamin K, 109, *110,* 126Q,
129E
deficiency, 121

Vitamins
amino acid metabolism, 264
as cofactors, 107–109, *108*
deficiencies and excesses,
120–121
dietary requirements, 4
fat-soluble, 109–110, *110*
TCA cycle, 113

W

Water (H$_2$O)
biochemistry of, 21
Weak acids
defined, 22
ionization, 39Q, 44E

X

Xanthine, purine degradation,
258
Xanthomas, 217
Xeroderma pigmentosum, DNA
damage and, 44, 88Q,
95E
Xylitol dehydrogenase, carbohydrate metabolism, 174

Z

Z form of DNA, 50
Zinc finger, 28
Zomogens, 234